BIOREMEDIATION OF WASTEWATER

Factors and Treatments

BIOREMEDIATION OF WASTEWATER

Factors and Treatments

Edited by
Olga Sánchez, PhD

Apple Academic Press Inc. | Apple Academic Press Inc.
3333 Mistwell Crescent | 9 Spinnaker Way
Oakville, ON L6L 0A2 | Waretown, NJ 08758
Canada | USA

©2015 by Apple Academic Press, Inc.

First issued in paperback 2021

Exclusive worldwide distribution by CRC Press, a member of Taylor & Francis Group

No claim to original U.S. Government works

ISBN 13: 978-1-77463-567-4 (pbk)
ISBN 13: 978-1-77188-162-3 (hbk)

Library and Archives Canada Cataloguing in Publication

Bioremediation of wastewater: factors and treatments / edited by Olga Sánchez, PhD.

Includes bibliographical references and index.
ISBN 978-1-77188-162-3 (bound)
1. Sewage--Purification--Biological treatment. 2. Bioremediation. 3. Microbial biotechnology. I. Sánchez, Olga, author, editor

TD755.B55 2015   628.3'5   C2015-902523-0

Library of Congress Cataloging-in-Publication Data

Bioremediation of wastewater : factors and treatments / Olga Sánchez, PhD., editor.

pages cm
Includes bibliographical references and index.
ISBN 978-1-77188-162-3 (alk. paper)
1. Water--Purification--Biological treatment. 2. Sewage--Purification--Biological treatment. 3. Water--Purification. I. Sánchez Martínez, Olga, editor.

TD475.B58 2015   628.1'62--dc23   2015013062

Apple Academic Press also publishes its books in a variety of electronic formats. Some content that appears in print may not be available in electronic format. For information about Apple Academic Press products, visit our website at **www.appleacademicpress.com** and the CRC Press website at **www.crcpress.com**

About the Editor

OLGA SÁNCHEZ, PhD

Olga Sánchez received her PhD degree in Biological Sciences from the Universitat Autònoma de Barcelona (Spain) in 1996. Her research began with the study of the physiology of photosynthetic bacteria and derived to the utilization of complex microbial biofilms in packed reactors for the treatment of contaminated effluents. In 2007 she got a fixed position as aggregate teacher at the Department of Genetics and Microbiology of the Universitat Autònoma de Barcelona, and presently, her investigation focuses on the application of molecular techniques for the characterization of the diversity of different natural microbial communities, including marine environments or wastewater treatment systems. These methodologies include clone libraries, FISH (Fluorescence In Situ hybridization), fingerprinting techniques such as DGGE (Denaturing Gradient Gel Electrophoresis) and next-generation sequencing technologies.

Contents

Acknowledgment and How to Cite .... *ix*

List of Contributors .... *xi*

Introduction .... *xvii*

Part I: Microbial Communities for Wastewater Treatment

1. **Bacterial Communities in Different Sections of a Municipal Wastewater Treatment Plant Revealed by 16S rDNA 454 Pyrosequencing** 3

   Lin Ye and Tong Zhang

2. **Taxonomic Precision of Different Hypervariable Regions of 16S rRNA Gene and Annotation Methods for Functional Bacterial Groups in Biological Wastewater Treatment** 25

   Feng Guo, Feng Ju, Lin Cai, and Tong Zhang

3. **The Choice of PCR Primers Has Great Impact on Assessments of Bacterial Community Diversity and Dynamics in a Wastewater Treatment Plant** 55

   Nils Johan Fredriksson, Malte Hermansson, and Britt-Marie Wilén

4. **Abundance and Diversity of Bacterial Nitrifiers and Denitrifiers and Their Functional Genes in Tannery Wastewater Treatment Plants Revealed by High-Throughput Sequencing** 101

   Zhu Wang, Xu-Xiang Zhang, Xin Lu, Bo Liu, Yan Li, Chao Long, and Aimin Li

5. **Assessing Bacterial Diversity in a Seawater-Processing Wastewater Treatment Plant by 454-Pyrosequencing of the 16S rRNA and amoA Genes** 125

   Olga Sánchez, Isabel Ferrera, Jose M González, and Jordi Mas

6. **Assessment of Bacterial and Structural Dynamics in Aerobic Granular Biofilms** 141

   David G. Weissbrodt, Thomas R. Neu, Ute Kuhlicke, Yoan Rappaz, and Christof Holliger

Part II: Environmental Factors

7.   **Abundance, Diversity and Seasonal Dynamics of Predatory Bacteria in Aquaculture Zero Discharge Systems** 177

Prem P. Kandel, Zohar Pasternak, Jaap van Rijn, Ortal Nahum, and Edouard Jurkevitch

8.   **Microbial Community Functional Structures in Wastewater Treatment Plants as Characterized by GeoChip** 205

Xiaohui Wang, Yu Xia, Xianghua Wen, Yunfeng Yang, and Jizhong Zhou

9.   **Key Design Factors Affecting Microbial Community Composition and Pathogenic Organism Removal in Horizontal Subsurface Flow Constructed Wetlands** 231

Jordi Morató, Francesc Codony, Olga Sánchez, Leonardo Martín Pérez, Joan García, and Jordi Mas

Part III: Treatments

10.   **An Application of Wastewater Treatment in a Cold Environment and Stable Lipase Production of Antarctic Basidiomycetous Yeast** *Mrakia blollopis* 259

Masaharu Tsuji, Yuji Yokota, Kodai Shimohara, Sakae Kudoh, and Tamotsu Hoshino

11.   **Revealing the Factors Influencing a Fermentative Biohydrogen Production Process Using Industrial Wastewater as Fermentation Substrate** 277

Iulian Zoltan Boboescu, Mariana Ilie, Vasile Daniel Gherman, Ion Mirel, Bernadett Pap, Adina Negrea, Éva Kondorosi, Tibor Bíró, and Gergely Maróti

12.   **Microbial Communities from Different Types of Natural Wastewater Treatment Systems: Vertical and Horizontal Flow Constructed Wetlands and Biofilters** 311

Bárbara Adrados, Olga Sánchez, Carlos Alberto Arias, Eloy Bécares, Laura Garrido, Jordi Mas, Hans Brix, and Jordi Morató

Author Notes 329

Index 335

Acknowledgment and How to Cite

The editor and publisher thank each of the authors who contributed to this book. The chapters in this book were previously published in various places in various formats. To cite the work contained in this book and to view the individual permissions, please refer to the citation at the beginning of each chapter. Each chapter was read individually and carefully selected by the editor; the result is a book that provides a nuanced look at the bioremediation of wastewater. The chapters included are broken into three sections, which describe the following topics:

- In Chapter 1, while investigating the bacterial diversity and abundance in a Hong Kong wastewater treatment plant for sewage, Ye and Zhang analyzed the influent, activated sludge, digestion sludge, and effluent. They found that the bacterial diversities in the four samples varied significantly, but what is significant is that they introduce a method for differentiating between "gene percantage" and "cell percentage" when analyzing bacterial genera.
- In Chapter 2, Guo and colleagues offer helpful methodology for more precisely assessing microbial ecology in activated sludge and other complex environmental samples.
- The results of Chapter 3, performed by Fredriksson and colleagues, demonstrate that experimental comparisons of universal 16S rRNA primers can reveal differences not detected by theoretical comparisons. As with both the Ye and Guo articles, these discussions contribute to a more thorough and detailed understanding of how we can analyze and measure microbial communities in wastewater.
- The work of Wang and colleagues in Chapter 4 helpfully expands our knowledge about the microbial nitrification and denitrification processes and their underlying biological mechanisms for industrial wastewater treatment.
- Sánchez and colleagues were the first to analyze the activated sludge of a seawater-processing plant in Chapter 5. We concluded that the bacterial diversity was as high as has been observed in conventional wastewater treatment plants. However, the composition of the bacterial community differed greatly from other plants. Our results suggest that only a few populations of specialized bacteria (likely with high transcription activity) are responsible for removal of ammonia in seawater systems.

- The research conducted by Weissbrodt and colleagues in Chapter 6 investigated the dynamics of bacterial communities and the structures of bioaggregates during transitions from activated sludge flocs to early-stage nuclei and to mature granular biofilms. Their findings are helpful regarding the mechanisms of bacterial selection, granule formation, and biofilm maturation in relation to the evolution of process variables.
- In Chapter 7, Kandel and colleagues. offer a useful study that assesses the impact of predator bacteria on microbial processes.
- In Chapter 8, Wang and colleagues, while finding distinct similarities between microbial functional communities among activated sludge samples from four wastewater treatment plants, also reach helpful conclusions regarding correlations between microbial functional potentials and water temperature, dissolved oxygen, ammonia concentrations, and the loading rate of chemical oxygen demand.
- The investigation in Chapter 9 by Morató and colleagues of wastewater treatment plant design factors—such as granular media, water depth, and season effect—clarify our understanding of variables that affect the removal of microbial indicators in constructed wetlands.
- In Chapter 10, Tsuji and colleagues address concerns relating to dairy wastewater, reporting the effects of both lipase and a particular yeast strain on milk fat.
- In Chapter 11, Boboescu and colleagues implement a useful experimental design to investigate the process variables involved in batch-mode biohydrogen production, and offer implications that point to a greater optimization of this methodology on a larger scale.
- The comparison in Chapter 12 by Adrados and colleagues regarding the composition of the microbial communities of three different types of domestic wastewater treatment systems enlarges microbial analysis by analyzing both the bacterial and archaeal populations. They focused on the possible influence of the water influent composition, the design, and the bed filling of the treatment systems. They found no relation between the influent and effluent bacterial communities inside the same treatment system.

List of Contributors

Bárbara Adrados
Health and Environmental Microbiology Laboratory, Optics and Optometry Department & AQUA-SOST – UNESCO Chair on Sustainability, Universitat Politècnica de Catalunya, Edifici Gaia, Pg. Ernest Lluch/Rambla Sant Nebridi, Terrassa 08222, Spain

Carlos Alberto Arias
Aarhus University, Department of Bioscience, Ole Worms Allé 1, Building 1135, 8000 Århus C, Denmark

Eloy Bécares
Facultad de Ciencias Biológicas y Ambientales, Universidad de León, Campus de Vegazana s/n, 24071 León, Spain

Tibor Bíró
Szent István University, Faculty of Economics, Agricultural and Health Studies, Szarvas, Hungary

Iulian Zoltan Boboescu
Polytechnic University of Timisoara, Timisoara, Romania and Seqomics Biotechnology Ltd, Szeged, Hungary

Hans Brix
Aarhus University, Department of Bioscience, Ole Worms Allé 1, Building 1135, 8000 Århus C, Denmark

Lin Cai
Environmental Biotechnology Laboratory, The University of Hong Kong, Hong Kong SAR, China

Francesc Codony
Laboratory of Health and Environmental Microbiology, Department of Optics and Optometry, Aqua-Sost-UNESCO Chair on Sustainability, Technical University of Catalonia, c/Violinista Vellsolà, 37, 08222 Terrassa, Barcelona, Spain

Laura Garrido
Departament de Genètica i Microbiologia, Universitat Autònoma de Barcelona, 08193 Bellaterra, Spain

Vasile Daniel Gherman
Polytechnic University of Timisoara, Timisoara, Romania

Feng Guo
Environmental Biotechnology Laboratory, The University of Hong Kong, Hong Kong SAR, China

Isabel Ferrera
Departament de Biologia Marina i Oceanografia, Institut de Ciències del Mar, ICM-CSIC, Barcelona, Spain

Nils Johan Fredriksson
Department of Civil and Environmental Engineering, Water Environment Technology, Chalmers University of Technology, Gothenburg, Sweden

Joan García
GEMMA-Group of Environmental Engineering and Microbiology, Department of Hydraulic, Maritime and Environmental Engineering, Universitat Politècnica de Catalunya-BarcelonaTech, c/Jordi Girona, 1–3, Building D1, E-08034 Barcelona, Spain

Jose M. González
Department of Microbiology, University of La Laguna, ES-38206, La Laguna, Tenerife, Spain

Malte Hermansson
Department of Chemistry and Molecular Biology, University of Gothenburg, Gothenburg, Sweden

Christof Holliger
Laboratory for Environmental Biotechnology, School for Architecture, Civil and Environmental Engineering, Ecole Polytechnique Fédérale de Lausanne, Lausanne, Switzerland

Tamotsu Hoshino
Biomass Refinery Research Center (BRRC), National Institute of Advanced Industrial Science and Technology (AIST), Kagamiyama, Higashihiroshima, Hiroshima, Japan and Graduate School of Life Science, Hokkaido University, Kita-ku, Sapporo, Hokkaido, Japan

Mariana Ilie
Polytechnic University of Timisoara, Timisoara, Romania

Feng Ju
Environmental Biotechnology Laboratory, The University of Hong Kong, Hong Kong SAR, China

Edouard Jurkevitch
Department of Plant Pathology and Microbiology, Faculty of Agriculture, Food and Environment, The Hebrew University of Jerusalem, Rehovot, Israel

Prem P. Kandel
Department of Plant Pathology and Microbiology, Faculty of Agriculture, Food and Environment, The Hebrew University of Jerusalem, Rehovot, Israel

Éva Kondorosi
Hungarian Academy of Sciences, Biological Research Centre, Temesvari krt. 62., Szeged 6726, Hungary

Sakae Kudoh
National Institute of Polar Research (NIPR), Midori-cho, Tachikawa, Tokyo, Japan

Ute Kuhlicke
Microbiology of Interfaces, Department of River Ecology, Helmholtz Centre for Environmental Research - UFZ, Magdeburg, Germany

Aimin Li
State Key Laboratory of Pollution Control and Resource Reuse, School of the Environment, Nanjing University, Nanjing, China

Yan Li
State Key Laboratory of Pollution Control and Resource Reuse, School of the Environment, Nanjing University, Nanjing, China

Bo Liu
State Key Laboratory of Pollution Control and Resource Reuse, School of the Environment, Nanjing University, Nanjing, China and Research Institute of Nanjing University in Lianyungang, Lianyungang, China

Chao Long
State Key Laboratory of Pollution Control and Resource Reuse, School of the Environment, Nanjing University, Nanjing, China

Xin Lu
State Key Laboratory of Pollution Control and Resource Reuse, School of the Environment, Nanjing University, Nanjing, China

Leonardo Martín Pérez
Laboratory of Health and Environmental Microbiology, Department of Optics and Optometry, Aqua-Sost-UNESCO Chair on Sustainability, Technical University of Catalonia, c/Violinista Vellsolà, 37, 08222 Terrassa, Barcelona, Spain and Rosario Chemical Institute (IQUIR-CONICET, UNR), Faculty of Biochemical and Pharmacological Sciences, National University of Rosario, Suipacha 531, 2000 Rosario, Santa Fe, Argentine

Gergely Maróti
Polytechnic University of Timisoara, Timisoara, Romania and Hungarian Academy of Sciences, Biological Research Centre, Temesvari krt. 62., Szeged 6726, Hungary

Jordi Mas
Departament de Genètica i Microbiologia, Universitat Autònoma de Barcelona, 08193, Bellaterra, Spain

Ion Mirel
Polytechnic University of Timisoara, Timisoara, Romania

Jordi Morató
Laboratory of Health and Environmental Microbiology, Department of Optics and Optometry, Aqua-Sost-UNESCO Chair on Sustainability, Technical University of Catalonia, c/Violinista Vellsolà, 37, 08222 Terrassa, Barcelona, Spain

Ortal Nahum
Department of Plant Pathology and Microbiology, Faculty of Agriculture, Food and Environment, The Hebrew University of Jerusalem, Rehovot, Israel

Adina Negrea
Polytechnic University of Timisoara, Timisoara, Romania

Thomas R. Neu
Microbiology of Interfaces, Department of River Ecology, Helmholtz Centre for Environmental Research - UFZ, Magdeburg, Germany

Bernadett Pap
Seqomics Biotechnology Ltd, Szeged, Hungary

Zohar Pasternak
Department of Plant Pathology and Microbiology, Faculty of Agriculture, Food and Environment, The Hebrew University of Jerusalem, Rehovot, Israel

Yoan Rappaz
Laboratory for Environmental Biotechnology, School for Architecture, Civil and Environmental Engineering, Ecole Polytechnique Fédérale de Lausanne, Lausanne, Switzerland

Olga Sánchez
Departament de Genètica i Microbiologia, Universitat Autònoma de Barcelona, 08193, Bellaterra, Spain

Kodai Shimohara
Hokkaido High-Technology College (HHT), Megunino-kita, Eniwa, Hokkaido, Japan

Masaharu Tsuji
Biomass Refinery Research Center (BRRC), National Institute of Advanced Industrial Science and Technology (AIST), Kagamiyama, Higashihiroshima, Hiroshima, Japan

Jaap van Rijn
Department of Animal Sciences, Faculty of Agriculture, Food and Environment, The Hebrew University of Jerusalem, Rehovot, Israel

Xiaohui Wang
Environmental Simulation and Pollution Control State Key Joint Laboratory, School of Environment, Tsinghua University, Beijing, China and Department of Environmental Science and Engineering, Beijing University of Chemical Technology, Beijing, China

Zhu Wang
State Key Laboratory of Pollution Control and Resource Reuse, School of the Environment, Nanjing University, Nanjing, China

David G. Weissbrodt
Laboratory for Environmental Biotechnology, School for Architecture, Civil and Environmental Engineering, Ecole Polytechnique Fédérale de Lausanne, Lausanne, Switzerland, Institute of Environmental Engineering, ETH Zurich, Zurich, Switzerland, and Department of Process Engineering, Eawag, Duebendorf, Switzerland

Xianghua Wen
Environmental Simulation and Pollution Control State Key Joint Laboratory, School of Environment, Tsinghua University, Beijing, China

Britt-Marie Wilén
Department of Civil and Environmental Engineering, Water Environment Technology, Chalmers University of Technology, Gothenburg, Sweden

Yu Xia
Environmental Simulation and Pollution Control State Key Joint Laboratory, School of Environment, Tsinghua University, Beijing, China

Yunfeng Yang
Environmental Simulation and Pollution Control State Key Joint Laboratory, School of Environment, Tsinghua University, Beijing, China

Lin Ye
Environmental Biotechnology Lab, Department of Civil Engineering, The University of Hong Kong, Pokfulam Road, Hong Kong SAR, China

Yuji Yokota
Bio-Production Research Institute, National Institute of Advanced industrial Science and Technology (AIST), Tsukisamu-higashi, Toyohira-ku, Sapporo, Hokkaido, Japan

Tong Zhang
Environmental Biotechnology Lab, Department of Civil Engineering, The University of Hong Kong, Pokfulam Road, Hong Kong SAR, China

Xu-Xiang Zhang
State Key Laboratory of Pollution Control and Resource Reuse, School of the Environment, Nanjing
University, Nanjing, China

Jizhong Zhou
Environmental Simulation and Pollution Control State Key Joint Laboratory, School of Environment,
Tsinghua University, Beijing, China, Institute for Environmental Genomics and Department of Micro-
biology and Plant Biology, University of Oklahoma, Norman, Oklahoma, United States of America,
and Earth Sciences Division, Lawrence Berkeley National Laboratory, Berkeley, California, United
States of America

Introduction

The quantity and quality of waste generated and discharged into natural water bodies is a topic of serious concern in our world today. Consequently, there is a need for different strategies to address wastewater treatment and subsequent reuse, especially in arid and semi-arid areas where water shortages are the rule. Biological treatment processes constitute crucial tools in the biodegradation of organic matter, transformation of toxic compounds into harmless products, and nutrient removal in wastewater microbiology.

Activated sludge in wastewater treatment plants (WWTPs), which contains a complex mixture of microbial populations dominated by bacteria, plays an essential role in the biological treatment in these facilities. Other natural wastewater treatment systems such as constructed wetlands, biological sand filters, and other decentralized solutions provide other valuable options to conventional methods due to their efficiency, low establishment costs, and reduced operation and management requirements. In all these systems, microbial processes are essential, since many reactions are microbiologically mediated.

The composition of the microbial community has been extensively studied in the past decades. Different methods have been used in order to deepen our understanding of the structure and functionality of the microorganisms involved. By applying culture-dependent techniques, many species have been isolated from systems, such as activated sludge, although most of the microorganisms cannot be obtained by these classical methods. However, molecular tecniques such as gene libraries, fingerprinting methodologies, fluorescence in situ hybridization, and the recently incorporated next-generation sequencing technologies have greatly expanded our knowledge on wastewater biodiversity, showing that a significant hidden diversity of unknown and uncultured microorganisms have the potential to act as degraders of environmental pollutants. Therefore, further attempts to isolate the key microorganisms involved will be essential in order to explore the degradation capacity of the microbial communities developed in biological treatment processes. The articles in this compendium were chosen to represent an overview of the most current research into these facets of wastewater bioremediation. Ongoing research is essential to the future of human health, as well as the over all health of our planet.

Olga Sánchez, PhD

In Chapter 1, Ye and Zhang successfully demonstrated that 454 pyrosequencing was a powerful approach for investigating the bacterial communities in the activated sludge, digestion sludge, influent, and effluent samples of a full scale wastewater treatment plant treating saline sewage. For each sample, 18,808 effective sequences were selected and utilized to do the bacterial diversity and abundance analysis. In total, 2,455, 794, 1,667, and 1,932 operational taxonomic units were obtained at 3 % distance cutoff in the activated sludge, digestion sludge, influent, and effluent samples, respectively. The corresponding most dominant classes in the four samples were *Alphaproteobacteria, Thermotogae, Deltaproteobacteria,* and *Gammaproteobacteria.* About 67 % sequences in the digestion sludge sample were found to be affiliated with the *Thermotogales* order. Also, these sequences were assigned into a recently proposed genus *Kosmotoga* by the Ribosomal Database Project classifier. In the effluent sample, the authors found high abundance of *Mycobacterium* and *Vibrio,* which are genera containing pathogenic bacteria. Moreover, in this study, they authors proposed a method to differentiate the "gene percentage" and "cell percentage" by using Ribosomal RNA Operon Copy Number Database.

High throughput sequencing of 16S rRNA gene leads us into a deeper understanding on bacterial diversity for complex environmental samples, but introduces blurring due to the relatively low taxonomic capability of short read. For wastewater treatment plants, only those functional bacterial genera categorized as nutrient remediators, bulk/foaming species, and potential pathogens are significant to biological wastewater treatment and environmental impacts. Precise taxonomic assignment of these bacteria at least at genus level is important for microbial ecological research and routine wastewater treatment monitoring. Therefore, the focus of Chapter 2, by Guo and colleagues, was to evaluate the taxonomic precisions of different ribosomal RNA (rRNA) gene hypervariable regions generated from a mix activated sludge sample. In addition, three commonly used classification methods including RDP Classifier, BLAST-based best-hit annotation, and the lowest common ancestor annotation by MEGAN were evaluated by comparing their consistency. Under an unsupervised method, analysis of consistency among different classification methods suggests there are no hypervariable regions with good taxonomic coverage for all genera. Taxonomic assignment based on certain regions of the 16S rRNA genes, e.g. the V1&V2 regions—provide fairly consistent taxonomic assignment for a relatively wide range of genera. Hence, it is recommended to use these regions for studying functional groups in activated sludge. Moreover, the inconsistency among methods also demonstrated that a specific method might not be suitable for identification of some bacterial genera using certain 16S rRNA gene regions. As a general rule, drawing conclusions based only on one sequencing region and one classification method should be avoided due to the potential false negative results.

Assessments of bacterial community diversity and dynamics are fundamental for the understanding of microbial ecology as well as biotechnological applications. In Chapter 3, Fredriksson and colleagues show that the choice of PCR primers has great impact on the results of analyses of diversity and dynamics using gene libraries and DNA fingerprinting. Two universal primer pairs targeting the 16S rRNA gene, 27F&1492R and 63F&M1387R, were compared and evaluated by analyzing the bacterial community in the activated sludge of a large-scale wastewater treatment plant. The two primer pairs targeted distinct parts of the bacterial community, none encompassing the other, both with similar richness. Had only one primer pair been used, very different conclusions had been drawn regarding dominant phylogenetic and putative functional groups. With 27F&1492R, *Betaproteobacteria* would have been determined to be the dominating taxa while 63F&M1387R would have described *Alphaproteobacteria* as the most common taxa. Microscopy and fluorescence in situ hybridization analysis showed that both *Alphaproteobacteria* and *Betaproteobacteria* were abundant in the activated sludge, confirming that the two primer pairs target two different fractions of the bacterial community. Furthermore, terminal restriction fragment polymorphism analyses of a series of four activated sludge samples showed that the two primer pairs would have resulted in different conclusions about community stability and the factors contributing to changes in community composition. In conclusion, different PCR primer pairs, although considered universal, target different ranges of bacteria and will thus show the diversity and dynamics of different fractions of the bacterial community in the analyzed sample. The authors also show that while a database search can serve as an indicator of how universal a primer pair is, an experimental assessment is necessary to evaluate the suitability for a specific environmental sample.

Biological nitrification/denitrification is frequently used to remove nitrogen from tannery wastewater containing high concentrations of ammonia. However, information is limited about the bacterial nitrifiers and denitrifiers and their functional genes in tannery wastewater treatment plants (WWTPs) due to the low-throughput of the previously used methods. In Chapter 4, Wang and colleagues used 454 pyrosequencing and Illumina high-throughput sequencing, combined with molecular methods, to comprehensively characterize structures and functions of nitrification and denitrification bacterial communities in aerobic and anaerobic sludge of two full-scale tannery WWTPs. Pyrosequencing of 16S rRNA genes showed that *Proteobacteria* and *Synergistetes* dominated in the aerobic and anaerobic sludge, respectively. Ammonia-oxidizing bacteria (AOB) *amoA* gene cloning revealed that *Nitrosomonas europaea* dominated the ammonia-oxidizing community in the WWTPs. Metagenomic analysis showed that the denitrifiers mainly included the genera of *Thauera, Paracoccus, Hyphomicrobium, Comamonas* and *Azoarcus*, which may greatly contribute to the nitrogen removal in the two WWTPs. It is interesting that AOB and ammonia-oxidizing archaea had low abundance although both WWTPs demonstrated high ammonium removal efficiency. Good correlation between the qPCR and metagenomic analysis is observed for the quantification of functional genes *amoA, nirK, nirS* and *nosZ*, indicating that the metagenomic approach may be a promising method used to comprehensively investigate the abundance of functional genes of nitrifiers and denitrifiers in the environment.

In Chapter 5, Sánchez and colleagues investigated the bacterial community composition of activated sludge from a wastewater treatment plant (Almería, Spain) with the particularity of using seawater by applying 454-pyrosequencing. The results showed that *Deinococcus-Thermus, Proteobacteria, Chloroflexi* and *Bacteroidetes* were the most abundant retrieved sequences, while other groups, such as *Actinobacteria, Chlorobi, Deferribacteres, Firmicutes, Planctomycetes, Spirochaetes* and *Verrumicrobia* were reported at lower proportions. Rarefaction analysis showed that very likely the diversity is higher than what could be described despite most of the unknown microorganisms probably correspond to rare diversity. Furthermore, the majority of taxa could not be classified at the genus level and likely represent novel members of these groups. Additionally, the nitrifiers in the sludge were characterized by pyrosequencing the *amoA* gene. In contrast, the nitrifying bacterial community, dominated by the genera *Nitrosomonas*, showed a low diversity and rarefaction curves exhibited saturation. These results suggest that only a few populations of low abundant but specialized bacteria are responsible for removal of ammonia in these saline wastewater systems.

Aerobic granular sludge (AGS) is based on self-granulated flocs forming mobile biofilms with a gel-like consistence. In Chapter 6, by Weissbrodt and colleagues, bacterial and structural dynamics from flocs to granules were followed in anaerobic-aerobic sequencing batch reactors (SBR) fed with synthetic wastewater, namely a bubble column (BC-SBR) operated under wash-out conditions for fast granulation, and two stirred-tank enrichments of *Accumulibacter* (PAO-SBR) and *Competibacter* (GAO-SBR) operated at steady-state. In the BC-SBR, granules formed within 2 weeks by swelling of *Zoogloea* colonies around flocs, developing subsequently smooth zoogloeal biofilms. However, *Zoogloea* predominance (37–79%) led to deteriorated nutrient removal during the first months of reactor operation. Upon maturation, improved nitrification (80–100%), nitrogen removal (43–83%), and high but unstable dephosphatation (75–100%) were obtained. Proliferation of dense clusters of nitrifiers, *Accumulibacter*, and *Competibacter* from granule cores outwards resulted in heterogeneous bioaggregates, inside which only low abundance *Zoogloea* (<5%) were detected in biofilm interstices. The presence of different extracellular glycoconjugates detected by fluorescence lectin-binding analysis showed the complex nature of the intracellular matrix of these granules. In the PAO-SBR, granulation occurred within two months with abundant and active *Accumulibacter* populations (56 ± 10%) that were selected under full anaerobic uptake of volatile fatty acids and that aggregated as dense clusters within heterogeneous granules. Flocs self-granulated in the GAO-SBR after 480 days during a period of over-aeration caused by biofilm growth on the oxygen sensor. Granules were dominated by heterogeneous clusters of *Competibacter* (37 ± 11%). *Zoogloea* were never abundant in biomass of both PAO- and GAO-SBRs. This study showed that *Zoogloea, Accumulibacter*, and *Competibacter* affiliates can form granules, and that the granulation mechanisms rely on the dominant population involved.

Standard aquaculture generates large-scale pollution and strains water resources. In aquaculture using zero discharge systems (ZDS), highly efficient fish growth and water recycling are combined. The wastewater stream is directed through compartments in which beneficial microbial activities induced by creating suitable environmental conditions remove biological and chemical pollutants, alleviating both problems. Bacterial predators, preying on bacterial populations in the ZDS, may affect their diversity, composition and functional redundancy, yet in-depth understanding of this phenomenon is lacking. In Chapter 7, Kandel and colleagues analyzed the dynamics of populations belonging to the obligate predators *Bdellovibrio* and like organisms (BALOs) in freshwater and saline ZDS over a 7-month period using QPCR targeting the *Bdellovibrionaceae*, and the *Bacteriovorax* and *Bacteriolyticum* genera in the *Bacteriovoracaeae*. Both families co-existed in ZDS compartments, constituting 0.13–1.4% of total Bacteria. Relative predator abundance varied according to the environmental conditions prevailing in different compartments, most notably salinity. Strikingly, the *Bdellovibrionaceae*, hitherto only retrieved from freshwater and soil, also populated the saline system. In addition to the detected BALOs, other potential predators were highly abundant, especially from the *Myxococcales*. Among the general bacterial population, *Flavobacteria, Bacteroidetes, Fusobacteriaceae* and unclassified Bacteria dominated a well mixed but seasonally fluctuating diverse community of up to 238 operational taxonomic units, as revealed by 16S rRNA gene sequencing.

Biological WWTPs must be functionally stable to continuously and steadily remove contaminants which rely upon the activity of complex microbial communities. However, knowledge is still lacking in regard to microbial community functional structures and their linkages to environmental variables. Therefore, in Chapter 8, Wang and colleagues aimed to investigate microbial community functional structures of activated sludge in wastewater treatment plants (WWTPs) and to understand the effects of environmental factors on their structure. Twelve activated sludge samples were collected from four WWTPs in Beijing. A comprehensive functional gene array named GeoChip 4.2 was used to determine the microbial functional genes involved in a variety of biogeochemical processes such as carbon, nitrogen, phosphorous and sulfur cycles, metal resistance, antibiotic resistance and organic contaminant degradation. High similarities of the microbial community functional structures were found among activated sludge samples from the four WWTPs, as shown by both diversity indices and the overlapped genes. For individual gene category, such as *egl, amyA, lip, nirS, nirK, nosZ, ureC, ppx, ppk, aprA, dsrA, sox* and *benAB*, there were a number of microorganisms shared by all 12 samples. Canonical correspondence analysis (CCA) showed that the microbial functional patterns were highly correlated with water temperature, dissolved oxygen (DO), ammonia concentrations and loading rate of chemical oxygen demand (COD). Based on the variance partitioning analyses (VPA), a total of 53% of microbial community variation from GeoChip data can be explained by wastewater characteristics (25%) and operational parameters (23%), respectively. This study provided an overall picture of microbial community functional structures of activated sludge in WWTPs and discerned the linkages between microbial communities and environmental variables in WWTPs.

Constructed wetlands constitute an interesting option for wastewater reuse since high concentrations of contaminants and pathogenic microorganisms can be removed with these natural treatment systems. In Chapter 9, by Morató and colleagues, the role of key design factors which could affect microbial removal and wetland performance, such as granular media, water depth and season effect was evaluated in a pilot system consisting of eight parallel horizontal subsurface flow (HSSF) constructed wetlands treating urban wastewater from Les Franqueses del Vallès (Barcelona,

Spain). Gravel biofilm as well as influent and effluent water samples of these systems were taken in order to detect the presence of bacterial indicators such as total coliforms (TC), *Escherichia coli*, fecal enterococci (FE), *Clostridium perfringens*, and other microbial groups such as *Pseudomonas* and *Aeromonas*. The overall microbial inactivation ratio ranged between 1.4 and 2.9 log-units for heterotrophic plate counts (HPC), from 1.2 to 2.2 log units for total coliforms (TC) and from 1.4 to 2.3 log units for *E. coli*. The presence of fine granulometry strongly influenced the removal of all the bacterial groups analyzed. This effect was significant for TC ($p = 0.009$), *E. coli* ($p = 0.004$), and FE ($p = 0.012$). Shallow HSSF constructed wetlands were more effective for removing *Clostridium* spores ($p = 0.039$), and were also more efficient for removing TC ($p = 0.011$) and *E. coli* ($p = 0.013$) when fine granulometry was used. On the other hand, changes in the total bacterial community from gravel biofilm were examined by using denaturing gradient gel electrophoresis (DGGE) and sequencing of polymerase chain reaction (PCR)-amplified fragments of the 16S rRNA gene recovered from DGGE bands. Cluster analysis of the DGGE banding pattern from the different wetlands showed that microbial assemblages separated according to water depth, and sequences of different phylogenetic groups, such as *Alpha, Beta* and *Delta-Proteobacteria, Nitrospirae, Bacteroidetes, Acidobacteria, Firmicutes, Synergistetes* and *Deferribacteres* could be retrieved from DGGE bands.

Milk fat curdle in sewage is one of the refractory materials for active sludge treatment under low temperature conditions. For the purpose of solving this problem by using a bio-remediation agent, in Chapter 10, Tsuji and colleagues screened Antarctic yeasts and isolated SK-4 strain from algal mat of sediments of Naga-ike, a lake in Skarvsnes, East Antarctica. The yeast strain showed high nucleotide sequence homologies (>99.6%) to *Mrakia blollopis* CBS8921T in ITS and D1/D2 sequences and had two unique characteristics when applied on an active sludge; i.e., it showed a potential to use various carbon sources and to grow under vitamin-free conditions. Indeed, it showed a biochemical oxygen demand (BOD) removal rate that was 1.25-fold higher than that of the control. The authors considered that the improved BOD removal rate by applying SK-4 strain was based on its lipase activity and characteristics. Finally, the authors purified the lipase from SK-4 and found that the enzyme was quite stable under wide ranges of temperatures and pH, even in the presence of various metal ions and organic solvents. SK-4, therefore, is a promising bio-remediation agent for cleaning up unwanted milk fat curdles from dairy milk wastewater under low temperature conditions.

Biohydrogen production through dark fermentation using organic waste as a substrate has gained increasing attention in recent years, mostly because of the economic advantages of coupling renewable, clean energy production with biological waste treatment. An ideal approach is the use of selected microbial inocula that are able to degrade complex organic substrates with simultaneous biohydrogen generation. Unfortunately, even with a specifically designed starting inoculum, there is still a number of parameters, mostly with regard to the fermentation conditions, that need to be improved in order to achieve a viable, large-scale, and technologically feasible solution. In Chapter 11, Boboescu and colleagues applied statistics-based factorial experimental design methods to investigate the impact of various biological, physical, and chemical parameters, as well as the interactions between them on the biohydrogen production rates. By developing and applying a central composite experimental design strategy, the effects of the independent variables on biohydrogen production were determined. The initial pH value was shown to have the largest effect on the biohydrogen production process. High-throughput sequencing-based metagenomic assessments of microbial communities revealed a clear shift towards a *Clostridium* sp.-dominated environment, as the responses of the variables investigated were maximized towards the highest H_2-producing potential. Mass spectrometry analysis suggested that the microbial consortium largely followed hydrogen-generating metabolic pathways, with the simultaneous degradation of complex organic compounds, and thus also performed a biological treatment of the beer brewing industry wastewater used as a fermentation substrate. Therefore, the authors have developed a complex optimization strategy for batch-mode biohydrogen production using a defined microbial consortium as the starting inoculum and beer brewery wastewater as the fermentation substrate. These results have the potential to bring us closer to an optimized, industrial-scale system which will serve the dual purpose of wastewater pre-treatment and concomitant biohydrogen production.

In Chapter 12, Adrados and colleagues analazed the prokaryotic microbial communities (*Bacteria* and *Archaea*) of three different systems operating in Denmark for the treatment of domestic wastewater (horizontal flow constructed wetlands (HFCW), vertical flow constructed wetlands (VFCW) and biofilters (BF)) using endpoint PCR followed by Denaturing Gradient Gel Electrophoresis (DGGE). Further sequencing of the most representative bacterial bands revealed that diverse and distinct bacterial communities were found in each system unit, being γ-*Proteobacteria* and *Bacteroidetes* present mainly in all of them, while *Firmicutes* was observed in HFCW and BF. Members of the *Actinobacteria* group, although found in HFCW and VFCW, seemed to be more abundant in BF units. Finally, some representatives of α, β and δ-*Proteobacteria, Acidobacteria* and *Chloroflexi* were also retrieved from some samples. On the other hand, a lower archaeal diversity was found in comparison with the bacterial population. Cluster analysis of the DGGE bacterial band patterns showed that community structure was related to the design of the treatment system and the organic matter load, while no clear relation was established between the microbial assemblage and the wastewater influent.

PART I

MICROBIAL COMMUNITIES FOR WASTEWATER TREATMENT

Bacterial Communities in Different Sections of a Municipal Wastewater Treatment Plant Revealed by 16S rDNA 454 Pyrosequencing

LIN YE AND TONG ZHANG

1.1 INTRODUCTION

Biological treatment processes are the most widely used approach for treating municipal and industrial wastewater in wastewater treatment plants (WWTPs) due to their high efficiency for various organic/nutrient matters removal and low operational cost. The microbial community, which is dominated by bacteria (Wagner et al. 2002), plays an essential role in the biological treatment reactors and has been studied for several decades by both isolation (Neilson 1978) and molecular methods, such as polymerase chain reaction (PCR)-denaturing gradient gel electrophoresis (Muyzer et al. 1993; Ye and Zhang 2010), terminal restriction fragment length polymorphism (Liu et al. 1997), cloning (Schuppler et al. 1995), and fluorescent in situ hybridization (Erhart et al. 1997). The culturing

Bacterial Communities in Different Sections of a Municipal Wastewater Treatment Plant Revealed by 16S rDNA 454 Pyrosequencing. © *Ye L and Zhang T.* Applied Microbiology and Biotechnology **97**,6 (2013), doi: 10.1007/s00253-012-4082-4. *Licensed under a Creative Commons Attribution License, http://creativecommons.org/licenses/by/3.0.*

methods have been a very direct and effective way to characterize the microbial community. However, most of the bacteria in the natural environment cannot be cultured in an artificial medium in the laboratory (Giovannoni et al. 1990; Hugenholtz et al. 1998) and the diversity of the uncultured bacteria is quite considerable (Whitman et al. 1998). The molecular methods greatly promoted our understanding of the microbial community. But for complex environmental samples (such as soil and activated sludge) with overwhelming genetic diversities, these methods are still far away from revealing the panorama of the bacterial community and can only investigate the most abundant population in these samples (Claesson et al. 2009).

The next generation sequencing technologies originated several years ago made the high throughput sequencing easy to be implemented with low cost (Glenn 2011). 454 Pyrosequencing is one of the popular high throughput sequencing systems, which can generate more than 400,000 effective reads with average read length up to several hundred base pairs and average quality of greater than 99.5 % accuracy rate (Droege and Hill 2008; Glenn 2011). By incorporating barcode sequences on primers, a certain number of DNA samples can be sequenced at the same time in one run. This technology has been successfully used in investigating microbial diversity and abundance in various samples, such as marine water (Qian et al. 2010), soil (Roesch et al. 2007), and human distal intestine (Claesson et al. 2009). There are also several studies applying pyrosequencing in exploring the microbiota in WWTP (Kim et al. 2011; McLellan et al. 2010; Ye et al. 2011); however, these results are very limited and preliminary.

In this study, to investigate the bacterial diversity and abundance in a full scale WWTP treating saline sewage in Hong Kong, we systematically analyzed 16S rRNA gene in the influent, activated sludge, digestion sludge, and effluent by using 454 high throughput pyrosequencing. It was found that the bacterial diversities in the four samples are quite different. Besides, in this study, we also introduced a method to consider and compare the difference of 16S rRNA gene percentage and bacteria cell number percentage in these samples. The results showed that some genera, especially the abundant genera, may be underestimated or overestimated when using 16S rRNA gene to reflect their abundances.

1.2 MATERIALS AND METHODS

1.2.1 WWTP DESCRIPTION AND SAMPLING

In this study, the activated sludge, digestion sludge, influent, and effluent samples were taken from Shatin WWTP, which is a full scale municipal wastewater treatment plant in Hong Kong. Because seawater has been used extensively in the toilet flushing system in Hong Kong, this WWTP treats saline sewage (salinity ~1.2 %) containing about 30 % seawater with activated sludge system. It treats about 216,000 m^3 wastewater per day with a COD concentration of 226 ~ 491 mg l^{-1}. The aeration tank was partitioned into two zones (anoxic zone and aerobic zone) for carbon and nitrogen removal but no special facilities for phosphate removal. The activated sludge sample analyzed in this study was taken from the aerobic zone. To reduce sludge volume, both the primary sludge and the surplus activated sludge are digested in mesophilic anaerobic digesters. Detailed information about the WWTP and the samples were summarized in Table S4 and Fig. S3. When taking samples, the activated sludge (mixed liquid containing both flocs and suspended bacteria in the aerobic zone of the aeration tank) and the digestion sludge were taken from the reactor and mixed thoroughly and then fixed on site by mixing with 100 % ethanol at a volume ratio of 1:1 and kept in an ice box for transportation and then stored in our laboratory at −20 °C before DNA extraction. Influent and effluent samples were kept in plastic containers and delivered to our laboratory within 3 h. Immediately after arriving at the laboratory, 12 ml influent was centrifuged at 4,000 rpm for 10 min at 4 °C and 400 ml effluent was filtrated using a 0.45-μm glass fiber filter to collect the bacteria cells. The collected residue was used for DNA extraction.

1.2.2 DNA EXTRACTION AND PCR

For all samples in this study, DNA was extracted using FastDNA® SPIN Kit for Soil (MP Biomedicals, Illkirch, France). Before pyrosequencing, the above DNA of each sample was amplified with a set of primers targeting the hypervariable V4 region of the 16S rRNA gene (RDP's Pyro-sequencing Pipeline: http://pyro.cme.msu.edu/pyro/help.jsp). The forward primer is 5'-AYTGGGYDTAAAGNG-3' and the reverse primers are the mixture of four primers, i.e., 5'-TACCRGGGTHTCTAATCC-3', 5'-TACCAGAGTATCTAATTC-3', 5'-CTACDSRGGTMTCTAATC-3', and 5'-TACNVGGGTATCTAATCC-3' (Claesson et al. 2009). Barcodes that allow sample multiplexing during pyrosequencing were incorporated between the 454 adaptor and the forward primers. The PCR amplification was conducted in a 100-μl reaction system using MightyAmp polymerase (TaKaRa, Dalian, China). The amplification was conducted in an i-Cycler (BioRad, Hercules, CA, USA) under the following conditions: 98 °C for 2 min, 28 cycles at 98 °C for 15 s, 56 °C for 20 s and 68 °C for 30 s, and a final extension at 68 °C for 10 min. The PCR products were purified by using PCRquick-spinTM PCR Product Purification Kit (iNtRON Biotechnology, South Korea) and then mixed equally before conducting pyrosequencing.

For clone library construction, the DNA of another digestion sludge sample taken from the sample anaerobic digester was amplified by PCR using universal bacterial 16S rRNA primer set EUB8F (5'-AGAGTTT-GATCMTGGCTCAG-3') (Heuer et al. 1997) and UNIV1392R (5'-AC-GGGCGGTGTGTRC-3') (Ferris et al. 1996). The vector used for ligation was pMD®18-T (TaKaRa). The purified plasmid was sequenced on an ABI 3730xl capillary sequencer (Applied Biosystems, Foster City, CA, USA).

1.2.3 HIGH THROUGHPUT PYROSEQUENCING

The PCR products of the V4 region of 16S rRNA gene were sequenced using the Roche 454 FLX Titanium sequencer (Roche, Nutley, NJ, USA). Samples in this study were individually barcoded to enable multiplex sequencing. The results are deposited into the NCBI short reads archive database (accession number: SRA026842.2).

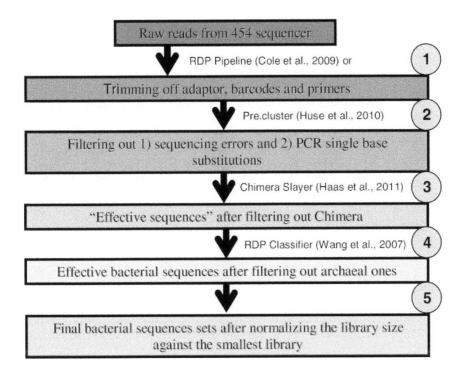

FIGURE 1: Sequences quality trimming flow chart

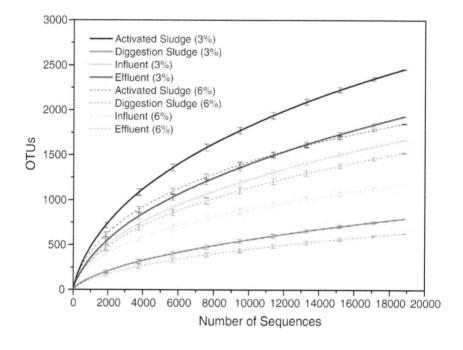

FIGURE 2: Rarefaction curves of the four samples at cutoff levels of 3 % (solid lines) and 6 % (dash lines) created by using RDP's pyrosequencing pipeline. The error bars show 95 % confidence of upper and lower limits

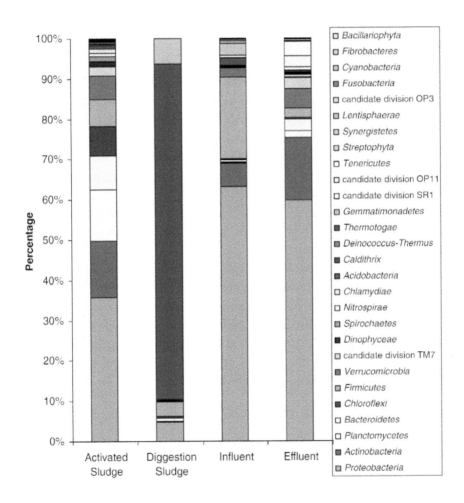

FIGURE 3: Relative abundances of different phyla in the four samples (the results were obtained by using BLASTN combined with MEGAN)

1.2.4 SEQUENCE PROCESSING
AND BACTERIAL POPULATION ANALYSIS

Following pyrosequencing, Python scripts were written to (1) remove sequences containing more than one ambiguous base ('N'), (2) check the completeness of the barcodes and the adaptor, and (3) remove sequences shorter than 150 bp.

454 Sequencing noises were removed by Pre.cluster (Huse et al. 2010) tool in Mothur package (Roeselers et al. 2011). Chimeras introduced in the PCR process were detected using ChimeraSlayer (Haas et al. 2011) in Mothur package. Because all sequences flagged as chimeras are not recommended to be discarded blindly (http://microbiomeutil.sourceforge. net/), so the reads flagged as chimeras were submitted to Ribosomal Database Project (RDP) classifier (Wang et al. 2007; Cole et al. 2009). Those being assigned to any known genus with 50 % confidence threshold were merged with the non-chimera reads to form the collection of "effective sequences" for each sample.

Although the primers used in this study are bacteria-specific primers, a few archaeal sequences might be obtained. The same situation was also observed in another study (Qian et al. 2010). To remove these archaeal sequences, the effective sequences of each sample were submitted to the RDP classifier again to identify the archaeal and bacterial sequences, and the archaeal sequences were filtered out using a self-written Python script. The average length of all effective bacterial sequences without the primers was 207 bp. The above quality trimming process was summarized in Fig. 1.

After that, the "RDP Align" tool in RDP's Pyrosequencing Pipeline was used to align the effective bacterial sequences of each sample. A cluster file was generated for each sample with "RDP Complete Linkage Clustering" tool. With the cluster file, the rarefaction curves were generated using the "RDP Rarefaction" tool.

All effective bacterial sequences obtained from pyrosequencing in this study were compared with Greengenes 16S rRNA gene database (DeSantis et al. 2006) annotated with NCBI taxonomy using NCBI's BLASTN tool (Altschul et al. 1990) and the default parameters except for the maximum hit number of 100 (Claesson et al. 2009). Then, the sequences were assigned to NCBI taxonomies with MEGAN (Huson et al. 2007) by using the lowest common ancestor algorithm and the default parameters except the BLAST bitscore, for which we found three values used in previous studies: 35 (Huson et al. 2007), 86 (Urich et al. 2008), and 250 (Claesson et al. 2009). In this study, the intermediate BLAST bitscore cutoff threshold 86 was applied.

After the sequences were assigned to NCBI taxonomies, the percentage of bacteria in each taxon could be calculated, which was defined as "gene percentage" in this study. From the Ribosomal RNA Operon Copy Number Database (rrnDB) (http://rrndb.mmg.msu.edu), we downloaded the data of rRNA operon copy numbers of each genus. According to these copy number data, the abovementioned "gene percentage" could be converted to "cell percentage". It was noted that a few genera found in this study have no rRNA operon copy numbers in rrnDB. For those genera, we used the average rRNA operon copy number values in the whole rrnDB. And if the species in a genus contains different rRNA operon copy numbers, we used the average value of all the species in this genus.

1.3 RESULTS

1.3.1 EFFECTIVENESS CHECK FOR THE RAW READS

In this study, we obtained 51,072, 47,072, 45,577, and 46,785 reads for the activated sludge, digestion sludge, influent, and effluent samples, respectively. After initial quality check mentioned in the "Materials and methods" section, the chimera and *Achaea* sequences were also checked and filtered. As shown in Table 1, 43–65 % of the raw reads met the quality and length criteria. By using ChimeraSlayer in Mothur package, 10–20 % of the reads were flagged as chimeras in the activated sludge and influent and effluent samples, while only 2.6 % reads in the digestion sludge sample were judged to be chimeras. Some of these so-called chimeras may represent naturally formed sequences that do not represent PCR artifacts and are not recommended to be blindly discarded (http://microbiomeutil.sourceforge.net/). So, in this study, all the chimeras picked by ChimeraSlayer were submitted to RDP classifier for further checking and it was found that about half (shown in Table 1) of the chimeras can be assigned to a known genus at 50 % confidence threshold. Hence, only those sequences that cannot be assigned into a genus were regarded as real chimeras and excluded in the downstream analysis. Although the primers used in this study were specific for bacteria, some sequences were assigned to *Archaea* by the RDP classifier. These sequences were filtered out in this step. At last, in order to do the comparison at the same sequencing depth, 18,808 effective bacterial sequences were extracted from each sample to do the downstream analysis.

1.3.2 BACTERIAL COMMUNITY COMPOSITION

In order to compare the bacterial species richness among these samples, operational taxonomic units (OTUs) were determined for each sample at distance levels of 3 and 6 % (Table 1). The OTU number in the activated sludge was the largest among the four samples, i.e., 2,455 and 1,852 at distance cutoff levels of 3 and 6 %, respectively. And the digestion sludge contained the least OTU amount. The bacterial phylotype richness levels can also be reflected using Shannon diversity index (Table 1) which also revealed that the activated sludge had the highest bacterial diversity among the four samples. The rarefaction curves of the four samples at distance cutoff levels of 3 and 6 %, as shown in Fig. 2, demonstrated clearly that the bacterial phylotype richness of the activated sludge was much higher than the other samples. The influent and effluent samples had moderate richness and the digestion sludge had the least richness. The rarefaction curves, especially those of the activated sludge, influent, and effluent samples, did not level off even at the sequencing depth of 18,808, suggesting that this sequencing depth was still not enough to cover the whole bacterial diversity and thus further sequencing would be valuable to detect more species.

The effective bacterial sequences in the four samples were all assigned to corresponding taxonomies by using BLAST combined with MEGAN. The relative abundances of different phyla in the four samples were shown in Fig. 3. From the phylum assignment result, it was found that the bacterial diversity in the digestion sludge was significantly different from the other three samples. Over 80 % of the sequences in the digestion sludge were assigned into *Thermotogae*, which is a phylum that may have much in common with species in *Archaea* (Cracraft and Donoghue 2004). In other samples, *Proteobacteria* were the dominant phylum, accounting for about 36, 63, and 60 % in the activated sludge, influent, and effluent samples, respectively. *Actinobacteria* were the secondary phylum in the activated sludge and effluent samples, corresponding to the percentages of about 14 and 15 %. However, in the influent sample, the secondary phylum was *Firmicutes* at a percentage of 20.26 %. The total phyla number in the activated sludge, digestion sludge, influent, and effluent samples were 23, 13, 19, and 23, respectively, suggesting that the bacterial diversity in the digestion sludge was lower than the other samples even at the phylum level. There were in total 12 phyla shared by the four samples, which were *Proteobacteria, Actinobacteria, Planctomycetes, Bacteroidetes, Chloroflexi, Firmicutes, Verrucomicrobia, candidate division TM7, Thermotogae,* and *Synergistetes*. The detailed comparison of these phyla in the samples was shown in Table S1.

Besides the phylum, bacterial diversity and abundance were also analyzed more specifically at other taxonomic units, i.e., class (Table S2, order (Table S3), and genus (Fig. 4). In any of the taxonomic units, the bacterial diversity of the digestion sludge was always found dramatically different from those of other samples. Only a few sequences assigned into the taxonomic units shared with the other samples. It is interesting to find that the most dominant classes in the activated sludge, digestion sludge, influent, and effluent are different, which were *Alphaproteobacteria, Thermotogae, Deltaproteobacteria,* and *Gammaproteobacteria*, respectively. At the order level, it was found that the top five dominant populations in the activated sludge samples were *Planctomycetales, Actinomycetales, Rhizobiales, Caldilineales,* and *Sphingobacteriales*, which were totally different from those dominant populations of the influent samples, i.e., *Desulfobacterales, Clostridiales, Desulfovibrionales, Lactobacillales,* and *Bifidobacteriales*. This probably indicated that some bacterial populations in the influent may not proliferate in the activated sludge. Figure 4 shows the comparison of the sequence assignment results on the genus level. It was found that at the genus level of MEGAN's cladogram, there were 11 nodes shared by the activated sludge and influent sample and 17 nodes shared by the activated sludge and the effluent samples. However, there were only five nodes shared by the activated sludge, influent, and effluent samples.

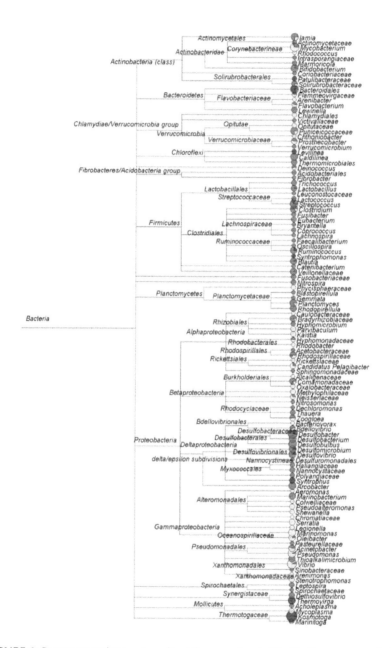

FIGURE 4: Sequences assignment results at the genus level. All effective sequences in the four samples were assigned into NCBI taxonomies by using BLASTN and MEGAN. Only nodes with over 30 sequences were shown in this figure. Pie charts indicate the relative sequence abundances of the corresponding nodes in the four samples.

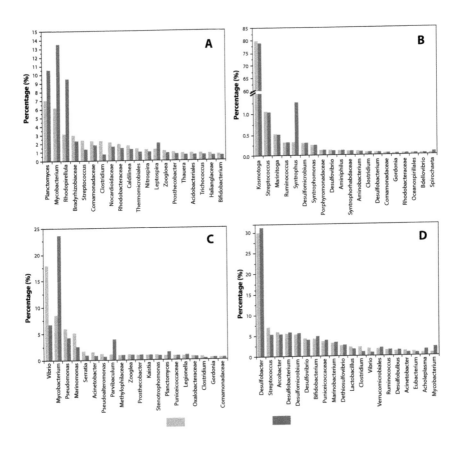

FIGURE 5: Top 20 genera in each sample showing the comparison of percentages of 16S rRNA gene number and cell number (a activated sludge, b digestion sludge, c effluent, d influent)

1.3.3 HIGH ABUNDANCE OF THERMOTOGALES IN THE DIGESTION SLUDGE

In this study, we observed extremely high abundance of *Thermotogales* that exists in the anaerobic digestion samples. According to the taxonomic analysis results of BLAST combined with MEGAN, there were 12,594 out of the totally 18,808 sequences assigned to the *Thermotogales* order, corresponding to a percentage of 67 % (Table S3). Using the RDP classifier, these high abundant sequences were further assigned into the Thermotogaceae family and *Kosmotoga* genus (Fig. S1) at a confidence threshold of 50 %. The high abundance of *Kosmotoga* in this digestion sludge was further confirmed by another independent sample from the same digester using the cloning library method with the universal bacterial 16S rRNA gene primers (EUB8F and UNIV1392R). Among 60 clones sequenced by the Sanger method, 24 of them were assigned to *Thermotogales* order and *Kosmotoga* genus by the RDP classifier (Fig. S2).

1.3.4 DIFFERENCES BETWEEN "GENE PERCENTAGE" AND "CELL PERCENTAGE"

In previous studies, 16S rRNA gene percentage was often used to reflect the bacterial abundance in the samples (Bibby et al. 2010; Park et al. 2008; Ye et al. 2011). However, the copy numbers of 16S rRNA gene in prokaryotic microorganisms are not equal and may vary from 1 to 15 (Klappenbach et al. 2001). So, the cell percentages may disagree with the 16S rRNA gene percentages. In this study, we converted the 16S rRNA gene numbers into cell number at the genus level referring to the rrnDB and linked the two percentages commonly used in environmental microbiology (Lee et al. 2009). Figure 5 shows the comparison of the percentage of 16S rRNA gene and cells of the four samples, i.e., activated sludge, digestion sludge, influent, and effluent. The result indicated that there were a few genera that could be significantly overestimated or underestimated. The 16S rRNA gene percentages of the three genera (*Planctomyces, Mycobacterium*, and *Rhodopirellula*) in the activated sludge sample, two genera (*Mycobacterium* and *Parvibaculum*) in the effluent sample, and one genus (*Syntrophus*) in the digestion sludge were lower than their corresponding cell percentage. Moreover, one genus (*Vibrio*) in the effluent sample had a higher 16S rRNA gene copy percentage than its cell percentage. In the influent sample, the percentage of 16S rRNA gene copy and the cell number matched with each other consistently for most of the genera. The largest difference observed in this study was the *Syntrophus* genus in the digestion sludge sample, which was up to 389 %.

1.4 DISCUSSION

In this study, we analyzed the bacterial diversity in a municipal WWTP. The results show that activated sludge harbored the highest diversity of bacteria. At a distance cutoff of 3 %, as shown in Table 1, 2,455 OTUs were obtained from 18,808 sequences, indicating that activated sludge is a highly species-rich ecosystem, which is more complicated than seawater (Qian et al. 2010) and comparable to soil (Roesch et al. 2007). The influent and effluent samples contained moderate amounts of bacterial populations. The OTU number of the influent in the present study was much smaller than a previous study conducted on a WWTP influent (McLellan et al. 2010), in which over 3,000 OTUs were identified with sequencing depths between 17,338 and 34,080 V6 region reads for each sample. The digestion sludge contained the least OTUs and most of the sequences (67 %) concentrated in the order *Thermotogales*.

Regarding the bacterial community in the activated sludge, so far, only very few studies have been conducted, especially with the activated sludge in the full scale municipal WWTPs. Snaidr et al. investigated the bacterial community structure of activated sludge of a large municipal WWTP by using cloning methods (Snaidr et al. 1997). In those days, due to the lack of a high throughput approach, their results were very preliminary. *Proteobacteria* was found to be the dominant phylum, which was consistent in this study. Their cloning result showed that *Betaproteobacteria* was the most abundant class in the phylum of *Proteobacteria*, followed by *Gammaproteobacteria*. However, the results in the present study suggested that *Alphaproteobacteria* and *Actinobacteria* were the top two classes in the activated sample (Table S2). Another recently published study analyzed the bacterial diversity of a full-scale integrated fixed-film activated sludge system using the high throughput pyrosequencing method (Kwon et al. 2010). Their results also showed that *Betaproteobacteria* and *Gammaproteobacteria* were the most abundant groups in the activated sludge. The inconsistence between the present study and other studies may be caused by the different conditions, especially the salinity of the wastewater, which was 1.2 % in the activated sludge system of the present study. In common domestic wastewater, as studied by Kwon et al. (2010), the salinity was usually lower than 0.05 %. The activated sludge in this study was a suspended system while in the study of Kwon et al (2010) the activated sludge was an integrated fixed-film system.

Consistent with the high abundance of *Thermotogales* bacteria in the digestion sludge observed in this study, other researchers also reported that there were about 53 % *Thermotogales* bacterium in an anaerobic digester based on cloning results (Sasaki et al. 2010). The *Thermotogales* order contains many different members, which are gram-negative and rod-shaped anaerobic thermopiles containing unique lipids (Huber et al. 1990). The novelty found in the present study was that nearly all of the *Thermotogales* bacteria were affiliated with *Kosmotoga* genus which was a recently proposed new genus (DiPippo et al. 2009) containing members isolated from oil production fluid (*Kosmotoga olearia*) (DiPippo et al. 2009) and from a shallow hydrothermal system occurring within a coral reef (*Kosmotoga arenicorallina*) (Nunoura et al. 2010). So far, this is the first time to report the extremely high abundance (~67 %). As a comparison, only about 0.05 % *Kosmotoga* species was found in another freshwater sludge digester operated under similar conditions. So, it could be very interesting to do further studies to investigate these *Kosmotoga* species in this seawater-containing digester and explore their roles in the sludge digestion.

The results of McLellan et al. (2010) on the bacterial diversity in a WWTP influent showed that *Actinobacteria*, *Bacteroidetes*, and *Firmicutes* were the most dominant three groups of bacteria in the influent, adding up to 37.5 % of the total bacteria. However, in the present study, as shown in Table S2, the top three classes of bacteria were *Deltaproteobacteria*, *Clostridia*, and *Gammaproteobacteria*. At the order level, the top three dominant groups were *Desulfobacterales*, *Clostridiales*, and

Desulfovibrionales, among which *Desulfobacterales* and *Desulfovibrio-nales* are the common sulfate-reducing bacteria. The high abundance of sulfate-reducing bacteria that existed in the influent was mainly because of seawater portion in the influent. During transportation of sewage in the pipeline, anaerobic condition led to the blooming of *Desulfobacterales* and *Desulfovibrionales*. Therefore, the dominant bacterial groups in this study were different from those of McLellan et al. (2010).

For the effluent, the present study is the first one which analyzed the whole bacterial diversity symmetrically and comprehensively using 454 pyrosequencing. An important phenomenon found in the present study was the high percentage of *Mycobacterium* and *Vibrio* genera (Fig. 4), which could be potentially pathogenic bacteria and harmful to humans (Bibby et al. 2010). Although it is difficult to tell whether these *Mycobacterium* and *Vibrio* bacteria in the effluent were pathogens since not all species in these genera are pathogenic bacteria (Ye and Zhang, 2011), it deserves further studies to explore the impacts of the bacteria in the effluent. Besides, a few bacterial diversity differences between the activated sludge sample and the effluent sample could be observed from Fig. 4. For example, *Caldilinea, Lewinella, Leptospira*, etc. were only detected in the activated sample. *Serratia, Shewanella, Marinomonas*, etc. were found in the effluent sample but not in the activated sample. This is an indication that the sedimentation process may change the abundance of bacteria. Because the bacteria in the activated sludge flocs were removed during the sedimentation process, some free-swimming bacteria populations still remained in the effluent.

To our best knowledge, this is the first study introducing a method to differentiate the "gene percentage" and "cell percentage" by refereeing to the rrnDB (Lee et al. 2009). This approach successfully demonstrated the underestimation or overestimation of some bacterial populations using 16S rRNA gene percentage only. Due to the incomplete data in the rrnDB, the average 16 S rRNA gene copy number was used to represent those data-not-available genera. Such practice may introduce some errors, but these errors could be corrected soon with the rapid development of rrnDB due to the more and more completion of whole genome projects of known bacterial species.

REFERENCES

1. Altschul S, Gish W, Miller W, Myers E, Lipman D. Basic local alignment search tool. J Mol Biol. 1990;215(3):403–410.
2. Bibby K, Viau E, Peccia J. Pyrosequencing of the 16S rRNA gene to reveal bacterial pathogen diversity in biosolids. Water Res. 2010;44(14):4252–4260. doi: 10.1016/j.watres.2010.05.039.
3. Claesson M, O'Sullivan O, Wang Q, Nikkila J, Marchesi J, Smidt H, De Vos W, Ross R, O'Toole P. Comparative analysis of pyrosequencing and a phylogenetic microarray for exploring microbial community structures in the human distal intestine. PLoS One. 2009;4(8):e6669. doi: 10.1371/journal.pone.0006669.
4. Cole J, Wang Q, Cardenas E, Fish J, Chai B, Farris R, Kulam-Syed-Mohideen A, McGarrell D, Marsh T, Garrity G. The ribosomal database project: improved alignments and new tools for rRNA analysis. Nucleic Acids Res. 2009;37(suppl 1):D141–D145. doi: 10.1093/nar/gkn879.
5. Cracraft J, Donoghue MJ. Assembling the tree of life. Oxford: Oxford University Press; 2004.
6. DeSantis TZ, Hugenholtz P, Larsen N, Rojas M, Brodie EL, Keller K, Huber T, Dalevi D, Hu P, Andersen GL. Greengenes, a chimera-checked 16S rRNA gene database and workbench compatible with ARB. Appl Environ Microbiol. 2006;72(7):5069–5072. doi: 10.1128/AEM.03006-05.
7. DiPippo JL, Nesbo CL, Dahle H, Doolittle WF, Birkland NK, Noll KM. *Kosmotoga olearia* gen. nov., sp. nov., a thermophilic, anaerobic heterotroph isolated from an oil production fluid. Int J Syst Evol Microbiol. 2009;59(12):2991–3000. doi: 10.1099/ijs.0.008045-0.
8. Droege M, Hill B. The Genome Sequencer FLX (TM) System—longer reads, more applications, straight forward bioinformatics and more complete data sets. J Biotechnol. 2008;136(1–2):3–10. doi: 10.1016/j.jbiotec.2008.03.021.
9. Erhart R, Bradford D, Seviour R, Amann R, Blackall L. Development and use of fluorescent in situ hybridization probes for the detection and identification of Microthrix parvicella in activated sludge. Syst Appl Microbiol. 1997;20(2):310–318. doi: 10.1016/S0723-2020(97)80078-1.
10. Ferris M, Muyzer G, Ward D. Denaturing gradient gel electrophoresis profiles of 16 S rRNA-defined populations inhabiting a hot spring microbial mat community. Appl Environ Microbiol. 1996;62(2):340–346.
11. Giovannoni SJ, Britschgi TB, Moyer CL, Field KG. Genetic diversity in Sargasso sea bacterioplankton. Nature. 1990;345:60–63. doi: 10.1038/345060a0.
12. Glenn TC. Field guide to next generation DNA sequencers. Mol Ecol Resour. 2011;15:759–769. doi: 10.1111/j.1755-0998.2011.03024.x.
13. Haas BJ, Gevers D, Earl AM, Feldgarden M, Ward DV, Giannoukos G, Ciulla D, Tabbaa D, Highlander SK, Sodergren E. Chimeric 16 S rRNA sequence formation and detection in Sanger and 454-pyrosequenced PCR amplicons. Genome Res. 2011;21(3):494–504. doi: 10.1101/gr.112730.110.
14. Heuer H, Krsek M, Baker P, Smalla K, Wellington E. Analysis of actinomycete communities by specific amplification of genes encoding 16 S rRNA and gel-electropho-

retic separation in denaturing gradients. Appl Environ Microbiol. 1997;63(8):3233–3241.

15. Huber R, Woese CR, Langworthy TA, Kristjansson JK, Stetter KO. Fervidobacterium islandicum sp. nov., a new extremely thermophilic eubacterium belonging to the "Thermotogales" Arch Microbiol. 1990;154(2):105–111. doi: 10.1007/BF00423318.

16. Hugenholtz P, Goebel BM, Pace NR. Impact of culture-independent studies on the emerging phylogenetic view of bacterial diversity. J Bacteriol. 1998;180(18):4765–4774.

17. Huse SM, Welch DM, Morrison HG, Sogin ML. Ironing out the wrinkles in the rare biosphere through improved OTU clustering. Environ Microbiol. 2010;12(7):1889–1898. doi: 10.1111/j.1462-2920.2010.02193.x.

18. Huson D, Auch A, Qi J, Schuster S. MEGAN analysis of metagenomic data. Genome Res. 2007;17(3):377–386. doi: 10.1101/gr.5969107.

19. Kim T, Kim H, Kwon S, Park H. Nitrifying bacterial community structure of a full-scale integrated fixed-film activated sludge process as investigated by pyrosequencing. J Microbiol Biotechnol. 2011;21(3):293–298.

20. Klappenbach J, Saxman P, Cole J, Schmidt T. rrndb: the ribosomal RNA operon copy number database. Nucleic Acids Res. 2001;29(1):181–184. doi: 10.1093/nar/29.1.181.

21. Kwon S, Kim T, Yu G, Jung J, Park H. Bacterial community composition and diversity of a full-scale integrated fixed-film activated sludge system as investigated by pyrosequencing. J Microbiol Biotechnol. 2010;20(12):1717–1723.

22. Lee Z, Bussema C, III, Schmidt T. rrnDB: documenting the number of rRNA and tRNA genes in bacteria and archaea. Nucleic Acids Res. 2009;37:D489–D493. doi: 10.1093/nar/gkn689.

23. Liu WT, Marsh TL, Cheng H, Forney LJ. Characterization of microbial diversity by determining terminal restriction fragment length polymorphisms of genes encoding 16 S rRNA. Appl Environ Microbiol. 1997;63(11):4516–4522.

24. McLellan S, Huse S, Mueller Spitz S, Andreishcheva E, Sogin M. Diversity and population structure of sewage derived microorganisms in wastewater treatment plant influent. Environ Microbiol. 2010;12(2):378–392. doi: 10.1111/j.1462-2920.2009.02075.x.

25. Muyzer G, De Waal E, Uitterlinden A. Profiling of complex microbial populations by denaturing gradient gel electrophoresis analysis of polymerase chain reaction-amplified genes coding for 16 S rRNA. Appl Environ Microbiol. 1993;59(3):695–700.

26. Neilson A. The occurrence of aeromonads in activated sludge: isolation of Aeromonas sobria and its possible confusion with Escherichia coli. J Appl Microbiol. 1978;44(2):259–264. doi: 10.1111/j.1365-2672.1978.tb00798.x.

27. Nunoura T, Hirai M, Imachi H, Miyazaki M, Makita H, Hirayama H, Furushima Y, Yamamoto H, Takai K. Kosmotoga arenicorallina sp. nov. a thermophilic and obligately anaerobic heterotroph isolated from a shallow hydrothermal system occurring within a coral reef, southern part of the Yaeyama Archipelago, Japan, reclassification of Thermococcoides shengliensis as Kosmotoga shengliensis comb. nov., and

emended description of the genus *Kosmotoga*. Arch Microbiol. 2010;192:811–819. doi: 10.1007/s00203-010-0611-7.

28. Park S, Park B, Rhee S. Comparative analysis of *archaea*l 16 S rRNA and amoA genes to estimate the abundance and diversity of ammonia-oxidizing *archaea* in marine sediments. Extremophiles. 2008;12(4):605–615. doi: 10.1007/s00792-008-0165-7.

29. Qian P, Wang Y, Lee O, Lau S, Yang J, Lafi F, Al-Suwailem A, Wong T. Vertical stratification of microbial communities in the Red Sea revealed by 16 S rDNA pyro-sequencing. ISME J. 2010;5:507–518. doi: 10.1038/ismej.2010.112.

30. Roesch L, Fulthorpe R, Riva A, Casella G, Hadwin A, Kent A, Daroub S, Camargo F, Farmerie W, Triplett E. Pyrosequencing enumerates and contrasts soil microbial diversity. ISME J. 2007;1:283–290.

31. Roeselers G, Mittge EK, Stephens WZ, Parichy DM, Cavanaugh CM, Guillemin K, Rawls JF. Evidence for a core gut microbiota in the zebrafish. ISME J. 2011;5:1595–1608. doi: 10.1038/ismej.2011.38.

32. Sasaki D, Hori T, Haruta S, Ueno Y, Ishii M, Igarashi Y. Methanogenic pathway and community structure in a thermophilic anaerobic digestion process of organic solid waste. J Biosci Bioeng. 2010;111(1):41–46. doi: 10.1016/j.jbiosc.2010.08.011.

33. Schuppler M, Mertens F, Schön G, Göbel UB. Molecular characterization of nocar-dioform actinomycetes in activated sludge by 16 S rRNA analysis. Microbiology. 1995;141(2):513–521. doi: 10.1099/13500872-141-2-513.

34. Snaidr J, Amann R, Huber I, Ludwig W, Schleifer KH. Phylogenetic analysis and in situ identification of bacteria in activated sludge. Appl Environ Microbiol. 1997;63(7):2884–2896.

35. Urich T, Lanzén A, Qi J, Huson D, Schleper C, Schuster S. Simultaneous assessment of soil microbial community structure and function through analysis of the meta-transcriptome. PLoS One. 2008;3(6):e2527. doi: 10.1371/journal.pone.0002527.

36. Wagner M, Loy A, Nogueira R, Purkhold U, Lee N, Daims H. Microbial community composition and function in wastewater treatment plants. Antonie Van Leeuwen-hoek. 2002;81(1):665–680. doi: 10.1023/A:1020586312170.

37. Wang Q, Garrity GM, Tiedje JM, Cole JR. Naive Bayesian classifier for rapid as-signment of rRNA sequences into the new bacterial taxonomy. Appl Environ Micro-biol. 2007;73(16):5261–5267. doi: 10.1128/AEM.00062-07.

38. Whitman WB, Coleman DC, Wiebe WJ. Prokaryotes: the unseen majority. Proc Natl Acad Sci U S A. 1998;95(12):6578–6583. doi: 10.1073/pnas.95.12.6578.

39. Ye L, Zhang T. Estimation of nitrifier abundances in a partial nitrification reactor treating ammonium-rich saline wastewater using DGGE. T-RFLP and mathemati-cal modeling. Appl Microbiol Biotechnol. 2010;88(6):1403–1412. doi: 10.1007/s00253-010-2837-3.

40. Ye L, Zhang T. Pathogenic bacteria in sewage treatment plants as revealed by 454 pyrosequencing. Environ Sci Technol. 2011;45(17):7173–7179. doi: 10.1021/es201045e.

41. Ye L, Shao MF, Zhang T, Tong AHY, Lok S. Analysis of the bacterial community in a laboratory-scale nitrification reactor and a wastewater treatment plant by 454-pyro-sequencing. Water Res. 2011;45(15):4390–4398. doi: 10.1016/j.watres.2011.05.028.

42.

There is one table and several supplemental files that are not available in this version of the article. To view this additional information, please use the citation on the first page of this chapter.

CHAPTER 2

Taxonomic Precision of Different Hypervariable Regions of 16S rRNA Gene and Annotation Methods for Functional Bacterial Groups in Biological Wastewater Treatment

FENG GUO, FENG JU, LIN CAI, AND TONG ZHANG

2.1 INTRODUCTION

The activated sludge process is a key process in wastewater treatment that removes organic pollutants and nutrients (N and P) by the action of certain bacteria. Such bacteria include heterotrophic organic-utilizing bacteria, ammonia oxidizing bacteria (AOB), nitrite oxidizing bacteria (NOB), denitrifiers, polyphosphate accumulating organisms (PAOs), and others [1]. Many of these bacteria perform very specific functions (like AOB and NOB), and their activity is positively correlated with treating efficiencies [2], [3], [4]. These organisms are not only crucial in wastewater treatment plants (WWTPs), but also playing key roles in biogeochemistry [5].

Taxonomic Precision of Different Hypervariable Regions of 16S rRNA Gene and Annotation Methods for Functional Bacterial Groups in Biological Wastewater Treatment. © Guo F, Ju F, Cai L, and Zhang T. PLoS ONE **8**,10 (2013). doi:10.1371/journal.pone.0076185. Licensed under a Creative Commons Attribution License, creativecommons.org/licenses/by/3.0/

This group of bacteria will hereafter be referred to as functional remediators. However, some other bacteria may cause operational problems to WWTPs. These include foaming and bulking bacteria, which may hamper solid-liquid separation [6], [7], [8]. This group of bacteria will hereafter be referred to as the operational group. In addition, the occurrence and abundance of bacterial pathogens in activated sludge is also a concern, since it may be highly correlated with occurrence of pathogens in effluent discharged into receiving water [9], [10], [11]. Generally, the performance and environmental impacts of WWTPs are significantly affected by the above three functional groups Identifying them at the sufficient taxonomic level (e.g., genus level) is significantly important for investigations on either microbial ecology of both full-scale plants and lab-scale reactors or linking microbial community with sewage treatment.

During the last two decades, molecular methods have enhanced analysis of bacterial communities and their functions in activated sludge. Various detection methods of bacterial identification that use the 16S rRNA gene have been developed. These include fluorescence in situ hybridization (FISH) [12], denaturing gradient gel electrophoresis (DGGE) [13], terminal-restriction fragment length polymorphism (T-RFLP) [14], clone library construction [15], quantitative PCR (qPCR) [16], and microarray [17]. Such methods have aided in revealing invaluable knowledge of the microbial ecology of activated sludge. However, these methods usually target certain bacterial groups or profile communities without much detail or resolution. For complex bacterial communities like activated sludge, traditional molecular methods could not provide sufficient resolution to adequately characterize the bacterial community function. For example, the abundances of AOB and NOB usually represent less than 1% of the total bacterial population in activated sludge from full-scale WWTPs, but they are extremely important for nitrogen removal [18].

Currently, high throughput sequencing techniques have proved their superiority for profiling complex microbial communities [19], [20]. For engineering applications, the profiling of the functional groups is more useful than enumerating the entire bacterial community. Monitoring those functional bacterial groups with high throughput sequencing may be served as routine analysis of WWTPs components, at reduced cost and greater efficiency. However, there are a number of limitations in analyzing 16S rRNA by current high throughput sequencing methods such as 454 pyrosequencing and Illumina paired-end sequencing. There are nine hypervariable regions separated by the conserved regions across the full length 16S rRNA gene. The sequence length, i.e. ~450 bp for 454 pyrosequencing and ~200 bp for Illumina paired-end sequencing could only cover two or three consecutive hypervariable regions and several single regions of 16S rRNA gene, respectively [21], [22]. On one hand, a number of reports had revealed that the selected primer sets targeting partial 16S rRNA gene suffered from biases among taxa due to their coverage [23], [24]. On the other hand, the short read length undoubtedly increases the uncertainty in taxonomic assignment [25]. Unfortunately, most specific functional microorganisms are required to be classified into genus level at least and the results at higher taxonomic levels are much less meaningful for the correlation of bacterial community and functions. Consequently it would be valuable to study which regions may be more suitable for identification of functional bacterial groups as precise as possible. So far, a number of studies had evaluated the coverage the primers for different hypervariable regions of 16S rRNA gene. Few of them analyzed the taxonomic precision for various regions [26]. Moreover, in silico datasets usually derived from the full length sequences are used for evaluation. These datasets may not exactly reflect the experimental results. Also, for specific and significant bacterial groups, such as the functional groups focused by the present study, there is no systematic evaluation to fully check the taxonomic precision of their partial 16S rRNA gene sequences.

In addition, the taxonomic annotation of sequences was realized by specific classification tools. The three classification methods commonly used in 16S rRNA gene-based prokaryotic taxonomic analysis are Ribosomal Database Project (RDP) Classifier based on the Bayesian algorithm, BLAST-based best-hit (BH) output that only considers the similarity between query and reference, and MEtaGenome ANalyzer (MEGAN) using BLAST-based lowest common ancestor (LCA) algorithm, which cares both the similarity between query and reference and the inconsistency of results with close or the same similarity [27]. In some cases, these three tools may give inconsistent results for some taxa, especially using short reads [18], [28], which will undoubtedly mislead some judgments. Thus they should be assessed for the identification of functional groups in activated sludge.

With an overall microbial community profiling, we recently reported that pyrosequencing of V3V4 amplicon of 16S rRNA gene obtained the highest bacterial diversity of a mixed AS sample [29]. However, in this study, the same pyrosequencing data were revisited to answer the following two questions: 1) what is the optimal region for precise taxonomy of those functional genera in AS samples? 2) How the biases are if using different classification methods? Since the real bacterial community in the sample is totally unknown, we chose to unsupervisedly compare the classification consistency among different classification methods to evaluate the taxonomic accuracy of different hypervariable regions. It is well-understood that the consistently assigned results are more likely correct. The results will be helpful not only for the methodology of studying on microbial ecology in activated sludge and other complex environmental samples by high throughput sequencing, but also for routine monitoring of the functional bacterial groups in activated sludge. More importantly, the results will be informative to alert known biases when using pyrosequencing to profile the bacterial community in activated sludge or other systems by different sequencing region or classification methods.

2.2 MATERIALS AND METHODS

2.2.1 LIST OF FUNCTIONAL GROUPS

The assemblage of species in each functional group was selected based on previous reports 1,10, as shown in Table 1. Bacteria in the remediator group include typical and well-known genera of AOB, NOB, denitrifiers, PAOs, glycogen accumulating organisms (GAOs) and hydrolysers. As to the bulking and foaming bacteria group, only the formally named genera were recruited. Members of the pathogenic group, were classified as 'potential pathogens', since in many cases the 16S rRNA gene can only be used for the identification to the genus level, while the pathogens should be identified to the species at least. In total, the list contains 73 genera, including three multifunctional genera, *Rhodococcus, Mycobacterium* and *Tetrasphaera* that were assigned to more than one group. Moreover, there are three candidate genera, *'Accumulibacter' 'Competibacter'* and *'Mi-*

crothrix', which could not be recognized by the RDP Classifier or the LCA method due to limitation of current databases. They can only be assigned by the BH method.

TABLE 1: List of functional groups in activated sludge.

Functional group	Subgroup	Genus[a]	Reference sequence number in database[b]
Nutrient remediators	AOB	*Nitrosomonas*	48/78 (61.5%)
		Nitrosospira	68/68 (100%)
		Nitrosococcus	19/19 (100%)
	NOB	*Nitrobacter*	73/80 (91.3%)
		Nitrospira	21/21 (100%)
	Denitrifier	*Acidovorax*	208/233 (89.3%)
		Arcobacter	92/92 (100%)
		Azoarcus	57/64 (89.0%)
		Comamonas	221/244 (90.6%)
		Curvibacter	12/14 (85.7%)
		Dechloromonas	31/34 (91.2%)
		Hyphomicrobium	61/65 (93.8%)
		Meganema	2/2 (100%)
		Methylobacillus	10/13 (76.2%)
		Methylophilus	26/29 (89.7%)
		Paracoccus	308/318 (96.9%)
		Rhodobacter	99/137 (72.3%)
		Rhodococcus	783/800 (97.9%)
		Thauera	61/61 (100%)
		Zoogloea	31/36 (86.1 %)
	PAO	*"Accumulibacter"*	0/12 (0%)
		Tetrasphaera	21/21 (100%)
	GAO	*"Competibacter"*	0/1 (0%)
		Defluviicoccus	2/2 (100%)
	Hydrolyser	*Saprospira*	4/11 (36.4%)
		Lewinella	19/19 (100%)
Bulking and foaming group	Bulking bacteria	*"Microthrix"*	18/18 (0%)
		Acinetobacter	1,677/1,690 (99.2%)

TABLE 1: *Cont.*

Functional group	Subgroup	Genus[a]	Reference sequence number in database[b]
		Aquaspirillum	4/11 (36.4%)
		Beggiatoa	10/26 (38.5%)
		Caldilinea	2/2 (100%)
		Flexibacter	22/74 (29.7%)
		Gordonia	223/225 (99.1 %)
		Haliscomenobacter	7/8 (87.5%)
		Leucothrix	4/5 (80%)
		Moraxella	84/123 (68.3%)
		Sphaerotilus	16/16 (100%)
		Thiothrix	70/72 (97.2%)
		Trichococcus	40/40 (100%)
	Foaming bacteria	*Isosphaera*	2/12 (16.7%)
		Mycobacterium	1,063/1,068 (99.5%)
		Nocardia	768/777 (98.8%)
		Rhodococcus	783/800 (97.9%)
		Skermania	12/12 (100%)
		Tetrasphaera	21/21 (100%)
Potential pathogen		*Aeromonas*	768/770 (99.7%)
		Bordetella	186/192(96.9%)
		Borrelia	322/322 (100%)
		Brucella	187/201 (93.0%)
		Campylobacter	450/452 (99.6%)
		Chlamydia	72/78 (92.3%)
		Chlamydophila	30/30 (100%)
		Clostridium	1,586/1,694 (93.6%)
		Corynebacterium	614/624 (98.4%)
		Escherichia	710/731 (97.1%)
		Enterococcus	843/856 (98.5%)
		Francisella	100/101 (99.0%)
		Haemophilus	882/964 (91.5%)
		Helicobacter	415/418 (99.3%)
		Klebsiella	517/ 806 (64.1%)
		Legionella	193/206 (93. 7%)

TABLE 1: *Cont.*

Functional group	Subgroup	Genus[a]	Reference sequence number in database[b]
		Leptospira	315/316 (99.7%)
		Listeria	190/190 (100%)
		Mycobacterium	1,063/1,068 (99.5%)
		Mycoplasma	356/533 (66.8%)
		Neisseria	1,229/1,235 (99.5%)
		Pseudomonas	5,819/6,057 (96.1 %)
		Rickettsia	164/164 (100%)
		Salmonella	339/365 (92.9%)
		Serratia	629/652 (96.5%)
		Shigella	288/288 (100%)
		Staphylococcus	1,207/1,214 (99.4%)
		Streptococcus	2,289/2,311 (99.0%)
		Treponema	1,074/1,089 (98.6%)
		Vibrio	1,533/1,705 (89.9%)
		Yersinia	311/315 (98.7%)
		Total	29,930/31,522 (94.9%)

[a]*Genus names with single quote marks were candidate genera that could not be classified by RDP classifier. They are kept in the final database.*
[b]*The first number indicates the sequences that could be identically classified in RDP at genus level (except for the candidate genera) and thus are kept in the final database. The second is the number of total downloaded sequences from Greengenes database. Percentage showed the portion of identically classified sequences.*

2.2.2 MODIFIED GREENGENES DATABASE USING VALIDATED SEQUENCES OF THE FUNCTIONAL GROUPS

Nearly full length 16S rDNA sequences (>1,250 bp and 1,444 bp in average) of all the above 73 genera were manually downloaded from Greengenes online database [30]. After that, the identity assigned by RDP classifier (at 80% confidence threshold cutoff) was compared with the original Greengene assignments of all these sequences. If the RDP results were inconsistent with the actual identity for one sequence, the set was removed from the downloaded dataset (about 5.1% sequences were eliminated). The remaining sequences were added to the pre-cleaned Greengenes database (released at May 9, 2011, totally 385,791 reference sequences with mean length of 1,405 bp), from which the 73 genera listed in Table 1 had been removed. This was done to produce a more comprehensive and updated database to use in the BLAST search in this study. After removing redundancies, the modified Greengenes database contained 415,721 sequences, with 29,930 sequences belonging to the selected functional groups (Table 1).

2.2.3 ACTIVATED SLUDGE SAMPLE PREPARATION, PYROSEQUENCING AND DATA PROCESSING

FastDNA® SPIN Kit for Soil (MP Biomedicals, France) was used to extract total DNA from the eleven activated sludge samples [29], according to manufacturer's instructions. We confirm that: i) the sampling sites were not privately-owned or protected in any way; and ii) the field studies did not involve endangered or protected species. This kit has been found to be the most efficient method for extracting DNA from activated sludge [31]. Equal masses of DNA from different samples were mixed for the following PCR amplification. The primers used in PCR and their coverage for domain *Bacteria* determined by RDP 'Probe Match' were listed in Table S1 in file S1. The details for barcoded-PCR and pyrosequencing can be found in reference [29]. We sequenced the same amplicon from both directions. For instance, two types of pyrotags, i.e. V12 (sequencing from V1 to V2) and V21 (sequencing from V2 to V1) were obtained for V1&V2 regions and distinguished by the differently barcoded forward and reverse primers. The two sub-datasets for the same amplicon hypervariable regions were analyzed as duplications below.

After the raw reads were obtained, they were quality-checked to remove all reads with any ambiguous bases. Then the barcodes and primers were trimmed from the reads in RDP Pyrosequencing Pipeline. The average sequencing length of reads (for those regions over 400 bp) was approximately 400~430 bp, most reads could fully cover the V1&V2 (~350 bp) and V5&V6 (~300 bp) amplicons, but reads only partially covered the V3&V4 (440~470 bp) and V7&V8&V9 (~420 bp) amplicons. To avoid bias caused by trimming with both primers, we trimmed the reads of V3&V4 and V7&V8&V9 only with the single primer (at the beginning of sequencing). Two mismatches for each primer were allowed during trimming and only trimmed tags over certain length (V5&V6, >200 bp; V1&V2, >250 bp; V3&V4 and V7&V8&V9, >300 bp) were kept for further analysis. Then the tags were further cleaned by Denoise [32] and Chimera Check [33] using the software package of MOTHUR 1.22 [34]. After the eight sets of cleaned pyrotags were obtained, they were randomly normalized at the same depth of 32,000 each, except for V987 at 24,000. The raw. sff files for this study were uploaded to the Sequence Read Archive of NCBI (SRR790735).

2.2.4 EVALUATION OF TAXONOMIC PRECISION OF DIFFERENT SEQUENCING REGIONS AND CLASSIFICATION METHODS

To evaluate taxonomic accuracies of the different variable regions, we compared the consistency of results from three widely used 16S-rDNA-based taxonomic assignment methods: 1) online RDP Classifier [25], 2) BH output, and 3) MEGAN -LCA output after BLAST against the modified Greengenes database. We assumed that higher consistency among different methods indicated higher precision of taxonomic assignment for that variable region. Moreover, the more consistently classified tags suggest the higher efficiency derived from both taxonomic capability and PCR recovery (i.e., primer coverage).

In most cases the confidence threshold for RDP classifier was set at 80% if not mentioned. For the BLAST against the modified Greengenes database, the expect-value (e-value) was set at 10^{-100} and the top 50 hits were kept in the output file. The e-value was approximately equal to 85% similarity for a 300 bp alignment length. To filter the low similarity hits, the results were edited by eliminating all hits with less than 95% (for LCA annotation) or 97% (for BH output) similarity. Because the IDs of the DNA sequences from Greengenes in the BLAST output file are Greengenes No., not names of taxa, a script was written in Python to transform the Greengenes IDs into the lowest taxon recorded in the modified Green-genes database. For the BH method, a Python script was used to extract the first output line (with the highest bit score) for each type of pyrotag, and the abundance of all functional genera was counted by another script. The LCA annotation by MEGAN (Version 4.70.4) was performed using the default parameters except that 'Min Support' (the hit number cutoff for output) was set at 1 instead of 5. The files containing read names and their corresponding taxonomic names were output at genus level for all the three methods. For each genus and variable region, the number of pyrotags was concordantly classified by all three methods, by two of the three meth-ods, or by only one of the methods were quantified using a Python script.

2.3 RESULTS

2.3.1 DATABASE CONSTRUCTION

Table 1 showed the functional groups (including remediators, operational and pathogenic groups) of bacteria in activated sludge. In total, 31,522 nearly full-length 16S rDNA sequences belonging to the 73 genera were downloaded from Greengenes database, therein 29,930 (94.9%) of them could be correctly identified by RDP Classifier at confidence threshold of 80%. For most genera, the eliminated sequences represented only a small portion of the total (<10%). However, For *Nitrosomonas, Beggiatoa, Klebsiella* and *Mycoplasma*, more than 20% were eliminated. For *Sapros-pira, Flexibacter, Aquaspirillum* and *Isosphaera*, over half of the reference sequences were eliminated.

2.3.2 EVALUATION OF HYPERVARIABLE REGIONS

The percentages of sequences that could be assigned into any certain ge-nus and all functional groups are shown in Fig. 1A and B, respectively. About 28.2%~46.6% reads could be assigned into a genus at 80% confi-dence threshold in RDP Classifier, while approximately 25.5%~35.3% and

26.3%~41.7% reads were assigned to a genus for BH and LCA methods, respectively (Fig. 1A). For functional groups, as shown in Fig. 1B, bacteria belonging to the functional groups accounted for 12.6%~25.8% of total bacteria based on the RDP results. Also approximately 42.5%~55.7% of the sequence tags were assigned to genus level. These results suggest that roughly half of the genera present did not belong to the listed functional groups, and their roles in activated sludge may need further exploration.

On average, analysis by all three methods indicated that genus-level assignment was the most effective when using the V12 and V21 sequences. This was true when including all species and when only using bacteria belonging to the functional groups. The V34 and V789 regions had the least reads assigned by BH and LCA methods, respectively. Interestingly, discrepancies between RDP and BLAST-based methods for functional groups were not as large as those for the total assemblage. This suggested that the three methods were more consistent when focusing on the functional groups than when including all species. It may be due to the fact that bacteria in the functional groups are usually well-characterized compared with most other bacterial genera, which makes them more likely to be classified correctly.

The percentages of members of the functional groups are displayed using a heat map in Fig. 2 according to 16S rRNA gene region and classification method used. Five genera, such as *Nitrococcus, 'Competibacter', Campylobacter, Chlamydophila* and *Helicobacter* were not detected for all regions and classification methods used (detection limit at 0.003%). Two candidate genera, *'Microthrix'* and *'Accumulibacter'* were only detected by the BH method because they were not included in the RDP taxonomic system and MEGAN-LCA annotation. There were 45 to 58 genera that belonged to the functional groups, depending on the 16S rRNA gene region and classification method used. Four genera with >1% average abundance were *Zoogloea* (3.53%), *Dechloromonas* (2.47%), *Nitrospira* (1.53%) and *Trichococcus* (1.52%). Because the mixed activated sludge samples were collected from eleven municipal WWTPs located in several countries, these highly abundant genera represented the common key groups in municipal activated sludge. For example, *Zoogloea*, one of the most common genera, is not only a denitrifier, but also a specific floc-forming bacterium, and may be important in the organization of activated sludge [35].

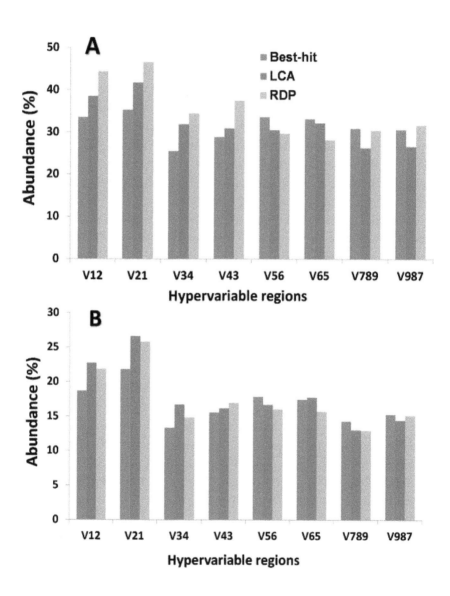

FIGURE 1: Percentages of pyrotags that could be assigned at genus level (A) and concerned functional genera (B) by three annotation methods of RDP Classifier, Best-hit and LCA.

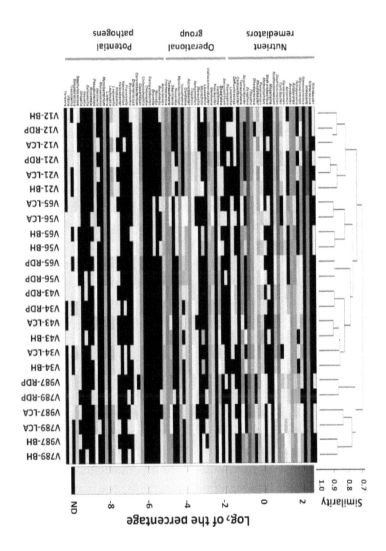

FIGURE 2: Percentages of pyrotags identified as concerned genera for various multi-variable-regions by three methods. The cluster among columns was based on the mean Bray-Curtis distances calculated by the percentages of each genus to total pyrotags. The outer, meddle and inner rings are corresponding to remediators, operational group and potential pathogens.

The results from cluster analysis of the classification results using different regions, sequencing direction, and annotation methods, is also shown in Fig. 2. These results indicate that sequences were clustered first according to regions. However, except for the RDP results, V56 and V65 clustered with the V34 region. Moreover, there was usually less than 80%~90% similarity among results determined by different classification methods of the same sequencing regions, indicating the classification methods could cause noticeable biases as well as the sequencing regions.

The classification consistencies using different multi-variable regions were evaluated by comparing results from the three classification methods and were summarized in Fig. 3. Results were classified according to 3 categories: 1) Those consistently assigned to the same genus by all three methods 2) those assigned to the same genus by two different methods and 3) those assigned to a genus by only one classification method. Obviously, sequence tags that were consistently classified tags by all three methods or by two methods are more likely correct than those exclusively assigned. Using this criterion, the V1&V2 (average of V12 and V21) and V3&V4 region is good for classifying all the functional group. And V1&V2 is slightly better than V3&V4 due to its high percentage of consistently assigned tags for remediators. On the other hand, the V7&V8&V9 region is not recommended for any of the functional groups, due to a large proportion of inconsistently-assigned tags.

Other than the overall presentation of Fig. 3, Fig. 4 summarized the most and the least efficient regions for moderately abundant genera (over 0.1% abundance as determined by at least one method), supporting by the abundance of consistently assigned tags (see the left box chart). It is obvious that no region could have a good coverage for all 25 genera since every region have the several lowest-classified genera. For different genera, abundances of consistently assigned tags fluctuated from 1.5 to 30.4 folds among different regions, if not considering the undetected genera in certain regions (*Nitrosonomas, Caldilinea, Curvibacter* and *Azoarcus*). It was suggested that quantification of those functional genera might be seriously biased with a bad choice of hypervariable regions. The variation of quantification results should be derived from either primer coverage (PCR efficiency) or taxonomic precision of different segments. The abundance of consistently assigned tags was affected by both primer coverage and taxonomic precision of the selected region. However, the consistencies among methods only potentially reflect the taxonomic precision of specific hypervariable regions. In general, agreed with Fig. 3, V1&V2 showed relatively high efficiency, regarding to both the abundance of consistently assigned tags (left box chart) and the consistency among methods (right box chart). Many genera (10 genera), especially for most of the high-ranked genera (six of top ten genera), got the best consistent assignment while only 4 (*Caldilinea, Curvibacter, Paracoccus* and *Haliscomenobacter*) got the poorest consistencies for V1&V2. On the contrary, it could be drawn that V7&V8&V9 is not recommended for profiling functional bacteria in activated sludge due to too many genera get the fewest consistently assigned tags (13 genera) in this region and the poorest assignment consistency (16 genera) for the 25 genera.

Further, we estimated how uncertain the inconsistent assigned results were, i.e. whether the sequences could be correctly assigned at higher taxonomic levels (e.g. family and order). The most inconsistently assigned genus (see Fig. 4), *Acidovorax* were only analyzed here. As illustrated in Fig. S1 in file S1, the tags that could only be solely assigned by LCA or BH were checked by RDP Classifier at the confidence threshold of 80%. In most cases, the assignments were correct at order levels except for the V3&V4 sequences that were solely assigned as *Acidovorax* by LCA. The V789 amplicons were often mis-annotated at family level. The V1&V2 and V5&V6 regions exhibited well accuracy at family level, suggesting the relatively higher taxonomic accuracy for this family on the other side.

2.3.3 EVALUATION OF DIFFERENT CLASSIFICATION METHODS

Firstly, since the tags solely assigned by RDP Classifier with 80% confidence threshold usually account for large portions of total assigned tags for all regions (Fig. 3), it is worth to figure out whether this method is precise enough under various confidence thresholds. We tested 50%, 80%, 95% and 100% thresholds and found that the consistency between RDP and other two methods was increased with the strictness (Fig. 5). Generally, about 30%~50% classified tags were filtered out if increasing the confidence threshold from 50% to 100%. Among all the tags assigned as concerned groups, 23.9%~45.2% of them were solely assigned by RDP at 50% confidence, while the ratios decreased to 5.3%~18.1% at 100% confidence. The consistencies at 100% confidence threshold are relatively high in V3&V4 and V1&V2. The region of V7&V8&V9 showed the worst performance because there was about 20% reads assigned to the functional groups could not be recognized by BH and LCA.

Even more importantly, this result also suggested that the current version RDP Classifier may have some biases because many tags that were identified very surely (at 100% confidence) could not be agreed by BLASTn-based methods, which exhibit the real phylogenetic distances between sequences. Although the overall error rate for RDP Classifier is not high, it may introduce significant biases for some important groups, such as *Nitrosomonas* [18]. Under this consideration, a further investigation on the detected moderately abundant genera (>0.1% abundance at least in one of four multi-variable regions) was conducted. Here, two indexes, one for potential underestimation and the other for potential overestimation, were proposed. The former is the percentage of tags that were exclusively classified by BH and/or LCA to total classified tags and the latter is percentage of tags that were exclusively classified by RDP Classifier to total tags. Larger indexes (typically over 50%) suggested the higher potential biases. As listed in Table 2, there were some genera that could be underestimated or overestimated for all multi-variable regions. *Caldilinea* were always potentially over-counted, while *Nitrosomonas, Rhodobacter, Curvibacter* and *Leptospira* tended to be seriously underestimated no matter which multi-variable region was selected. It agreed with our previous report about mistakenly assigned the AOB genus *Nitrosomonas* by RDP [18]. Table 1 also showed *Nitrosomonas* and *Rhodobacter* got many inconsistent taxonomic assignments between Greengenes database and RDP Classifier even using full length of 16S rRNA gene. Further, we found that at stricter confidence thresholds the potential overestimation decreases dramatically, but the situation of underestimation becomes reasonably more serious, and vice versa for looser confidence threshold.

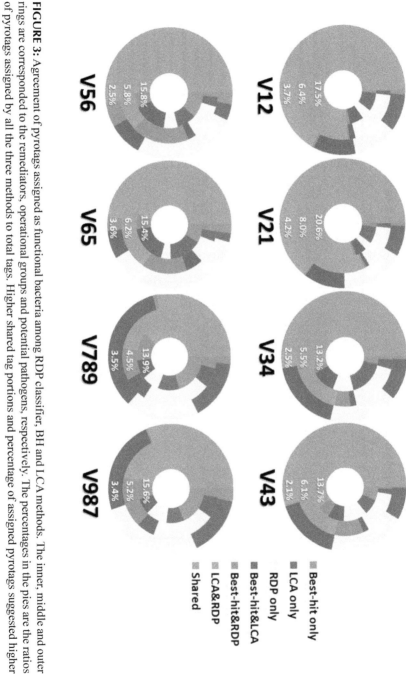

FIGURE 3: Agreement of pyrotags assigned as functional bacteria among RDP classifier, BH and LCA methods. The inner, middle and outer rings are corresponded to the remediators, operational groups and potential pathogens, respectively. The percentages in the pies are the ratios of pyrotags assigned by all the three methods to total tags. Higher shared tag portions and percentage of assigned pyrotages suggested higher efficiencies of the variable regions.

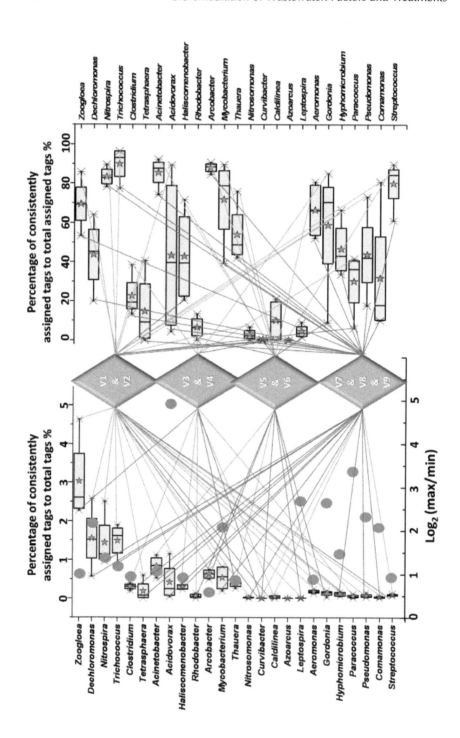

FIGURE 4: Comparison of consistently assigned tags by 3 methods among different sequencing regions for the moderately abundant genera. In the left box chart, abundance is the percentage of consistently assigned tags for 4 regions (V1&V2, V3&V4, V5&V6 and V7&V8&V9, abundance of V1&V2 is the mean value of V12 and V21). Only 25 genera which were over 0.1% abundance as determined by at least one method were listed. *Azoarcus* had no consistently assigned tag although its abundance is high as determined by LCA annotation. Dots showed the logarithmetics of the ratios of maximum identified tags to minimum ones among different regions. No dots for *Nitrosonomas, Caldilinea, Curvibacter* and *Azoarcus* because they get at least one 0-hit region. Percentages of 3-method consistently assigned tags to total assigned tags for each region were compared in the right box chart. Shown are the minimum, 25% quantile, median, 75% quantile and maximum. Stars indicated the average value. For each genus, colored lines link to the most abundant/consistently assigned region and the least abundant/consistently assigned region. More links of lines indicated the priority of the regions, such as V1&V2, while others suggested the poor performance of the region (like V7&V8&V9) for these abundant genera.

Further, we also compared the divergence between BH and LCA for different multi-variable regions. The results were presented in Table S2 in file S1. The two BLAST-based methods also exhibited divergences. The pyrotags that could be assigned by BH but not by LCA should be resulted from the share of similar partial sequence of references among different taxa. Interestingly, in most cases this type of inconsistency existed in certain regions. For instance, V1&V2 exhibited high capability in classification of *Tetrasphaera*, while other regions were mostly LCA-negative. On the other hand, there are many tags only could be assigned by LCA. It indicated many tags are over 95% but less than 97% similarity to the references. Their phylogenetic position is still hard to be evaluated. They may be the uncharacterized species or genera closely related to the references.

2.4 DISCUSSION

For factors affecting on bacterial consortia profiling by high throughput sequencing, current methodological investigations mainly focused on DNA extraction [31] and primer coverage [29], [36]. The shortages from short read length are not often mentioned partially because the technique unprecedentedly refined the resolution of community profile and quantitative capability. However, in order to achieve the best taxonomic resolution for certain groups, the selection of sequencing regions and classification methods should be evaluated. Therefore, the real purpose for our study is to avoid the wrong or insufficient taxonomic assignments for those sequences that were truly generated from certain genera as possible and to get pre-evaluation of the potential biases for the concerned genera. Noticeably, these biases had been luckily amended in some studies and may be undermined in the others [18], [28].

Many studies had examined the classification capability of different variable region [23], [26], [37]. However, most of them analyzed the consistency between the artificially cut partial segments and their mother full-length sequences, which are not really generated by PCR. More importantly, all the previously reports targeted the overall classification without considering the weight of those significant functional groups. Recently, a report surveying 16S rRNA gene sequences from many environments showed that there was no perfect primer sets to cover all the examined sources [38]. The authors proposed that validation of primer sets (or regions) must be pre-evaluated for specific samples. However, besides matching of primers, obviously the different regions of 16S rRNA gene sequences and annotation methods have different validities for specific genera. Although our study only focused on the highly concerned functional genera in wastewater treatment systems, the results suggested that even for a specific sample, it is still hard, even impossible to get unbiased taxonomic information due to primer coverage, PCR amplification, taxonomic precision of the various sequencing regions and different annotation methods. To get pre-evaluated information for the highly concerned bacterial taxa rather than sample types is an alternative way to avoid significant biases.

For the selection of pyrosequencing region to analyze major functional bacterial community in activated sludge, based on both the abundance of consistently classified tags and the consistencies among methods, V1&V2 is the preferred region and V7&V8&V9 is the worst choice of 16S rRNA gene pyrosequencing. V1&V2 is very suitable in length for current pyrosequencing (around 350 bp amplicon). This primer set had been widely adopted in many previous studies among various sample types [39], [40].

Nevertheless, we also found that several functional genera would be underestimated or got the poorest taxonomic consistency in this region, such as *Caldilinea, Curvibacter, Hyphomicrobium* and *Streptococcus*.

The result is thought to be somewhat contradictive with our previous report that V3&V4 region is the most powerful for bacterial diversity profiling of activated sludge [29]. Thus, it was raised a concerning on whether there was a perfect selection of variable regions for pyrosequencing of 16S rRNA gene for different intentions, such as investigations on overall community or specific groups. Our results suggested that the selection would be never perfect even within very limited groups (genera). It also reminded us that the classification results may be 'fake' or 'blurred'. To get pre-evaluation for the most important bacteria will be helpful for judgments.

Another interesting result is that the V1&V2 region has the most classified pyrotags at genus-level for overall community and functional groups. We found that the used primer set for this region is the only one without any degenerated base. Using degenerated base for primer will expand their coverage undoubtedly. However, whether it affects the quantification need further tests.

The precision and accuracy of classification methods are highly dependent on their databases. It should be noted that the RDP Classifier and LCA algorithm will be very precise if their databases are perfect since they are tree-based and supervised, respectively. However, the Best-Hit method is only dependent on the similarity unsupervisedly, which could hardly be consistently correct by using a single cutoff for all bacteria. RDP classifier is very powerful in dealing with large datasets due to its simple algorithm and the high precision for long sequences under high confidence threshold. Regarding to the speed, online RDP Classifier could get the classification of the 32,000 pyrotags within a few minutes, while several days is needed for BLAST-based methods (against Greengenes database, with a single CPU of 2.8 G Hz). However, the RDP might make wrong assignments for short reads [25]. In this study, we found that RDP Classifier overestimated and/or underestimated several genera in some regions or even all regions. It should be highly careful to avoid drawing conclusions from the negative results based on the absence of specific genera, which are investigated by amplifying the inefficient region or using biased annotation method. The false negative results were much more misleading since significantly false positivity is hardly possible for the complex community like activated sludge. Thus, the problem of underestimation or even missing of certain genera using RDP Classifier, which could only be realized and solved by applying other methods, is more serious than overestimation. Moreover, it was found that some genera were innately less distinguishable by RDP Classifier no matter what regions were selected, such as Nitrosomonas and Leptospira (see Table 2).

On the other hand, BLAST-based methods also generated potential errors [28], [41]. For taxonomic assignment of the short 16S rDNA segments, these bugs are mainly derived from highly similar reference sequences (at specific variable regions) that belong to different taxa, even at genus or higher levels. The LCA algorithm that based on the agreement of qualified 'voters' was invented to solve this confuse [27]. Moreover, the reliability of BLAST-based taxonomy elementarily relies on the accuracy and integrity of database and suitable parameter setup. For example, the top 10% score-cutoff (default setting) for the LCA methods in MEGAN may require to be evaluated for the annotation on short tags.

In summary, high throughput sequencing of the segments of 16S rRNA genes is like taking high-resolution images for microbial community. However, current short read length that only cover a couple of hypervariable regions may make some 'pixels' defective due to PCR bias and low taxonomic precision. For studies on some specific groups, it may introduce large and unexplainable biases although these "pixels" may be insignificant in the studies about overall microbial communities. Thus pre-evaluation is needed to select the suitable regions and annotation methods for the concerned specific groups. This study focusing on the highly concerned bacterial genera in the activated sludge demonstrated an example for the pre-evaluation on different regions and classification methods. The results showed that there was no perfect region for all concerning abundant functional genera, except for some regions with less bias, i.e. V1&V2, which could be the preferred region of functional groups in activated sludge. On the other hand, the results also demonstrated the divergences among different classification methods, which may cause misleading results. To avoid drawing wrong conclusion from negative results based on only one region or method, cross-checking using multiple regions and annotation methods should be proposed for taxonomic assignment of pyrosequences of short 16S rRNA gene amplicons.

TABLE 2: Evaluation of taxonomic precision of RDP Classifier for the moderately abundant genera by referring to Best-Hit and the lowest common ancestor methods.

Genus	Abundance[a] %	Index of potential overestimation[b] %								Index of potential underestimation[c] %							
		V12	V21	V34	V43	V56	V65	V789	V987	V12	V21	V34	V43	V56	V65	V789	V987
Zoogloea	4.28	2.9	3.9	9.5	8.2	1.7	1.6	3.9	3.3	5.7	6.1	14.8	7.0	20.0	28.6	31.0	19.9
Dechloromonas	3.41	1.7	1.5	2.1	4.0	0.3	0.3	2.0	1.5	20.5	21.7	43.2	33.5	**52.0**	**59.1**	**56.4**	**60.0**
Nitrospira	1.68	2.6	1.1	8.8	8.3	8.6	10.9	3.3	7.7	10.3	3.6	4.6	1.7	4.4	5.8	2.8	3.5
Trichococcus	1.62	2.8	2.2	2.8	1.8	0.4	1.7	8.7	2.2	0.2	0.0	4.0	1.6	0.9	3.5	9.6	10.5
Clostridium	1.44	22.9	30.5	37.4	42.7	**62.2**	**72.1**	43.6	**58.7**	9.0	7.7	22.2	16.4	6.7	4.6	31.9	23.5
Tetrasphaera	1.04	6.0	6.6	22.7	28.2	2.3	2.1	14.9	20.5	0.0	0.5	9.5	6.6	3.0	4.5	21.9	9.6
Acinetobacter	0.96	2.0	4.3	1.0	4.7	0.0	0.2	0.0	0.0	2.4	1.4	4.5	2.3	5.9	9.2	23.1	21.4
Acidovorax	0.94	2.7	1.5	1.0	0.0	**53.2**	24.9	6.0	3.0	6.4	6.2	36.8	22.8	14.4	24.6	**63.7**	**75.7**
Haliscomenobacter	0.83	**70.5**	**63.4**	42.7	39.6	23.1	32.7	**71.3**	67.6	0.0	0.0	0.0	0.4	0.0	0.3	0.0	0.0
Rhodobacter	0.81	0.7	0.0	5.6	9.6	22.7	19.7	29.3	22.8	**82.5**	**83.2**	**79.2**	**56.3**	**76.8**	**80.3**	40.8	36.6
Arcobacter	0.71	6.4	3.1	2.5	5.1	6.8	10.5	1.6	7.0	5.3	3.1	4.5	5.1	0.5	0.7	2.9	6.2
Mycobacterium	0.70	0.0	0.6	0.0	0.0	0.0	0.3	0.0	0.0	16.9	12.7	26.2	23.1	9.3	11.2	**63.5**	58.0
Thauera	0.70	1.5	4.5	7.7	9.4	0.0	2.8	5.7	2.4	6.7	5.0	16.8	10.0	6.9	10.6	40.9	36.1
Nitrosomonas	0.55	0.0	0.9	0.0	5.1	4.6	1.2	0.0	0.0	**89.9**	**85.9**	**93.9**	**93.9**	**89.4**	**92.2**	**100.0**	**98.9**
Curvibacter	0.47	61.1	39.2	3.2	2.6	0.8	2.4	2.5	19.9	32.9	**51.2**	**96.8**	**97.4**	**98.9**	**97.6**	**96.9**	**71.7**

TABLE 2: *Cont.*

Genus	Abundance[a] %	Index of potential overestimation[b] %								Index of potential underestimation< %							
		V12	V21	V34	V43	V56	V65	V789	V987	V12	V21	V34	V43	V56	V65	V789	V987
Caltilinea	0.32	**85.9**	**93.8**	**70.1**	**61.1**	**62.3**	48.3	**51.0**	**52.6**	14.1	6.3	3.7	2.8	23.9	45.8	4.9	3.9
Azoarcus	0.31	0.0	N.D.	**69.4**	**77.5**	0.0	0.0	**70.6**	**65.3**	**50.0**	N.D.	24.1	18.6	**100.0**	**100.0**	26.5	13.3
Leptospira	0.28	0.0	0.0	3.0	7.3	0.0	0.0	0.0	0.0	**94.8**	**94.7**	**95.5**	**85.5**	**94.1**	**87.5**	**97.8**	**98.9**
Aeromonas	0.28	0.0	4.3	0.0	1.6	0.0	0.0	0.0	1.4	11.2	19.6	15.6	17.7	18.2	7.9	41.7	23.3
Gordonia	0.27	2.5	6.5	0.0	0.0	0.0	0.0	0.0	0.0	19.0	5.6	7.7	14.0	21.7	29.5	36.0	22.4
Hyphomicrobium	0.25	3.8	3.6	3.8	1.3	0.0	0.0	12.5	7.0	20.5	18.1	**54.8**	49.7	**50.8**	**51.1**	39.3	**53.5**
Paracoccus	0.19	28.3	25.8	2.8	12.0	0.0	2.9	0.0	0.0	13.2	27.4	48.6	**50.0**	**61.5**	**68.6**	**54.5**	**60.0**
Pseudomonas	0.16	9.1	10.3	0.0	0.0	0.0	0.0	0.0	0.0	6.1	8.8	45.9	40.9	42.9	57.1	**76.9**	**81.4**
Comamonas	0.14	14.3	0.0	8.3	3.4	21.2	14.3	2.3	0.0	10.7	3.3	**58.3**	**62.1**	**57.6**	46.9	**85.1**	**90.0**
Streptococcus	0.13	10.7	16.0	0.0	0.0	0.0	0.0	0.0	0.0	3.6	0.0	6.8	7.1	11.1	9.4	40.0	34.5

[a]Twenty-five genera accounted over 0.1 % average abundance determined by all three methods were listed.

[b]Index of potential overestimation is the percentage of tags that were exclusively classified by RDP classifier to total classified tags for certain genus. The confidence threshold for RDP classifier is 80%. The boldface number stressed those seriously underestimated or overestimated regions with index over 50%.

[c]Index of potential underestimation is the percentage of tags that were exclusively classified by BH and/or LCA to total classified tags.

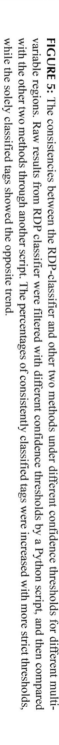

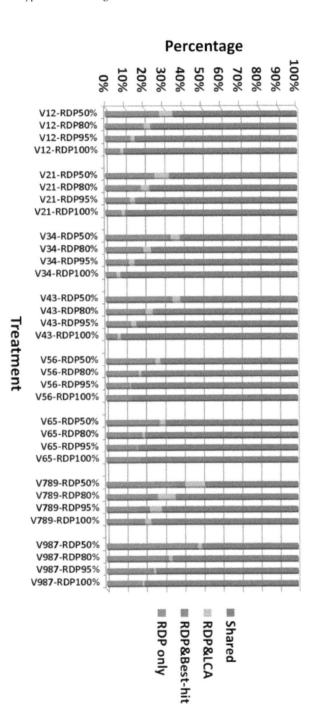

FIGURE 5: The consistencies between the RDP-classifier and other two methods under different confidence thresholds for different multi-variable regions. Raw results from RDP classifier were filtered with different confidence thresholds by a Python script, and then compared with the other two methods through another script. The percentages of consistently classified tags were increased with more strict thresholds, while the solely classified tags showed the opposite trend.

REFERENCES

1. Seviour RJ, Nielsen PH (2010) Microbial communities in activated sludge plants. In: Seviour RJ, Nielsen PH, editors. Microbial Ecology of Activated Sludge. London: IWA Publishing. pp. 95–126.
2. Seviour RJ, Mino T, Onuki M (2003) The microbiology of biological phosphorus removal in activated sludge systems. FEMS Microbiology Reviews 27(1): 99–127. doi: 10.1016/s0168-6445(03)00021-4
3. Yang Q, Peng YZ, Liu XH, Zeng W, Mino T, et al. (2007) Nitrogen removal via nitrite from municipal wastewater at low temperatures using real-time control to optimize nitrifying communities. Environmental Science & Technology 41(23): 8159–8164. doi: 10.1021/es070850f
4. Zhang T, Ye L, Tong AHY, Shao MF, Lok S (2011) Ammonia-oxidizing Archaea and Ammonia-oxidizing Bacteria in six full-scale wastewater treatment bioreactors. Applied Microbiology and Biotechnology 91(4): 1212–1225. doi: 10.1007/s00253-011-3408-y
5. Daims H, Michael WT, Wagner M (2006) Wastewater treatment: a model system for microbial ecology. Trends in Biotechnology 24(11): 483–489. doi: 10.1016/j.tibtech.2006.09.002
6. Soddell JA, Seviour RJ (1990) Microbiology of foaming in activated-sludge plants. Journal of Applied Bacteriology 68(2): 145–176. doi: 10.1111/j.1365-2672.1990.tb01506.x
7. Martins AMP, Pagilla K, Heijnen JJ, van Loosdrecht MCM (2004) Filamentous bulking sludge – a critical review. Water Research 38(4): 793–817. doi: 10.1016/j.watres.2003.11.005
8. Guo F, Zhang T (2012) Profiling bulking and foaming bacteria in activated sludge by high throughput sequencing. Water Research 46(8): 2772–2782. doi: 10.1016/j.watres.2012.02.039
9. Chen LH, Xiong ZH, Sun LL, Yang J, Jin Q (2012) VFDB 2012 update: toward the genetic diversity and molecular evolution of bacterial virulence factors. Nucleic Acids Research 40: 641–645. doi: 10.1093/nar/gkr989
10. Ye L, Zhang T (2011) Pathogenic bacteria in sewage treatment plants as revealed by 454 pyrosequencing. Environmental Science & Technology 45(17), 7173–7179.
11. Redman JA, Grant SB, Olson TM, Estes MK (2001) Pathogen filtration, heterogeneity, and the potable reuse of wastewater. Environmental Science & Technology 35(9): 1798–1805. doi: 10.1021/es0010960
12. Wagner M, Amann R, Lemmer H, Schleifer K (1993) Probing activated sludge with oligonucleotides specific for proteobacteria: Inadequacy of culture-dependent methods for describing microbial community structure. Applied Environmental Microbiology 59: 1520–1525.
13. Muyzer G, Dewaal EC, Uitterlinden AG (1993) Profiling of complex microbial-populations by denaturing gradient gel-electrophoresis analysis of polymerase chain reaction-amplified genes coding for 16S ribosomal RNA. Applied Environmental Microbiology 59(3): 695–700.

14. Liu WT, Marsh TL, Cheng H, Forney LJ (1997) Characterization of microbial diversity by determining terminal restriction fragment length polymorphisms of genes encoding 16S rRNA. Applied Environmental Microbiology 63(11): 4516–4522.

15. Zhang T, Jin T, Yan Q, Shao M, Wells G, et al. (2009) Occurrence of ammonia-oxidizing Archaea in activated sludges of a laboratory scale reactor and two wastewater treatment plants. Journal of Applied Microbiology 107: 970–977. doi: 10.1111/j.1365-2672.2009.04283.x

16. Zhang T, Fang HHP (2006) Applications of real-time polymerase chain reaction for quantification of microorganisms in environmental samples. Applied Microbiology and Biotechnology 70: 281–289. doi: 10.1007/s00253-006-0333-6

17. Xia SQ, Duan LA, Song YH, Li JX, Piceno YM, et al. (2010) Bacterial community structure in geographically distributed biological wastewater treatment reactors. Environmental Science and Technology 44: 7391–7396. doi: 10.1021/es101554m

18. Ye L, Shao MF, Zhang T, Tong AHY, Lok S (2011) Analysis of the bacterial community in a laboratory-scale nitrification reactor and a wastewater treatment plant by high-throughput pyrosequencing. Water Research 45(15): 4390–4398. doi: 10.1016/j.watres.2011.05.028

19. Hamady M, Lozupone C, Knight R (2010) Fast UniFrac: facilitating high-throughput phylogenetic analyses of microbial communities including analysis of pyrosequencing and PhyloChip data. ISME Journal 4(1): 17–27. doi: 10.1038/ismej.2009.97

20. Zhang T, Shao MF, Ye L (2012) 454 Pyrosequencing reveals bacterial diversity of activated sludge from 14 sewage treatment plants. ISME Journal. 6(6): 1137–1147. doi: 10.1038/ismej.2011.188

21. Kircher M, Kelso J (2010) High-throughput DNA sequencing-concepts and limitations. Bioessays 32(6): 524–536. doi: 10.1002/bies.200900181

22. Bartram AK, Lynch MDJ, Sterns JC, Moreno-Hagelsieb G, Neufeld JD (2011) Generation of multimillion-sequence 16S rRNA gene libraries from complex microbial communities by assembling paired-end Illumina reads. Applied and Environmental Microbiology 77(11): 3846–3852. doi: 10.1128/aem.02772-10

23. Kumar PS, Brooker MR, Dowd SE, Camerlengo T (2011) Target region selection is a critical determinant of community fingerprints generated by 16S Pyrosequencing. PLoS ONE 6(6): e20956. doi: 10.1371/journal.pone.0020956

24. Wang Y, Qian PY (2009) Conservative fragments in bacterial 16S rRNA genes and primer design for 16S ribosomal DNA amplicons in metagenomic studies. PLoS ONE 4(10): e7401. doi: 10.1371/journal.pone.0007401

25. Wang Q, Garrity GM, Tiedje JM, Cole JR (2007) Naive Bayesian classification for rapid assignment of rRNA sequences into the new bacterial taxonomy. Applied and Environmental Microbiology 73(16): 5261–5267. doi: 10.1128/aem.00062-07

26. Liu ZZ, DeSantis TZ, Andersen GL, Knight R (2008) Accurate taxonomy assignments from 16S rRNA sequences produced by highly parallel pyrosequencers. Nucleic Acids Research 36(18): e120. doi: 10.1093/nar/gkn491

27. Huson DH, Auch AF, Qi J, Schuster SC (2007) MEGAN analysis of metagenomic data. Genome Research 17(3): 377–386. doi: 10.1101/gr.5969107

28. Kan JJ, Clingenpeel S, Macur RE, Inskeep WP, Lovalvo D, et al. (2011) Archaea in Yellowstone Lake. ISME Journal 5(11): 1784–1795. doi: 10.1038/ismej.2011.56

29. Cai L, Ye L, Tong AHY, Lok S, Zhang T (2013) Biased diversity metrics revealed by bacterial 16S pyrotags derived from different primer sets. PLoS ONE 8(1): e54649 doi:10.1371/journal.pone.0053649.

30. DeSantis TZ, Hugenholtz P, Larsen N, Rojas M, Brodie EL, et al. (2006) Greengenes, a chimera-checked 16S rRNA gene database and workbench compatible with ARB. Applied and Environmental Microbiology 72(7): 5069–5072. doi: 10.1128/aem.03006-05

31. Guo F, Zhang T (2013) Biases during DNA extraction of activated sludge samples revealed by high throughput sequencing. Applied Microbiology and Biotechnology: DOI 10.1007/s00253-012-4244-4.

32. Huse SM, Welch DM, Morrison HG, Sogin ML (2010) Ironing out the wringkles in the rare biosphere through improved OTU clustering. Environmental microbiology 12: 1889–1898. doi: 10.1111/j.1462-2920.2010.02193.x

33. Haas BJ, Gevers D, Earl AM, Feldgarden M, Ward DV, et al. (2011) Chimeric 16S rRNA sequence formation and detection in Sanger and 454-pyrosequenced PCR amplicons. Genome Research 21: 494–504. doi: 10.1101/gr.112730.110

34. Schloss PD, Westcott SL, Ryabin T, Hall JR, Hartmann M, et al. (2009) Introducing mothur: Open-source, platform-independent, community-supported software for describing and comparing microbial communities. Applied and Environmental Microbiology 75: 7537–7541. doi: 10.1128/aem.01541-09

35. McKinney RE, Weichlein RG (1953) Isolation of floc-producing bacteria from activated sludge. Applied and Environmental Microbiology 1(5): 259–261.

36. Claesson MJ, Wang QO, O'Sullivan O, Greene-Diniz R, Cole JR, et al. (2010) Comparison of two next-generation sequencing technologies for resolving highly complex microbiota composition using tandem variable 16S rRNA gene regions. Nucleic Acids Research 38(22): e200. doi: 10.1093/nar/gkq873

37. Vasileiadis S, Puglisi E, Arena M, Cappa F, Cocconcelli PS, et al. (2012) Soil bacterial diversity screening using single 16S rRNA gene V regions coupled with multi-million read generating sequencing technologies. PLoS ONE 7(8): e42671. doi: 10.1371/journal.pone.0042671

38. Soergel DAW, Dey N, Knight R, Brenner SE (2012) Selection of primers for optimal taxonomic classification of environmental 16S rRNA gene sequences. ISME Journal 6(7): 1440–1444. doi: 10.1038/ismej.2011.208

39. Engelbrektson A, Kunin V, Wrighton KC, Zvenigorodsky N, Chen F, et al. (2010) Experimental factors affected PCR-based estimates of microbial species richness and evenness. ISME Journal 4(5): 642–646. doi: 10.1038/ismej.2009.153

40. Hamady M, Walker JJ, Harris JK, Gold NJ, Knight R (2008) Error-correcting barcoded primer for pyrosequencing hundreds of samples in multiplex. Nature Methods 5: 235–237. doi: 10.1038/nmeth.1184

41. Tripp HJ, Hewson I, Boyarsky S, Stuart JM, Zehr JP (2011) Misannotations of rRNA can now generated 90% false positive protein matches in metatranscriptomic studies. Nucleic Acids Research 39(20): 8792–8802. doi: 10.1093/nar/gkr576

There are several supplemental files that are not available in this version of the article. To view this additional information, please use the citation on the first page of this chapter.

The Choice of PCR Primers Has Great Impact on Assessments of Bacterial Community Diversity and Dynamics in a Wastewater Treatment Plant

NILS JOHAN FREDRIKSSON, MALTE HERMANSSON, AND BRITT-MARIE WILÉN

3.1 INTRODUCTION

In many environments bacterial communities are complex, with high number of individuals and high diversity. For example, estimates for bacterial communities in soil are in the range of 10^7–10^{10} bacterial cells [1,2] of 10^3–10^5 different taxa [2,3]. It is well established that only a fraction of this immense diversity can be described by the isolation and cultivation of single bacterial species (e.g. [4]) and microbial communities are therefore studied by cultivation-independent methods, typically using PCR targeting the 16S rRNA gene.

The 16S rRNA gene has several conserved regions which are common to a large number of bacterial species, and variable regions, which are

The Choice of PCR Primers Has Great Impact on Assessments of Bacterial Community Diversity and Dynamics in a Wastewater Treatment Plant. © *Fredriksson NJ, Hermansson M, and Wilén B-M.* PLoS ONE *8,10 (2013). doi:10.1371/journal.pone.0076431. Licensed under a Creative Commons Attribution License, http://creativecommons.org/licenses/by/4.0/.*

shared by fewer species. The conserved regions are used for the design of PCR primer pairs when the aim is to amplify as many bacterial species as possible. These primers are often referred to as universal primers implying that the target sequence is universally distributed. However, no universal primer pair can target all bacteria ([5,6]), and different universal primer pairs may amplify different fractions of a community. Evaluations and comparisons of universal primers are therefore necessary when 16S rRNA genes are used to assess bacterial community structure.

Both fast comparisons and thorough evaluations of universal primers can be made using on-line tools such as those available through the Microbial Community Analysis (MiCA) web site [7], the SILVA ribosomal RNA gene database project [8] or the Ribosomal Database Project (RDP) [9]. For example, in an extensive study, Klindworth et al. [6] used the SILVA ribosomal RNA gene database project [8] to evaluate the overall coverage of 512 primer pairs. However, when such tools are used the analysis is based on all deposited 16S rRNA genes in a database, regardless of environmental origin. Such comparisons may not be entirely adequate to evaluate the suitability of a primer pair for a specific environment. In addition, even when specific databases are used, the specificity predicted by the database comparison can be different from the observed specificity in an actual experiment [10]. Empirical comparisons of universal primers are therefore required.

Many different universal primer pairs have been compared using samples from a range of different environments and the fact that different primer pairs amplify different fractions of a microbial community have been illustrated by differences in DNA fingerprint patterns (e.g. Sipos et al.—rhizoplane [11], Fortuna et al.—soil [12]) and composition of gene libraries (e.g. Hong et al.—marine sediments [13], Lowe et al.—pig tonsils [14]). How different the amplified fractions are is highly dependent on which primer pairs that are compared and on which environment that is sampled, i.e. the composition of the sampled community. In some studies there are only minor differences between primer pairs [15], while other studies show larger differences [13,14].

Although the choice of primer pair clearly will affect which bacterial species that are detected, it is still common practice to only use one primer pair in environmental surveys. It is accepted that the resulting description of the bacterial community is not complete and, for example, by calculating the coverage of a 16S rRNA gene library, it is estimated how representative the description is [16]. However, estimations of community richness and gene library coverage are only based on the observed taxa, i.e. the community targeted by the primer pair, and does not reveal if there are other taxa in the true community that are not targeted by the primer pair. Without an experimental evaluation of the primer pair that is used, the accuracy of the resulting data can only be assumed. However, this assumption may lead to incomplete or false conclusions when the microbial community composition data is analyzed together with environmental parameters, because factors of importance for non-targeted bacterial groups will be missed and parameters affected by these groups will not be identified. In this study we show that the fraction of bacteria that is not targeted by a universal primer pair can be non-trivial, both in terms of phylogenetic and functional groups, and that this affects the interpretation of the observed community dynamics.

An increased understanding, and ultimately management, of the microbial community composition and dynamics is regarded as fundamental for the improvement of biotechnological processes for wastewater treatment [17–20]. Wastewater treatment plants (WWTPs) and reactors can also be regarded as model systems for microbial ecology [21] and as such, be used for analyses of the formation [22], diversity [23] and dynamics [24,25] of complex microbial communities. Since the use of microbial community data from WWTPs goes far beyond mere descriptions of community composition, knowing the limitations of the methods we use for identification and diversity estimations is fundamental.

The primer pair 27F&1492R [26], or variants targeting the same regions of the 16S rRNA gene, is common in surveys of full-scale activated sludge WWTPs [23,27–30]. This primer pair was also determined by Klindworth et al. [6] to be the best primer pair for amplification of nearly full-length 16S rRNA sequences. However, it is likely that a considerable fraction of the sequences in the 16S rRNA databases have been generated with 27F&1492R, since it is one of the most common primer pairs. As pointed out by Klindworth et al. [6], this may increase the coverage of 27F&1492R compared to other, less common, primer pairs in theoretical primer evaluations. Experimental comparisons are therefore valuable to consolidate the findings of theoretical evaluations. However, we have found few experimental evaluations of the 27F&1492R primer pair. By comparison with a fluorescence in situ hybridization (FISH) analysis it was found that Gram-positive bacteria in activated sludge were not properly represented in a gene library generated using the primer pair 8F & 1492R [30], where the forward primer 8F targets the same region as 27F. A comparison has also been made between the primer pairs HK12 & HK13 (a variant of 27F&1492R) and JCR15 & JCR14 using activated sludge samples, but only minor differences in composition of the different targeted communities were found [15]. In this study we compare the primer pairs 27F&1492R and 63F&M1387R. The latter is an adjusted version of the 63F & 1387R primer pair which was previously evaluated using strains of all major bacterial groups, including Gram-positive bacteria, and was found to be more successful than the primer pair 27F & 1392R [31]. Gram-positive bacteria were also found in abundance in the analysis of an activated sludge sample where primer pair 63F & 1390R was used [32]. In addition, the 63F&M1387R primers were found to successfully amplify 16S rRNA genes from environmental samples where the primer pairs 27F & 1392R and 27F&1492R had failed [31]. The indication from these two studies that the primer pairs 63F & 1387R and 63F & 1390R successfully target bacterial groups missed by the more common primer pairs 27F&1492R and 27F & 1392R motivates a detailed comparison. Furthermore, the target sites for 63F&M1387R are both located in regions different from the target sites of 27F&1492R, which might enable amplification of sequences not targeted by the latter.

Activated sludge is particularly suitable for evaluations of methods aiming to describe bacterial diversity as it harbors complex microbial communities including a wide range of bacterial taxa (e.g. [33]). In this study we compare the composition, richness, evenness and temporal dynamics of the bacterial communities targeted by primer pairs 27F&1492R and 63F&M1387R in the activated sludge of a large-scale WWTP, using terminal restriction fragment length polymorphism (T-RFLP), FISH and sequence analysis. We show that both primer pairs miss a substantial part of different phylogenetic and functional groups in the activated sludge, resulting in different descriptions of community composition and dynamics. We also compare the two primer pairs using a general and an environment specific database showing that the results of theoretical comparisons of primer pairs do not necessarily match the results of empirical comparisons.

3.2 RESULTS

3.2.1 ACTIVATED SLUDGE COMMUNITY COMPOSITION

16S rRNA gene libraries were generated from an activated sludge sample using the primer pairs 27F&1492R and 63F&M1387R. There was a big difference in the composition between the two gene libraries (Figure 1). Sequences of class *Betaproteobacteria* dominated the 27F&1492R library while *Alphaproteobacteria* was the most frequent class in the 63F&M1387R library. There was little overlap in the communities described by the two gene libraries. A combined division of the sequences in both libraries into operational taxonomic units (OTUs) based on DNA similarities of 98.7% (species level) resulted in a total of 90 OTUs, but only 5 of these included sequences from both libraries. The sequences in the five common OTUs were only a small fraction of the total: 10% and 22% of the sequences in the 27F&1492R library and the 63F&M1387R library, respectively. The common OTUs were identified as bacteria of families *Holophagaceae* (*Acidobacteria*), *Beijerinckiaceae* (*Alphaproteobacteria*) and *Comamonadaceae* (*Betaproteobacteria*) (three OTUs). To get an overall estimation of the ratios between different taxa in the activated sludge, the number of sequences of all OTUs was related to the number of sequences in the common OTU of phyla *Acidobacteria*. With the combined data from the two primer pairs, the three most abundant taxa were *Betaproteobacteria* (45%), *Alphaproteobacteria* (25%) and *Firmicutes* (12%). As a comparison, the activated sludge sample used for gene library construction was also analyzed by FISH using probes specific for the taxa *Alphaproteobacteria*, *Betaproteobacteria* and *Gammaproteobacteria*. The combined relative abundance of *Alphaproteobacteria*, *Betaproteobacteria* and *Gammaproteobacteria* was lower in the FISH analysis, 45% compared to 73% and 78% in the 27F&1492R and 63F&M1387R library, respectively. However, the FISH analysis resulted in a ratio between *Alphaproteobacteria* and *Betaproteobacteria* of 1 to 2, equal to the ratio in the combined analysis of the clone library data (Figure 2).

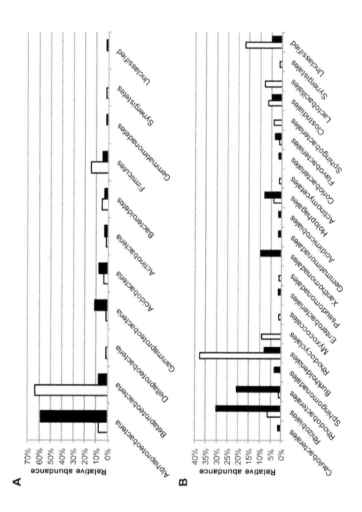

FIGURE 1: Composition of 16S rRNA gene libraries. Distribution of sequences in 16S rRNA gene libraries from an activated sludge sample generated with primer pairs 27F&1492R (77 sequences, white bars) and 63F&M1387R (63 sequences, black bars). Sequences were grouped at the level of phyla/class (panel A) and order (panel B) based on the classification by the RDP Classifier.

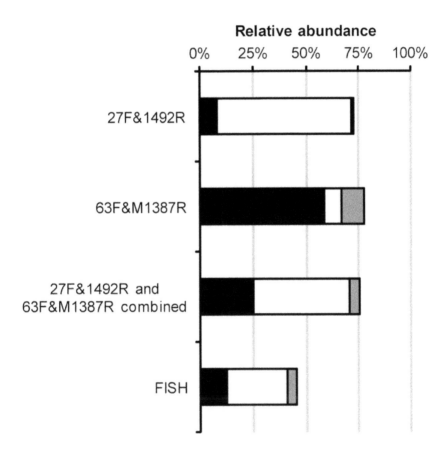

FIGURE 2: Comparison of different assessments of community composition. Comparison of the relative abundance of *Alphaproteobacteria* (black bars), *Betaproteobacteria* (white bars) and *Gammaproteobacteria* (gray bars) in an activated sludge sample. The relative abundances of the classes were derived from 16S rRNA gene libraries generated with primer pairs 27F&1492R and 63F&M1387R, analyzed separately and combined, and from FISH analysis using class-specific probes.

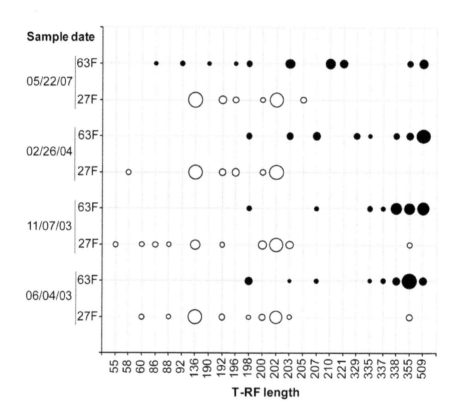

FIGURE 3: T-RFLP analysis of four activated sludge samples. T-RF profiles generated with 27F&1492R (white circles, marked as 27F on the Y-axis) and 63F&M1387R (black circles, marked as 63F on the Y-axis) using restriction enzyme HhaI. To allow for alignment of the T-RFs, 35 bases was added to the lengths of all T-RFs in the 63F&M1387R profiles. The size of the circles corresponds to the relative abundance of the T-RF, i.e. the peak height divided by the sum of all peak heights in the profile.

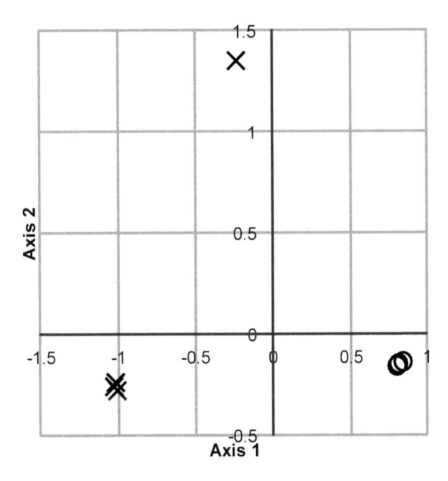

FIGURE 4: NMDS analysis of T-RF profiles. The T-RF profiles generated with 27F&1492R (circles) and 63F&M1387R (crosses) were analyzed by non-metric multidimensional scaling. The best 2-d configuration of 250 iterations is shown.

LIBSHUFF [34] was used to evaluate if the two libraries were significantly different. Figure S1 shows the homologous and heterologous coverage curves for the 63F&M1387R library compared with the 27F&1492R library. The difference in shape between the curves indicates that the two samples are different. The p-value was 0.001 which means that the difference between the homologous and heterologous coverage curves was bigger for the original samples than for any of the 999 randomly generated samples. The same results (homologous and heterologous coverage curves of different shapes, p-value 0.001) were obtained for both the data set of complete and 5' end sequences and the data set of complete and 3' end sequences and independent if the analysis was carried out by comparing the sequences in the 63F&M1387R library with the 27F&1492R library or vice versa. The two libraries were thus determined to be significantly different.

The bacterial community composition was also analyzed in four additional activated sludge samples using T-RFLP and the two primer pairs. In all four samples there were big differences between the T-RF profiles generated with the two primer pairs, with only three or less shared T-RFs per sample (Figure 3). In an ordination analysis the T-RF profiles generated with 27F&1492R clustered together, clearly separated from the T-RF profiles generated with 63F&M1387R (Figure 4). The ordination analysis also suggested that were greater differences among the 63F&M1387R profiles than among the 27F&1492R profiles, as one 63F&M1387R profile was separated from the others. To test if the differences between the profiles generated with different primers were significant a non-parametric analysis of similarity (ANOSIM) was applied. This analysis compares differences between groups, here the two groups of T-RF profiles generated with different primer pairs, with differences within groups. The test statistic R was 1, which is the highest possible value, indicating that there were differences between the T-RF profiles generated with different primer pairs. The differences were determined to be significant as the p-value was 0.026 and the hypothesis of no significant difference between the groups was rejected.

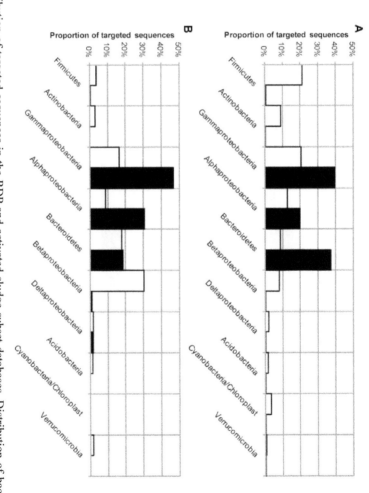

FIGURE 5: Distribution of targeted sequences in the RDP and activated sludge subset databases. Distribution of targeted sequences in the RDP database (panel A) and in the activated sludge subset of the RDP database (panel B) targeted by 27F&1492R (white bars) and 63F&M1387R (black bars) allowing 1 mismatch between primer and target sequence. The phyla and classes included in the figure are the ten phyla and classes with the highest number of genera in the RDP database.

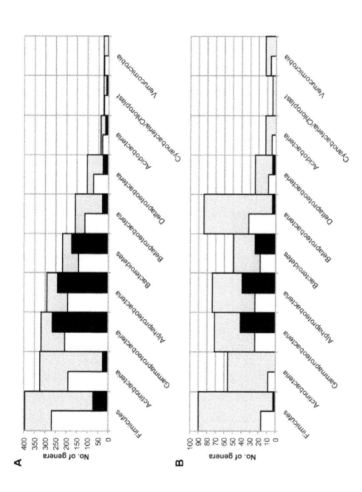

FIGURE 6: Richness of targeted phyla and classes in the RDP and activated sludge subset databases. Genus richness of different phyla and classes in the RDP database (panel A) and in the activated sludge subset of the RDP database (panel B) targeted by 27F&1492R (white bars) and 63F&M1387R (black bars) allowing 1 mismatch between primer and target sequence. The gray bars indicate the number of genera for each phylum or class in the RDP database (panel A) and in the activated sludge subset of the RDP database (panel B). The phyla and classes included in the figure are the ten richest in the RDP database.

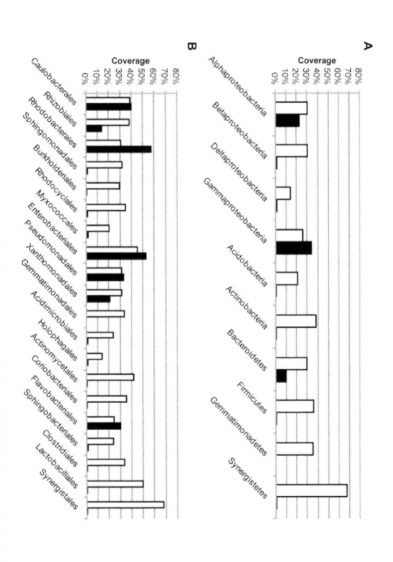

FIGURE 7: Coverage of the SILVA SSU Ref NR (release 114) database. Coverage of the SILVA SSU Ref NR (release 114) database for the primer pairs 27F&1492R (white bars) and 63F&M1387R (black bars). The coverage is the proportion of sequences long enough that match the primers with no mismatches. The coverage is shown for the phyla and taxa (panel A) and orders (panel B) that were observed in the gene libraries.

Order	Acc. No.	Source	27F[a] AGAGTTTGATCMTGGCTCAG	1492R[b] AAGTCGTAACAAGGTARCCGTA
Rhodobacterales				
	GU437447[c]	Microbial mat	--------------C--	---------------A--AAG
	HQ118543[c]	Sand	--------------C--	---------------A--AAG
	JF794560[c]	Marine sludge	--------------C--	---------------A--AAT
	AF234745[d]	Activated sludge	------------TC----	----------G------
	EU802227[e]	Estuary	--------------C--	------A--------G-----
Rhizobiales				
	HM066528[c]	Well water	--------------C--	---------------A--AAG
	HQ119006[c]	Soil	--------------C--	---------------A--AAG
	HQ119071[c]	Soil	--------------C--	---------------A--AAG
	AF234713[f]	Activated sludge	-----------TA-----	----------G-A---
	JQ996656[f]	Activated sludge	-----------TA-----	----------G-A---
Xanthomonadales				
	EU589302[c]	Soil	--------------C--	---------------A--AAT
	JX575910[c]	Pit mud	--------------C--	---------------A--AAT
	JX271974[c]	Activated sludge	--------------C--	---------------A--AAT
	HQ119763[f]	Sand	--------------C--	---------------A--AAG
	JX566536[f]	Soil	--------------C--	---------------A--AAT

FIGURE 8: Examples of mismatches between primer pair 27F&1492R and sequences targeted by 63F&M1387R. A dash (-) indicates the same base as in the primer. a) The degenerative base M is equal to bases A and C. b) The degenerative base R is equal to bases A and G. c) Sequence in the RDP database matching the 63F&M1387R primers but not the 27F&1492R primers. d) Sequence found in the BLAST search. 98.8% sequence similarity with a sequence from the 63F&M1387R library. e) Sequence found in the BLAST search. 97.5% sequence similarity with a sequence from the 63F&M1387R library. f) Sequence found in the BLAST search. Over 99% sequence similarity with sequences from the 63F&M1387R library.

			63F	M1387R[a]
Order	**Acc. No.**	**Source**	CAGGCCTAACACATGCAAGTC	GYCTTGTACACACCGCCC
Burkholderiales	KC633498[b]	Rya WWTP	T--T---T--T--------------	-T---------A------
	KC633510[b]	Rya WWTP	--T---T------T-----------	-T----------A-----
	KC633512[b]	Rya WWTP	--T---T-T----------------	-T----------A-----
	AB039335[c]	Compost	G--T----T--------T-------	-T-------A-----
	AJ412672[c]	Bioreactor	--T---T------T-----------	-T----------A-----
	AJ585992[c]	Activated sludge	--T---T-T--T-------------	-T-----------A-----
Rhodocyclales	KC633497[b]	Rya WWTP	A--T---T-T--T------------	-T-----------A-----
	KC633534[b]	Rya WWTP	--T---T----T-------------	-T-----------A-----
	KC633535[b]	Rya WWTP	--T---T-T---------------	-T-----------A-----
	AF395431[c]	Digester sludge	--T---T-T----------------	-T-----------A-----
	CR555306[c]	Freshwater mud	--T---T-T----------------	-T-----------A-----
	GQ388944[c]	Water distr. system	T--T---T-T-----------G---	-C-----------A-----
Lactobacillales	KC633477[b]	Rya WWTP	--GT------T--------G-----	-C-----------A-----
	AF308147[c]	Chicken feces	TGT----T-------T---------	-C------------A----
	AY167959[c]	Swine manure	-GT------T----------A----	-C------------A----
	AY741386[c]	Silkworm intestine	-GT---------T------------	-C------------A-T----

FIGURE 9: Examples of mismatches between primer pair 63F&M1387R and sequences targeted by 27F&1492R. A dash (-) indicates the same base as in the primer. a) The degenerative base Y is equal to bases C and T. The degenerative base W is equal to bases A and T. b) Sequence in the 27F&1492R gene library that does not match the 63F&M1387R primers. c) Sequence in the RDP database matching the 27F&1492R primers but not the 63F&M1387R primers.

3.2.2 THEORETICAL PRIMER EVALUATIONS

As in the analysis of the activated sludge, there was an apparent difference in the composition and distribution of sequences in the RDP database
targeted by 27F&1492R and 63F&M1387R (Figures 5 and 6). Three taxa,
Alphaproteobacteria, Gammaproteobacteria and *Bacteroidetes*, make up
97% of all sequences targeted by 63F&M1387R while the sequences targeted by 27F&1492R have a more even distribution (Figure 5). The six
most abundant taxa targeted by 27F&1492R: *Firmicutes, Gammaproteobacteria, Alphaproteobacteria, Actinobacteria, Bacteroidetes* and *Betaproteobacteria*, each represents between 8 and 21% of the total number
of targeted sequences. The sequences targeted by 27F&1492R were also
more evenly distributed in terms of number of different genera within each
taxa (Figure 6). The richness of the sequences targeted by 27F&1492R
was 1414 genera in 39 taxa and for 63F&M1387R the richness was 905
genera in 29 taxa. 27F&1492R covered 67% of all genera and 93% of
all phyla/classes in the RDP database. The coverage by 63F&M1387R
was lower: 43% of all genera and 69% of all phyla/classes in the RDP
database. However, more genera of the *Alphaproteobacteria, Gammaproteobacteria* and *Bacteroidetes* were targeted by the 63F&M1387R primer
pair than the 27F&1492R primer pair (Figure 6).

Although activated sludge contain diverse bacterial communities all
taxa in the RDP database cannot be expected to be found. An activated
sludge specific database was therefore generated to complement the theoretical evaluation of the two primer pairs. A search in the NCBI Nucleotide database for sequences longer than 600 bases and with any field containing the term "activated sludge" returned 12844 sequences. Most of
them, 10890 sequences, were also present in the RDP database and 10878
sequences were classified as bacterial sequences. The sequences in the
activated sludge subset of the RDP database (AS dataset) were distributed differently from the sequences in the complete RDP database, both
in terms of number of sequences and number of genera within each taxa
(Figures S2 and S3). The most common taxa in the AS dataset were, in de-
scending order, *Betaproteobacteria, Gammaproteobacteria, Bacteroidetes, Alphaproteobacteria* and *Firmicutes* whereas in the complete RDP
database the order was *Firmicutes, Gammaproteobacteria, Bacteroidetes*,

Actinobacteria and *Betaproteobacteria* (Figure S2). In terms of number of genera, the three richest taxa were *Firmicutes, Betaproteobacteria* and *Alphaproteobacteria* in the AS dataset while *Firmicutes, Actinobacteria* and *Gammaproteobacteria* were the richest in the complete database (Figure S3). The total richness in the AS dataset was low as it only included 527 of 2104 genera in 30 of 42 phyla/classes present in the complete database.

The primer pairs 27F&1492R and 63F&M1387R were matched against the AS dataset. As for the searches in the complete RDP database, the distribution of the targeted sequences was different for the 27F&1492R and 63F&M1387R primer pairs (Figures 5 and 6). For 63F&M1387R, 96% of the targeted sequences were classified as *Alphaproteobacteria, Gammaproteobacteria* and *Bacteroidetes* (Figure 5). The sequences targeted by 27F&1492R were distributed more evenly, with the three most abundant taxa, *Betaproteobacteria, Bacteroidetes* and *Gammaproteobacteria*, together only representing 65% of the sequences. The richness of the sequence sets were 161 genera in 19 phyla/classes for sequences targeted by 27F&1492R while 63F&M1387R only targeted 119 genera in 10 phyla/classes. The targeted sequences included 31% and 23% of all genera and 63% and 33% of all phyla/classes in the AS dataset, for the 27F&1492R and 63F&M1387R primer pairs, respectively. As in the analysis using the RDP database, the total number of genera targeted by the 63F&M1387R primer pair was lower than for the 27F&1492R primer pair, but for some taxa, e.g. *Alphaproteobacteria, Gammaproteobacteria* and *Bacteroidetes*, the 63F&M1387R primer pair targeted more genera than the 27F&1492R primer pair (Figure 6). Ten phyla in the AS dataset were not covered at all by either primer, hence the low percentages of total number of phyla/classes. However, these 10 phyla only represented 1% of the total number of sequences in the AS dataset.

3.2.3 EXPLORATION AND EXPLANATION OF THE DIFFERENCES IN TARGETED TAXA

Both in the analyzed activated sludge sample and in the database searches the two primer pairs targeted different sets of sequences. The 27F&1492R primer pair targeted more *Betaproteobacteria* and *Firmicutes* while the

63F&M1387R primer pair targeted more *Alphaproteobacteria* and *Gammaproteobacteria*. The TestPrime tool of the SILVA ribosomal RNA gene database project [8] was used to further evaluate the differences in sequence sets targeted by the two primer pairs. The coverage for the two primers, i.e. the percentage of the sequences long enough to include the primer sites that match the primers, of the taxonomic divisions observed in the gene libraries are shown in Figure 7. Overall, the 27F&1492R primer pair has a much better coverage than the 63F&M1387R primer pair. For the latter, the two classes with the best coverage, *Alphaproteobacteria* and *Gammaproteobacteria*, are also the most abundant in the gene library (Figure 1). However, at the order level, a high coverage does not correspond to a high abundance in the gene libraries. For example, 27F&1492R has a greater coverage for *Rhizobiales* sequences than 63F&M1387R, but fewer sequences were observed in the library. Likewise, 27F&1492R covers the *Xanthomonadales* sequences in the database better than 63F&M1387R, but no *Xanthomonadales* sequences were observed in the 27F&1492R library. To further explore the differences between the two primer pairs, sequences from the orders *Burkholderiales*, *Rhodocyclales* and *Bacillales*, which were the three most abundant in the 27F&1492R library, and from the orders *Rhodobacterales*, *Rhizobiales* and *Xanthomonadales*, which were the three most abundant in the 63F&M1387R library, were inspected. Very few of the sequences from the BLAST search that matched the 63F&M1387R library sequences of the orders *Rhodobacterales*, *Rhizobiales* and *Xanthomonadales* were long enough for a comparison to be made with either the 27F or the 1492R primer, let alone both of them. Among the sequences that were long enough for a comparison with the 27F&1492R primer pair there were both matching and non-matching sequences (see Figure 8 for examples of mismatches). An evaluation was also made of the sequences in the RDP database that matched the 63F&M1387R primer pair but not the 27F&1492R primer pair. Here it could be seen that the primer pair 27F&1492R does not match some sequences from the dominant orders in the 63F&M1387R library mainly due to mismatches with the 1492R primer (Figure 8). It should be noted that the analyzed sequences from the RDP database were not highly similar to the sequences in the 63F&M1387R library, the maximum similarity between the RDP sequences and the library sequences was 97.7%, 97.3% and 96.5% for *Rho-*

dobacterales, *Rhizobiales* and *Xanthomonadales*, respectively. However, most library sequences of the orders *Rhizobiales* and *Rhodobacterales* (11 of 19 and 8 of 13, respectively) were more similar to the RDP sequences targeted by only 63F&M1387R than to the RDP sequences targeted by both primer pairs. For *Xanthomonadales* the library sequences were equally similar to the RDP sequences targeted by only 63F&M1387R as to the RDP sequences targeted by both primers. In essence, we cannot conclude that the 27F&1492R primer pair failed to amplify more of the *Rhodobacterales*, *Rhizobiales* and *Xanthomonadales* in the activated sludge because of the same mismatches as seen in the RDP sequences or in the sequences from the BLAST search. However, we can see that the coverage by the 27F&1492R primer pair of these orders is not complete and based on the RDP sequences, it is mainly due to mismatches with the reverse primer. This is in contrast with the screening of all sequences in the SILVA database where the majority of the mismatches were due to mismatches with the forward primer (Table 1). The 63F&M1387R primer pair does not target sequences from the dominant orders in the 27F&1492R library almost exclusively because of mismatches with the 63F primer (Figure 9). Here too, this is different from the screening of all sequences in the SILVA database where 98% of the sequences not targeted by the primer pair had mismatches with the M1387R primer (Table 2). While 27F&1492R did target sequences in the RDP database of the *Rhodobacterales*, *Rhizobiales* and *Xanthomonadales* orders, including some not targeted by 63F&M1387R, 63F&M1387R only targeted a few *Burkholderiales* sequences, but no *Rhodocyclales* or *Bacillales* sequences.

3.2.4 IMPACT OF PCR PRIMER CHOICE ON THE OBSERVED DIVERSITY

The richness of the two gene libraries were similar, regardless if counts were based on phylogenetic classification (26 genera in 9 phyla/classes and 28 genera in 9 phyla/classes, for the 27F&1492R and 63F&M1387R library, respectively) or DNA similarities, approximating phyla and genera with 80% and 95% similarity, respectively (38 genera in 13 phyla and 34 genera in 12 phyla, for the 27F&1492R and 63F&M1387R library, respectively). However, for the 27F&1492R library the estimated richness was much lower at the level of species and genera, resulting in a greater estimated coverage (Figure 10). The sequences in the 63F&M1387R library appeared to be distributed more evenly than the sequences in the 27F&1492R library. In the evenness analysis using Pareto-Lorenz curves the Fo index was calculated to be 68% for the 63F&M1387R library and 75% for the 27F&1492R library, the lower index indicating a more even distribution (Figure S4).

TABLE 1: Distribution of sequences with mismatches with the 27F&1492R primer pair.

	SILVA Ref NR[a]	*Rhodobacterales*[b]	*Rhizobiales*[b]	*Xanthomonadales*[b]
Total no.	43561	107	62	20
Mismatch only with 27F[c]	64%	11%	8%	20%
Mismatch only with 1492R[d]	8%	73%	85%	70%
Mismatch with both[e]	28%	16%	7%	10%

a) Number of sequences in the SILVA SSU Ref NR (release 114) database that do not match the 27F&1492R primers. b) RDP database sequences of the given order that matched the 63F&M1387R primers but not the 27F&1492R primers. c) The proportion of the total number of analyzed sequences that only had mismatches with the 27F primer. d) The proportion of the total number of analyzed sequences that only had mismatches with the 1492R primer. e) The proportion of the total number of analyzed sequences that had mismatches with both the 27F and the 1492R primer.

3.2.5 IMPACT OF PCR PRIMER CHOICE ON THE OBSERVED COMMUNITY DYNAMICS

The composition of the T-RF profiles generated with 27F&1492R were found to be significantly different from the T-RF profiles generated with 63F&M1387R (Figure 3). In addition, the observed dynamics were also very different for the two primer pairs. The stability of a community over time can be analyzed by comparing all T-RF profiles with the first profile in a series of samples. While the 27F&1492R T-RF profiles showed a constant similarity of around 75% with the first T-RF profile in the series, suggesting a fairly stable community, the 63F&M1387R profiles showed a steady decrease in similarity from the first, indicating a steady deviation from the original community (Figure 11, panel A). By plotting the similarity between all consecutive T-RF profiles the times where the greatest changes in community composition occurred can be identified. The lowest similarity between two consecutive T-RF profiles was observed between November 2003 and February 2004 in the 27F&1492R analysis, while in the 63F&M1387R analysis the lowest similarity was observed between the T-RF profiles of February 2004 and May 2007 (Figure 11, panel B).

TABLE 2: Distribution of sequences with mismatches with the 63F&M1387R primer pair.

	SILVA Ref NR[a]	*Burkholderiales*[b]	*Rhodocyclales*[b]	*Lactobacillales*[b]
Total no.	192488	86	23	115
Mismatch only with 27F[c]	2%	97%	100%	98%
Mismatch only with 1492R[d]	52%	0%	0%	0%
Mismatch with both[e]	46%	3%	0%	2%

a) Number of sequences in the SILVA SSU Ref NR (release 114) database that do not match the 63F&M1387R primers. b) RDP database sequences and gene library sequences of the given order that matched the 27F&1492R primers but not the 63F&M1387R primers. c) The proportion of the total number of analyzed sequences that only had mismatches with the 63F primer. d) The proportion of the total number of analyzed sequences that only had mismatches with the M1387R primer. e) The proportion of the total number of analyzed sequences that had mismatches with both the 63F and the M1387R primer.

3.3 DISCUSSION

3.3.1 THE PRIMER PAIRS 27F&1492R AND 63F&M1387R DESCRIBE DIFFERENT FRACTIONS OF THE BACTERIAL COMMUNITY

The fact that primer pairs target different fractions of a community has been demonstrated in a number of studies by applying different primer pairs to a single sample [11–15]. However, the extent of primer bias and discrimination varies between different primers and environments and may be hard to predict without experimental data. The present investigation is the first one to report a significant primer bias of common universal 16S rRNA primers in the description of WWTP communities. Identification of limitations of common 16S rRNA primers is valuable because management of the diversity and dynamics of the bacterial communities in WWTPs is regarded as a possible, and perhaps even necessary, way to improve the function of the WWTPs [17–20] and to identify the factors that shape bacterial communities 16S rRNA gene sequence data is often used (e.g. [35–38]).

16S rRNA gene libraries were generated from an activated sludge sample using the two universal primer pairs 27F&1492R and 63F&M1387R, and very different descriptions of the bacterial community were obtained. Using the 27F&1492R primer pair the activated sludge community would have been described as dominated by *Betaproteobacteria* while the 63F&M1387R primer pair would have led us to believe that the activated sludge was dominated by *Alphaproteobacteria*. Different conclusions regarding the distribution of putative functional groups would also have been drawn. The sequences of the order *Rhizobiales*, the most abundant order in the 63F&M1387R library, were classified as *Beijerinckiaceae*, *Hyphomicrobiaceae* and *Methylocystaceae* which are heterotrophs [39], methylotrophs [40] and methanotrophs [41]. The *Burkholderiales* sequences, which were the most abundant in the 27F&1492R library, were almost all classified as different genera of the family *Comamonadaceae*, many of which are heterotrophs capable of denitrification [42]. Thus, with the 63F&M1387R primer pair methanotrophs and methylotrophs would have been determined to be abundant along with heterotrophic bacteria while with the 27F&1492R primer pair denitrifying heterotrophs would have been determined to be very abundant.

A previous study indicated that the primer pair 8F & 1492R [30] may fail to amplify Gram-positive bacteria in activated sludge, while another study did find Gram-positive bacteria in a gene library from activated sludge generated using 63F & 1390R [32]. In this study more Gram-positive sequences were found in the 27F&1492R library than in the 63F&M1387R library, 12% and 6% of all retrieved sequences, respectively. Of these sequences, only one from each library was of the same family, suggesting that both primer pairs do target Gram-positive bacteria, but different groups. Thus, depending on the community composition both primer pairs may appear to either fail or succeed in amplifying 16S rRNA gene sequences of Gram-positive bacteria.

There were many phyla represented in the activated sludge data set (Figure S2) that were not observed in the gene libraries. However, the phyla that were present in the gene libraries have been shown to be the most abundant in activated sludge of WWTPs and bioreactors world-wide [23,30,33]. The low number of observed phyla in the gene libraries are likely due to the relatively small library sizes. If a higher number of sequences had been analyzed less abundant phyla may also have been detected.

To further evaluate the accuracy of the descriptions of the bacterial community by the two primer pairs a comparison was made with a FISH analysis of the same activated sludge sample (Figure 2). Of course, results obtained by the FISH method may also be biased and erroneous since FISH probes, just as PCR primers, may not be as specific or inclusive as intended. Even so, the comparison can be used to highlight two aspects of the observed distribution of different taxa in the gene libraries. The comparison with the distribution obtained by FISH analysis showed that both primer pairs may overestimate the relative abundance of *Proteobacteria*, possibly because they fail to detect some other bacterial groups. However, the ratio between the *Alphaproteobacteria* and the *Betaproteobacteria*, is similar in the combined analysis of the gene library data and in the FISH analysis. This could be an indication that together, the two primer pairs describe the *Proteobacteria* accurately, at least in terms of abundance of the different classes within the phyla.

That the two primer pairs amplify distinct parts of the microbial community in the activated sludge is consistent with the results of other experimental evaluations of primer pairs. Hong et al. used marine sediment samples to compare not only two primer pairs (27F&1492R and 8F&1542R), but also two DNA extraction techniques, and found that the different methods each produced distinct results [13]. As in this study, the most abundant phyla were detected by both primer pairs, but in different proportions. Although the two primer pairs used in this study are universal in the sense that they amplified sequences from a wide range of taxa, each primer pair showed a clear bias towards certain taxa. This was also reported by Lowe et al [14] who compared gene libraries from pig tonsils generated by 27F & 1389R and 63F & 1389R. Consistent with the results of this study, the 63F primer generated a higher number of sequences of class *Gammaproteobacteria* and 27F a higher number of *Firmicutes*. These and other differences in the range of sequences targeted by the two primer pairs were also seen in the database searches (Figures 5-7). However, the results of database comparisons and theoretical evaluations of primer pairs can be misleading. In essence, it does not matter if one primer pair has a 75% coverage and another a 10% coverage of a certain taxa if the bacteria present in the sample of interest belong to the 10% that the second primer pair targeted. For example, the 27F&1492R primer pair was shown to have a greater coverage than 63F&M1387R for most orders, including the *Rhizobiales* and the *Xanthomonadales*. Despite this, sequences of these two orders were much more abundant in the 63F&M1387R library than in the 27F&1492R library. This illustrates that a high coverage of a taxa does not guarantee detection of sequences from that taxa. An evaluation of the sequences in the RDP database targeted by the two primer pairs showed that the 27F&1492R primers did match sequences of both *Rhizobiales* and *Xanthomonadales*, but that a fraction of these two orders were missed due to mismatches with the 1492R primer (Table 1). That the general conclusions from database evaluations of complete databases can differ from specific comparisons was also seen in the evaluation of the mismatches. While the majority of the mismatches between the 27F&1492R primer pair and the sequences in the SILVA SSU Ref NR (release 114) database were due to mismatches with the forward primer, the reverse was observed in the manual evaluation of three specific orders.

Based on the database searches and evaluations in this study the primer pair 27F&1492R appeared to be a better choice for assessment of bacterial diversity than 63F&M1387R since it targeted a wider range of taxa and had a much better coverage. In an extensive theoretical evaluation of primer pairs by Klindworth et al. [6], 27F&1492R was also determined to be the best choice for amplification of nearly full-length sequences. However, the experimental comparison presented here showed that for the activated sludge that was analyzed none of the two primer pairs was necessarily better than the other. Both primer pairs generated gene libraries with similar richness, both including taxa not present in the other. Thus, if only one of these two primer pairs is to be used, which of the two that is the most suitable depends on the aim of the analysis. If the focus of the analysis is on *Betaproteobacteria*, then 27F&1492R would be a better choice than 63F&M1387R since a higher number of *Betaproteobacteria* sequences was found in the 27F&1492R gene library. However, if *Alphaproteobacteria* or *Gammaproteobacteria* are of interest, 63F&M1387R would be a better choice since more sequences of these phyla were found in the 63F&M1387R gene library.

3.3.2 THE PRIMER PAIRS 27F&1492R AND 63F&M1387R DESCRIBE DIFFERENT DYNAMICS OF THE BACTERIAL COMMUNITY

If two primer pairs target different fractions of a community it implicitly follows that they may also describe different community dynamics but this is rarely discussed or shown. By analyzing four activated sludge samples with the primer pairs 27F&1492R and 63F&M1387R we show that the community dynamics can be described in very different ways depending on the primer pair used. While the T-RF profiles generated with the 27F&1492R primer pair showed a fairly stable community, the 63F & M1387 T-RF profiles showed a community that steadily deviated from the initial composition (Figure 11, panel A). This result stresses that the observation of a stable bacterial community, as indicated by the 27F&1492R T-RF profiles, may be misleading.

Studies of bacterial community dynamics are often done to investigate the effect of different environmental parameters on the community composition (e.g. [24,43]). In this study we show that depending on the primer pair being used different parameters may appear to have the greatest effect. In the 27F&1492R analysis the greatest change in community composition occurred between samples two and three (collected in November 2003 and February 2004, respectively) while in the 63F&M1387R analysis the greatest change was observed between samples three and four (collected in February 2004 and May 2007, respectively) (Figure 11, panel B). Consequently, for the community targeted by 27F&1492R, changes in environmental parameters between samples two and three would seem more important than any changes occurring between samples three and four, while for the community targeted by 63F&M1387R, the T-RFLP analysis would suggest the opposite. For the four samples included in this study the primer pair 63F&M1387R detected more changes in community composition than 27F&1492R. However, differences in the described dynamics between primer pairs are likely to depend on the samples that are analyzed. As for evaluations of community composition, a primer pair that it is suitable for one set of samples may not be so for another sample set.

3.4 CONCLUSIONS

In the present study we show that the universal 16S rRNA gene primers 27F&1492R and 63F&M1387R target different parts of the bacterial community in activated sludge samples and would have resulted in distinct conclusions regarding the structure, function and dynamics of the community. The results demonstrate that experimental comparisons of universal 16S rRNA primers can reveal differences not detected by theoretical comparisons, because while database comparisons indicated that primer pair 27F&1492R would be a better choice than 63F&M1387R, the empirical comparison showed that none of the two primer pairs was better than the other. We also conclude that different dynamics can be expected with different primers and if only one primer pair is used, which is common practice, the absence of change in the observed community composition does not necessarily indicate a stable community. Combining the results of several surveys with different universal primer pairs may therefore be necessary for a more complete description of community diversity and dynamics.

3.5 MATERIALS AND METHODS

3.5.1 ETHICS STATEMENT

Permission to enter the Rya WWTP and to collect activated sludge samples were granted by Gryaab AB (owner and operator of the WWTP).

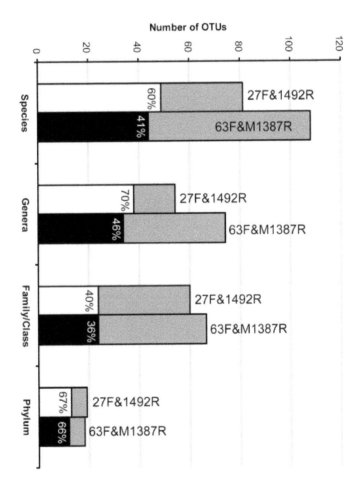

FIGURE 10: Observed and estimated richness of the 27F&1492R and 63F&M1387R 16S rRNA gene libraries. Gray columns represent richness as estimated by the Chao1 estimator. Black and white columns represent observed richness. The ratio observed: Estimated, i.e. the coverage, is given within each column. The taxonomic levels were approximated by DNA similarities of 98.7%, 95%, 90% and 80%, for species, genera, family/class and phylum, respectively.

Bioremediation of Wastewater: Factors and Treatments

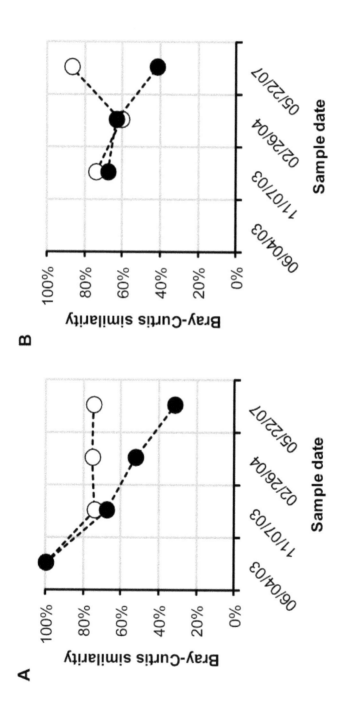

FIGURE 11: Community stability and rate of change. Community stability (panel A) Bray-Curtis similarity between the T-RF profile of 06/04/03 and all other profiles generated with 27F&1492R (white circles) and 63F&M1387R (black circles). Rate of change (panel B) Bray-Curtis similarity between subsequent T-RF profiles generated with 27F&1492R (white circles) and 63F&M1387R (black circles).

3.5.2 SAMPLE COLLECTION AND DNA EXTRACTION

Samples were collected at the end of the aerated basins at the Rya WWTP, a WWTP treating both industrial and municipal wastewater [44]. 50 mL of sample were centrifuged and the resulting pellet was stored at -20°C within 1.5 h from collection. DNA was extracted using Power Soil DNA Extraction Kit (MoBio Laboratories). The frozen sludge pellets were thawed, 15 mL sterile water were added and the samples were homogenized by 6 min of mixing in a BagMixer 100 MiniMix (Interscience). Water was removed by centrifugation and DNA was extracted from 0.25 g of homogenized sludge pellet according to the manufacturer's instructions. Samples collected 06/04/03, 11/07/03, 02/26/04 and 05/22/07 were used for T-RFLP analysis. The sample collected 07/15/04 was used for generation of 16S rRNA gene libraries and FISH analysis.

3.5.3 PCR FOR T-RFLP

16S rRNA genes were amplified using HotStarTaqPlus PCR kit (Qiagen) according to the manufacturer's instructions. Bacteria-specific primer pairs used were 27F (AGAGTTTGATCMTGGCTCAG) and 1492R (TACG-GYTACCTTGTTACGACTT) [26] and 63F (CAGGCCTAACACATG-CAAGTC) and M1387R (GGGCGGWGTGTACAAGRC). The primer pair 63F&M1387R was based on the previously published sequences 63F and 1387R [31]. The primer 1387R has a mismatch for some bacterial sequences at position 1388 [31] and was therefore modified, which increased the number of targeted sequences in the RDP database slightly (Table S1). The primers 27F and 63F were 5'-labeled with the fluorescent dye 6 – carboxyfluorescein. PCR reactions were carried out in the provided PCR buffer with 0.5 U HotStarTaqPlus, 200μM dNTP mix, 0.1 μM of each primer and 2-5 ng DNA. The PCR started with 5 min at 95°C for Taq polymerase activation followed by 35 cycles of denaturation at 94°C for 1 min, annealing at 55°C or 60°C for the 27F&1492R and 63F&M1387R primer pairs, respectively, for 30 s and elongation at 72°C for 1 min. The reactions were ended with a final elongation step at 72°C for 7 min. To evaluate the effect of annealing temperature on the T-RF profiles PCR was also done with the primer pair 63F&M1387R and annealing temperature 55°C for the sample collected 05/22/07. Two PCR reactions were prepared for each combination of primer pair, annealing temperature and restriction enzyme.

3.5.4 T-RFLP

The PCR products were purified using the Agencourt AMPure system (Beckman Coulter) and digested with 10 units of restriction enzyme *HhaI* or *RsaI* at 37°C for at least 16 hours. The restriction digests were purified and analyzed by capillary gel electrophoresis (3730 DNA Analyzer, Applied Biosystems). The size standard LIZ1200 (Applied Biosystems) was used for fragment size determination. The software GeneMapper (Applied Biosystems) was used to quantify the electropherogram data and to generate the terminal restriction fragment (T-RF) profiles. Peaks from fragments of size 50-1020 bases with a height above 100 fluorescent units were analyzed. The total fluorescence of a sample was defined as the sum of the heights of all the peaks in the profile and was interpreted as a measure of the amount of DNA that was loaded on the capillary gel. The T-RFs of the two profiles for each primer/enzyme combination were normalized as described by Dunbar et al [45], aligned using a moving average procedure [46] and then checked manually for errors. The two profiles were combined to a single consensus profile by taking the average size, height and areas of the fragments present in both. Consensus profiles that were compared were also normalized and aligned in the same way as the two replicate profiles. To allow for comparisons of the T-RF profiles generated with 27F&1492R and 63F&M1387R, 35 bases was added to the lengths of all T-RFs in the 63F&M1387R profiles. The relative abundance of a T-RF was calculated as the peak height of that T-RF divided by the sum of all peak heights in the profile.

3.5.5 ORDINATION ANALYSIS

Ordination analysis of all T-RF profiles was carried out using Bray-Curtis distances (described in [47]) calculated from relative abundance data. The

Bray-Curtis distance coefficient is a semi-metric distance measure, i.e. not strictly metric, and therefore it cannot be used for principal coordinate analysis unless a correction for negative eigenvalues is carried out [47]. It can however be used for ordination by non-metric multidimensional scaling (NMDS). NMDS of Bray-Curtis distance matrices was carried out using the software Primer 6 (Primer-E). The analysis was performed using 250 repetitions, Kruskal stress formula number 1 and a minimum stress of 0.01.

3.5.6 ANOSIM

To test if there was a significant difference between the T-RF profiles generated with 27F&1492R and 63F&M1387R, an analysis of similarity (ANOSIM) was carried out using the software PAST [48]. ANOSIM is a nonparametric multivariate procedure to test the significance of differences between groups of samples. The distances between all samples are converted to ranks and the ranks of the distances between the groups are compared with the ranks of the distances within the groups. A test statistic R is calculated which can have values between -1 and 1, where large positive values signify dissimilarity between the groups. The significance of the R-value is then calculated by Monte Carlo permutations where the samples are randomly assigned to the groups. The ANOSIM analysis was carried out using Bray-Curtis distances calculated from relative abundance data, 1000 Monte Carlo permutations and the T-RF profiles separated in two groups: profiles generated using 27F&1492R and profiles generated using 63F&M1387R.

3.5.7 CLONING AND SEQUENCING

16S rRNA gene libraries were generated from an activated sludge sample collected 07/15/04. For both primer pair 27F&1492R and primer pair 63F&M1387R, 16S rRNA genes were amplified in six replicate reactions as described above, with the exception that the forward primers were not labeled. The six replicate PCR-products were pooled and purified using

Qiagen QiaQuick PCR Purification Kit (Qiagen). 10 ng of purified PCR product were ligated into the plasmid vector pCR 4 TOPO (Invitrogen). One Shot DH5alpha-T1R competent Escherichia coli cells (Invitrogen) were transformed with the vector construct according to the manufacturer's instructions. Transformed cells were spread on LB-agar plates with 50 µg/ml Kanamycin and incubated at 37°C for 18 hours. For each library, 96 cloned sequences were amplified directly from transformed single colonies by PCR using the vector specific primers M13forward (GTA-AAACGACGGCCAG) and M13reverse (CAGGAAACAGCTATGAC). To amplify the cloned sequences, the bacterial cells were lyzed by 5 min incubation at 94°C, Taq polymerase was activated by 5 min at 95°C followed by 30 cycles of denaturation at 94°C for 45 s, annealing at 55°C for 45 s and elongation at 72°C for 1 min 45 s. The reactions were ended with a final elongation step at 72°C for 7 min. Sequencing was done using both M13forward and M13reverse as sequencing primers by Macrogen Inc. (South Korea).

3.5.8 SEQUENCE ANALYSIS

3.5.8.1 SEQUENCE PROCESSING

DNA Baser (v2.91.5) was used to remove vector sequences, to trim the sequences according to quality and to assemble sequences. In the cases were the 3' and 5' ends of the sequences could not be assembled the partial sequences were analyzed separately.

The sequences were checked for anomalies or chimeras in three ways:

1. The sequences were aligned using ClustalW 1.83 [49] with default settings. The alignment was used as input to Bellerophon [50]. Sequences marked as chimeric were removed from the alignment and the remaining sequences were analyzed again. This was repeated until no chimeric sequences were detected.
2. The sequences were aligned using the greengenes web application [51] with default settings. The aligned sequences were used as input to the greengenes implementation of Bellerophon. Here each

sequence is checked not only against the sequences in the clone library but also against the greengenes database of non-chimeric sequences. The similarity threshold was set to 99% and the divergence ratio was set to 1.

3. The sequences were aligned together with an Escherichia coli sequence (accession number U00096) using ClustalW 1.83 [49] with default settings. The alignment was then used as input to the analysis tool Mallard [52]. Sequences marked as possibly anomalous were further checked following the anomaly confirmation protocol suggested by Ashelford et al [53]. In brief, a possible anomalous sequence is analyzed together with reference sequences retrieved by BLAST using Pintail [53].

Sequences marked as chimeric or anomalous in any of the three analyses were removed.

After removal of chimeric sequences and sequences shorter than 450 bases, a total of 77 and 63 sequences were analyzed in the 27F and the 63F library, respectively. Of these 41 and 57 were near full-length assembled sequences. The remaining sequences were either only 5'-end or 3'-end sequences or 5' and 3'-ends from the same clone that were too short to be assembled. The sequences are available in GenBank under accession numbers KC633451-KC633553 (sequences amplified by 27F&1492R) and KC633554-KC633617 (sequences amplified by 63F&M1387R).

3.5.8.2 RICHNESS ANALYSIS

The non-chimeric sequences were aligned using ClustalW with default settings. Alignment of all non-chimeric sequences at the same time resulted in incorrect alignment of the 3'-end sequences and the sequences where therefore aligned in two separate sets: 1) the 5'-end sequences together with the assembled sequences, and 2) the 3'-end sequences together with the assembled sequences. In the latter the sequences were first converted to reverse complement, or anti-sense, sequences, so that they started with the reverse primer sequence. The alignments were used as input to Dnadist (the Phylip package [54]) and analyzed using the F84 distance and standard settings. The distance matrix produced by Dnadist was then converted to a similarity matrix. There were slight differences between the similarities generated from the alignment of assembled and 5'-end sequences and the similarities from the alignment of the assembled and 3-end sequences. For the assembled sequences, which were included in both data sets, the differences were due to small differences in the alignments. The unassembled 5' and 3'-ends from the same clone showed differences in similarity with the assembled sequences, because the similarity was based on different sections of the gene (the 5' end and 3' end). For all clones with sequences included in both data sets, i.e. either an assembled sequence or both a 5' and a 3'-end sequence, the similarity with the other clones was recalculated as the average similarity of the similarities from both the 5'-end alignment and the 3'-end alignment. For example: In the 5'-end alignment the 5'–end sequence of clone D59 was determined to be 97.1% similar to the assembled sequence of clone D40 and in the 3'-end alignment the 3'-end sequence of clone D59 was determined to be 95.1% similar to clone D40. The similarity between clone D59 and D40 was then calculated to be 96.1%, the average of 97.1% and 95.1%. The similarity between a clone with only a 3'-end sequence and a clone with only a 5'-end sequence (or vice versa) was set to 0. After calculation of similarity values the clones were grouped in OTUs based on a similarity threshold of 98.7%—representing species [55], 95%—representing genus, 90%—representing family/class and 80%—representing phylum [56]. The observed frequencies of the OTUs were used as input to the program SPADE [57] and the richness of the community was estimated. The Chao1 estimator was used as a lower bound estimate of the richness.

3.5.8.3 LIBSHUFF

The sequences of the two gene libraries were aligned in two separate sets: 1) 5'-end alignment, including assembled sequences and 5'-end sequences, and 2) 3'-end alignment, including assembled sequences and 3'-end sequences. In the latter the sequences were first converted to reverse complement, or anti-sense, sequences, so that they started with the reverse primer sequence. The aligned sequences were analyzed using Dnadist (the

Phylip package [54]) with the F84 distance and standard settings and used as input to LIBSHUFF [34]. LIBSHUFF compares two samples, or sequence libraries, by calculating differences between homologous coverage curves, and heterologous coverage curves. The coverage C is calculated by counting the number of unique sequences at a given evolutionary distance threshold D and a coverage curve is generated by calculating the coverage for a range of different evolutionary distances. To calculate the homologous coverage, C_x, the number of unique sequences is counted by comparing each sequence with the other sequences in the same sample. To calculate the heterologous coverage, C_{XY}, the number of unique sequences is counted by comparing each sequence with the sequences in the other sample. Similar homologous and heterologous coverage curves are an indication that the two samples are similar. In addition, LIBSHUFF pools the two samples and randomly separates the sequences into two new samples of the same size as the original samples. This is done 999 times and the differences between the samples in each pair of randomly generated samples are compared with the difference between the two original samples to determine if the latter is significant.

3.5.8.4 CLASSIFICATION.

The sequences were classified using the RDP classifier [58]. For additional identification the sequences in the gene libraries were compared with sequences in GenBank using BLAST [59]. The BLAST searches were done 11/12/2012 and 11/13/2012.

3.5.8.5 CLONE LIBRARY COMPARISONS AND COMBINATIONS

To compare the two libraries the non-chimeric sequences from both clone libraries were aligned together and analyzed and divided into OTUs as described above. To combine the two libraries and get overall ratios of different taxa and phylogenetic groups, the number of sequences of all OTUs (at OTU division threshold 98.7%) was related to the number of sequences in the common OTU of phyla *Acidobacteria*. For the OTUs of

Alphaproteobacteria and *Betaproteobacteria* that were common to both libraries the average of the new ratios was used.

The complementarities of the sequences in the 27F1492R library with the 63F and M1387R primers were also analyzed. Only assembled sequences and sequences with both the 5' and 3'-ends, with sequence data starting before the 63F site and ending after the M1387R site were evaluated (see Material S1 for results).

3.5.8.6 PARETO-LORENZ EVENNESS CURVES

A Pareto-Lorenz evenness curve (see for example 60,61 for explanation and usage) was used to illustrate and quantify the evenness of the different sequence sets. The sequences were divided in OTUs based on phyla and the *Proteobacteria* classes and the OTUs were ranked from high to low, based on their abundance. The cumulative proportion of OTU abundances (Y) was then plotted against the cumulative proportion of OTUs (X) resulting in a concave curve starting at (X, Y) = (0%, 0%) and ending in (X, Y) = (100%, 100%). The functional organization (Fo) index is the horizontal y-axis projection on the intercept with the vertical 20% x-axis line, i.e. the combined relative abundance of 20% of the OTUs. In a community with high evenness all or most OTUs are equally abundant which results in a Pareto-Lorenz curve close to a straight line of 45°. The Fo index for such a community is close to 20%. Specialized communities with one or a few dominating OTUs generate concave curves with high Fo indices.

3.5.9 COMPLETE RDP DATABASE SEARCH

The RDP tool Probe Match ([62], accessed 09/28/12) was used to compare the primer pairs 27F&1492R, 63F & 1387R and 63F&M1387R. The search was restricted to the domain Bacteria and but with no restriction on region, i.e. sequences of all lengths were searched. The resulting dataset was refined using the following dataset options: Both type and non type strains, both uncultured and isolates, both sequences longer and shorter than 1200 bases and only good sequences (low quality sequences were removed). The total number of sequences included in the search and the number of matches allowing 0, 1, 2 and 3 mismatches were noted. For each number of allowed mismatches (0, 1, 2 and 3), the following procedure was carried out:

1. A list of the targeted sequences was downloaded as a text file.
2. From the text files, the RDP IDs were extracted and a list of the RDP IDs were saved as a new text file.
3. Lists with the combinations (intersection, complement, unique) of the RDP ID lists for the different primers were constructed using a Perl script with the Compare::List module (available from corresponding author). The number of sequences in each of the combination lists was noted.

Subsequently, for the datasets generated by allowing 1 mismatch, the following procedure was carried out:

1. The RDP ID lists were uploaded to Sequence Cart in RDP and the corresponding sequences were retrieved.
2. The sequences in Sequence Cart were classified in RDP Classifier.
3. The hierarchy and the list of sequences from the RDP classifier was downloaded at confidence level 95%.

3.5.10 GENERATION OF A DATASET WITH ONLY ACTIVATED SLUDGE SEQUENCES

A search in the NCBI Nucleotide database (http://www.ncbi.nlm.nih. gov/nucleotide, accessed 10/03/12) was done using the search term ((600:2000[Sequence Length]) AND "activated sludge"). The search result was saved as a list of accession numbers and uploaded to RDP Sequence Cart. The resulting dataset of retrieved sequences could not be refined like the datasets generated by Probe Match and thus included sequences of low quality. The retrieved sequences were classified using RDP Classifier, and the hierarchy was downloaded at confidence level 95%. The list of RDP IDs of the activated sludge sequences was compared with the

lists of the sequences in the complete RDP database matching primer pairs
27F&1492R and 63F&M1387R as described above.

3.5.11 EVALUATION OF THE PRIMERS USING THE SILVA TESTPRIME TOOL

The tool TestPrime, version 1.0, (http://www.arb-silva.de/search/testprime/,
accessed 06/02/2013) which is a part of the SILVA ribosomal RNA gene
database project [8], was used to evaluate the primer pairs 27F&1492R
and 63F&M1387R. The SSU Ref NR database (release 114) was used allowing no mismatches.

3.5.12 INSPECTION OF SEQUENCES AND EVALUATION OF MISMATCHES

The sequences retrieved in the BLAST search that was used for classification were evaluated. Sequences that were more than 97% similar to a
sequence of the order *Rhodobacterales*, *Rhizobiales* or *Xanthomonadales*
from the 63F&M1387R library were considered, and if long enough, used
for comparison with the 27F&1492R primer pair. The primer sites were
located manually using the software BioEdit [63] and the mismatches
were identified.

The RDP Probe Match tool and RDP Classifier were used a second
time to retrieve sequences that were targeted by only one of the primer
pairs or both (database accessed 06/02/2013). For *Burkholderiales*, *Rhodocyclales* and *Bacillales*, sequences only targeted by 27F&1492R and
for *Rhodobacterales*, *Rhizobiales* and *Xanthomonadales*, sequences only
targeted by 63F&M1387R. The same procedure as above was used but the
search was restricted to the domain Bacteria and sequences with data from
E. coli position 6 to 1515. No mismatches were allowed and all sequences
were included (both type and non type strains, both uncultured and isolates, both sequences longer and shorter than 1200 bases and both high and
low quality sequences). After retrieval of accession numbers using RDP
and the procedure described above, the sequences were obtained from the

Nucleotide database (http://www.ncbi.nlm.nih.gov/nucleotide/, accessed 06/02/2013).

The primer sites were located manually using the software BioEdit [63] and the mismatches were identified. For the sequences from the 27F&1492R library, the same mismatches were seen as in the RDP database sequences. However, the sequences in the 63F&M1387R library are too short to include the target sites of the primer pair 27F&1492R and the same comparison could not be made. An analysis was carried out to determine if the 63F&M1387R library sequences were more similar to the RDP sequences that were targeted only by 63F&M1387R or to the sequences targeted by both 63F&M1387R and 27F&1492R. The sequences of the orders *Rhodobacterales*, *Rhizobiales* and *Xanthomonadales* matching either only 63F&M1387R or both 63F&M1387R and 27F&1492R were retrieved as described above. For each order, the two sets of sequences were aligned with the 63F&M1387R library sequences from that order using ClustalW with standard settings. The aligned sequences were then analyzed using Dnadist (the Phylip package [54]) with the F84 distance and standard settings to obtain the similarities between the sequences.

3.5.13 FLUORESCENCE IN SITU HYBRIDIZATION

Activated sludge samples were fixed in 4% paraformaldehyde as previously described [64]. After fixation, 5 mL of sample were filtered onto 0.2 μm pore size membrane filter and washed with 1X PBS directly onto the filter placed in the filter holder. The samples were hybridized as described by Amann [65] for 1.5 h. Oligonucleotide probes were synthesized and 5' labeled with the fluorochrome fluorescein isothiocyanate (FITC) or one of the sulfoindocyanine dyes Cy3 and Cy5 (Thermohybaid Interactiva, Ulm, Germany). The supernatant samples were hybridized directly on the filter. All bacteria were detected by hybridizing with a mixture of EUB338, EUB338 II and EUB338 III (called EUBMIX) [66,67]. *Alphaproteobacteria*, *Betaproteobacteria* and *Gammaproteobacteria* were detected by the probes ALF1b, BET42a and GAM42a [64]. The FISH slides were viewed with a BioRad Radiance 2000 CLSM equipped with 60x inverted objective (oil immersion Nikon Eclipse TE300 Corp, Tokyo, Japan). Excitation of FITC, Cy3 and Cy5 were done at 488 nm (Ar laser), 543 nm (HeNe laser) and 637 nm (red diode laser), respectively. Emissions were collected with filters 515-530 nm BP(HQ) for FITC, 590-570 nm BP(HQ) for Cy3 and 660 nm LP for Cy5. The collected images were finally processed using Adobe Photoshop (Adobe Systems Inc., USA). For quantification at least 10 z-series with 1 µm (6-24 sections) steps at no zoom applied were made for each sludge suspension sample and at least 10 images were taken of each supernatant sample on filters. The surface coverage of probe-positive cells was analyzed with the software COMSTAT [68] with the threshold set manually. The percentage coverage in relation to cells binding to EU-BMIX was calculated and used as an estimate of the relative abundance of the probe-defined bacterial groups.

REFERENCES

1. Whitman WB, Coleman DC, Wiebe WJ (1998) Prokaryotes: The unseen majority. Proc Natl Acad Sci USA 95: 6578-6583. doi:10.1073/pnas.95.12.6578. PubMed: 9618454.
2. Dykhuizen DE (1998) Santa Rosalia revisited: Why are there so many species of bacteria? Antonie Van Leeuwenhoek 73: 25-33. doi:10.1023/A:1000665216662. PubMed: 9602276.
3. Curtis TP, Sloan WT, Scannell JW (2002) Estimating prokaryotic diversity and its limits. Proc Natl Acad Sci USA 99: 10494-10499. doi:10.1073/pnas.142680199. PubMed: 12097644.
4. Hugenholtz P (2002) Exploring prokaryotic diversity in the genomic era. Genome Biol 3: reviews0003. 0001 - reviews0003. 0008. PubMed: 11864374.
5. Baker GC, Smith JJ, Cowan DA (2003) Review and re-analysis of domain-specific 16S primers. J Microbiol Methods 55: 541-555. doi:10.1016/j.mimet.2003.08.009. PubMed: 14607398.
6. Klindworth A, Pruesse E, Schweer T, Peplies J, Quast C et al. (2013) Evaluation of general 16S ribosomal RNA gene PCR primers for classical and next-generation sequencing-based diversity studies. Nucleic Acids Res 41: e1. doi:10.1093/nar/gks1297. PubMed: 22933715.
7. Shyu C, Soule T, Bent SJ, Foster JA, Forney LJ (2007) MiCA: A Web-Based Tool for the Analysis of Microbial Communities Based on Terminal-Restriction Fragment Length Polymorphisms of 16S and 18S rRNA Genes. Microb Ecol 53: 562-570. doi:10.1007/s00248-006-9106-0. PubMed: 17406775.
8. Quast C, Pruesse E, Yilmaz P, Gerken J, Schweer T et al. (2013) The SILVA ribosomal RNA gene database project: improved data processing and web-based tools. Nucleic Acids Res 41: D590-D596. doi:10.1093/nar/gks1219. PubMed: 23193283.

9. Cole JR, Chai B, Marsh TL, Farris RJ, Wang Q et al. (2003) The Ribosomal Database Project (RDP-II): previewing a new autoaligner that allows regular updates and the new prokaryotic taxonomy. Nucleic Acids Res 31: 442-443. doi:10.1093/nar/gkg039. PubMed: 12520046.

10. Morales SE, Holben WE (2009) Empirical Testing of 16S rRNA Gene PCR Primer Pairs Reveals Variance in Target Specificity and Efficacy Not Suggested by In Silico Analysis. Appl Environ Microbiol 75: 2677-2683. doi:10.1128/AEM.02166-08. PubMed: 19251890.

11. Sipos R, Székely AJ, Palatinszky M, Révész S, Márialigeti K et al. (2007) Effect of primer mismatch, annealing temperature and PCR cycle number on 16S rRNA gene-targeting bacterial community analysis. FEMS Microbiol Ecol 60: 341-350. doi:10.1111/j.1574-6941.2007.00283.x. PubMed: 17343679.

12. Fortuna AM, Marsh TL, Honeycutt CW, Halteman WA (2011) Use of primer selection and restriction enzymes to assess bacterial community diversity in an agricultural soil used for potato production via terminal restriction fragment length polymorphism. Appl Microbiol Biotechnol 91: 1193-1202. doi:10.1007/s00253-011-3363-7. PubMed: 21667276.

13. Hong SH, Bunge J, Leslin C, Jeon S, Epstein SS (2009) Polymerase chain reaction primers miss half of rRNA microbial diversity. Isme J 3: 1365-1373. doi:10.1038/ismej.2009.89. PubMed: 19693101.

14. Lowe BA, Marsh TL, Isaacs-Cosgrove N, Kirkwood RN, Kiupel M et al. (2010) Microbial communities in the tonsils of healthy pigs. Vet Microbiol 147: 346-357. PubMed: 20663617.

15. Bramucci M, Kane H, Chen M, Nagarajan V (2003) Bacterial diversity in an industrial wastewater bioreactor. Appl Microbiol Biotechnol 62: 594-600. doi:10.1007/s00253-003-1372-x. PubMed: 12827322.

16. Kemp P, [!(surname)!] , Aller J, [!(surname)!] (2004) Estimating prokaryotic diversity: When are 16S rDNA libraries large enough? Limnol Oceanogr Methods 2: 114-125. doi:10.4319/lom.2004.2.114.

17. McMahon KD, Martin HG, Hugenholtz P (2007) Integrating ecology into biotechnology. Curr Opin Biotechnol 18: 287-292. doi:10.1016/j.copbio.2007.04.007. PubMed: 17509863.

18. Curtis TP, Sloan WT (2006) Towards the design of diversity: stochastic models for community assembly in wastewater treatment plants. Water Sci Technol 54: 227-236. doi:10.2166/wst.2006.391. PubMed: 16898156.

19. Yuan Z, Blackall LL (2002) Sludge population optimisation: a new dimension for the control of biological wastewater treatment systems. Water Res 36: 482-490. doi:10.1016/S0043-1354(01)00230-5. PubMed: 11827354.

20. Nielsen PH, Mielczarek AT, Kragelund C, Nielsen JL, Saunders AM et al. (2010) A conceptual ecosystem model of microbial communities in enhanced biological phosphorus removal plants. Water Res 44: 5070-5088. doi:10.1016/j.watres.2010.07.036. PubMed: 20723961.

21. Daims H, Taylor MW, Wagner M (2006) Wastewater treatment: a model system for microbial ecology. Trends Biotechnol 24: 483-489. doi:10.1016/j.tibtech.2006.09.002. PubMed: 16971007.

22. Sloan WT, Lunn M, Woodcock S, Head IM, Nee S et al. (2006) Quantifying the roles of immigration and chance in shaping prokaryote community structure. Environ Microbiol 8: 732-740. doi:10.1111/j.1462-2920.2005.00956.x. PubMed: 16584484.

23. Figuerola EL, Erijman L (2007) Bacterial taxa abundance pattern in an industrial wastewater treatment system determined by the full rRNA cycle approach. Environ Microbiol 9: 1780-1789. doi:10.1111/j.1462-2920.2007.01298.x. PubMed: 17564611.

24. Wells GF, Park HD, Eggleston B, Francis CA, Criddle CS (2011) Fine-scale bacterial community dynamics and the taxa-time relationship within a full-scale activated sludge bioreactor. Water Res 45: 5476-5488. doi:10.1016/j.watres.2011.08.006. PubMed: 21875739.

25. Saikaly PE, Oerther DB (2004) Bacterial Competition in Activated Sludge: Theoretical Analysis of Varying Solids Retention Times on Diversity. Microb Ecol 48: 274-284. doi:10.1007/s00248-003-1027-6. PubMed: 15116279.

26. Lane DJ (1991) 16S/23S rRNA sequencing. In: E. StackebrandtM. Goodfellow. Nucleic acid techniques in bacterial systematics. New York, N.Y.: John Wiley & Sons, Inc.. pp. 115-176.

27. Yang C, Zhang W, Liu RH, Li Q, Li BB et al. (2011) Phylogenetic Diversity and Metabolic Potential of Activated Sludge Microbial Communities in Full-Scale Wastewater Treatment Plants. Environ Sci Technol 45: 7408-7415. doi:10.1021/es2010545. PubMed: 21780771.

28. Layton AC, Karanth PN, Lajoie CA, Meyers AJ, Gregory IR et al. (2000) Quantification of Hyphomicrobium Populations in Activated Sludge from an Industrial Wastewater Treatment System as Determined by 16S rRNA Analysis. Appl Environ Microbiol 66: 1167-1174. doi:10.1128/AEM.66.3.1167-1174.2000. PubMed: 10698787.

29. Jin DC, Wang P, Bai ZH, Wang XX, Peng H et al. (2011) Analysis of bacterial community in bulking sludge using culture-dependent and -independent approaches. J Environ Sci China 23: 1880-1887. doi:10.1016/S1001-0742(10)60621-3. PubMed: 22432314.

30. Kong Y, Xia Y, Nielsen JL, Nielsen PH (2007) Structure and function of the microbial community in a full-scale enhanced biological phosphorus removal plant. Microbiology 153: 4061-4073. doi:10.1099/mic.0.2007/007245-0. PubMed: 18048920.

31. Marchesi JR, Sato T, Weightman AJ, Martin TA, Fry JC et al. (1998) Design and Evaluation of Useful Bacterium-Specific PCR Primers That Amplify Genes Coding for Bacterial 16S rRNA. Appl Environ Microbiol 64: 795-799. PubMed: 9464425.

32. Chouari R, Le Paslier D, Daegelen P, Ginestet P, Weissenbach J et al. (2005) Novel predominant archaeal and bacterial groups revealed by molecular analysis of an anaerobic sludge digester. Environ Microbiol 7: 1104-1115. doi:10.1111/j.1462-2920.2005.00795.x. PubMed: 16011748.

33. Zhang T, Shao MF, Ye L (2012) 454 Pyrosequencing reveals bacterial diversity of activated sludge from 14 sewage treatment plants. Isme J 6: 1137-1147. doi:10.1038/ismej.2011.188. PubMed: 22170428.

34. Singleton DR, Furlong MA, Rathbun SL, Whitman WB (2001) Quantitative Comparisons of 16S rRNA Gene Sequence Libraries from Environmental Samples.

Appl Environ Microbiol 67: 4374-4376. doi:10.1128/AEM.67.9.4374-4376.2001. PubMed: 11526051.

35. Akarsubasi AT, Eyice O, Miskin I, Head IM, Curtis TP (2009) Effect of sludge age on the bacterial diversity of bench scale sequencing batch reactors. Environ Sci Technol 43: 2950-2956. doi:10.1021/es8026488. PubMed: 19475976.

36. Nadarajah N, Allen DG, Fulthorpe RR (2007) Effects of transient temperature conditions on the divergence of activated sludge bacterial community structure and function. Water Res 41: 2563-2571. doi:10.1016/j.watres.2007.02.002. PubMed: 17448516.

37. Pholchan MK, Baptista JD, Davenport RJ, Curtis TP (2009) Systematic study of the effect of operating variables on reactor performance and microbial diversity in laboratory-scale activated sludge reactors. Water Res 44: 1341-1352. doi: 10.1016/j. watres.2009.11.005

38. Valentín-Vargas A, Toro-Labrador G, Massol-Deyá AA (2012) Bacterial Community Dynamics in Full-Scale Activated Sludge Bioreactors: Operational and Ecological Factors Driving Community Assembly and Performance. PLOS ONE 7: 12. PubMed: 22880016.

39. Dedysh SN, Ricke P, Liesack W (2004) NifH and NifD phylogenies: an evolutionary basis for understanding nitrogen fixation capabilities of methanotrophic bacteria. Microbiology 150: 1301-1313. doi:10.1099/mic.0.26585-0. PubMed: 15133093.

40. Urakami T, Sasaki J, Suzuki K-I, Komagata K (1995) Characterization and Description of Hyphomicrobium denitrificans sp. nov. Int J Syst Bacteriol 45: 528-532. doi:10.1099/00207713-45-3-528.

41. Wartiainen I, Hestnes AG, McDonald IR, Svenning MM (2006) Methylocystis rosea sp. nov., a novel methanotrophic bacterium from Arctic wetland soil, Svalbard, Norway (78° N). Int J Syst Evol Microbiol 56: 541-547. doi:10.1099/ijs.0.63912-0. PubMed: 16514024.

42. Khan ST, Horiba Y, Yamamoto M, Hiraishi A (2002) Members of the Family Comamonadaceae as Primary Poly(3-Hydroxybutyrate-co-3-Hydroxyvalerate)-DegradingDenitrifiers in Activated Sludge as Revealed by a Polyphasic Approach. Appl Environ Microbiol 68: 3206-3214. doi:10.1128/AEM.68.7.3206-3214.2002. PubMed: 12088996.

43. Saikaly PE, Stroot PG, Oerther DB (2005) Use of 16S rRNA Gene Terminal Restriction Fragment Analysis To Assess the Impact of Solids Retention Time on the Bacterial Diversity of Activated Sludge. Appl Environ Microbiol 71: 5814-5822. doi:10.1128/AEM.71.10.5814-5822.2005. PubMed: 16204492.

44. Wilén BM, Lumley D, Mattsson A, Mino T (2010) Dynamics in Flocculation and Settling Properties Studied at a Full-Scale Activated Sludge Plant. Water Environ Res 82: 155-168. doi:10.2175/106143009X426004. PubMed: 20183982.

45. Dunbar J, Ticknor LO, Kuske CR (2001) Phylogenetic Specificity and Reproducibility and New Method for Analysis of Terminal Restriction Fragment Profiles of 16S rRNA Genes from Bacterial Communities. Appl Environ Microbiol 67: 190-197. doi:10.1128/AEM.67.1.190-197.2001. PubMed: 11133445.

46. Smith CJ, Danilowicz BS, Clear AK, Costello FJ, Wilson B et al. (2005) T-Align, a web-based tool for comparison of multiple terminal restriction fragment length

polymorphism profiles. FEMS Microbiol Ecol 54: 375-380. doi:10.1016/j.femsec.2005.05.002. PubMed: 16332335.

47. Legendre P, Legendre L (1998) Numerical Ecology. Amsterdam: Elsevier Science BV..

48. Hammer Ö, Harper DAT, Ryan PD (2001) PAST: Paleontological Statistics Software Package for Education and Data Analysis. Palaeontol Electron 4.

49. Thompson JD, Higgins DG, Gibson TJ (1994) Clustal-W - Improving the Sensitivity of Progressive Multiple Sequence Alignment through Sequence Weighting, Position-Specific Gap Penalties and Weight Matrix Choice. Nucleic Acids Res 22: 4673-4680. doi:10.1093/nar/22.22.4673. PubMed: 7984417.

50. Huber T, Faulkner G, Hugenholtz P (2004) Bellerophon: a program to detect chimeric sequences in multiple sequence alignments. Bioinformatics 20: 2317-2319. doi:10.1093/bioinformatics/bth226. PubMed: 15073015.

51. DeSantis TZ, Hugenholtz P, Larsen N, Rojas M, Brodie EL et al. (2006) Greengenes, a Chimera-Checked 16S rRNA Gene Database and Workbench Compatible with ARB. Appl Environ Microbiol 72: 5069-5072. doi:10.1128/AEM.03006-05. PubMed: 16820507.

52. Ashelford KE, Chuzhanova NA, Fry JC, Jones AJ, Weightman AJ (2006) New Screening Software Shows that Most Recent Large 16S rRNA Gene Clone Libraries Contain Chimeras. Appl Environ Microbiol 72: 5734-5741. doi:10.1128/AEM.00556-06. PubMed: 16957188.

53. Ashelford KE, Chuzhanova NA, Fry JC, Jones AJ, Weightman AJ (2005) At Least 1 in 20 16S rRNA Sequence Records Currently Held in Public Repositories Is Estimated To Contain Substantial Anomalies. Appl Environ Microbiol 71: 7724-7736. doi:10.1128/AEM.71.12.7724-7736.2005. PubMed: 16332745.

54. Felsenstein J (2005) PHYLIP (Phylogeny Inference Package). 3.6 ed: Distributed by the author. Seattle: Department of Genome Sciences, University of Washington.

55. Stackebrandt E, Ebers J (2006) Taxonomic parameters revisited: tarnished gold standards Microbiology. Today: 152-155.

56. Schloss PD, Handelsman J (2004) Status of the Microbial Census. Microbiol Mol Biol Rev 68: 686-691. doi:10.1128/MMBR.68.4.686-691.2004. PubMed: 15590780.

57. Chao A, Shen TJ (2003) Program SPADE (Species Prediction And Diversity Estimation).

58. Wang Q, Garrity GM, Tiedje JM, Cole JR (2007) Naive Bayesian Classifier for Rapid Assignment of rRNA Sequences into the New Bacterial Taxonomy. Appl Environ Microbiol 73: 5261-5267. doi:10.1128/AEM.00062-07. PubMed: 17586664.

59. Altschul SF, Gish W, Miller W, Myers EW, Lipman DJ (1990) Basic Local Alignment Search Tool. J Mol Biol 215: 403-410. doi:10.1016/S0022-2836(05)80360-2. PubMed: 2231712.

60. Marzorati M, Wittebolle L, Boon N, Daffonchio D, Verstraete W (2008) How to get more out of molecular fingerprints: practical tools for microbial ecology. Environ Microbiol 10: 1571-1581. doi:10.1111/j.1462-2920.2008.01572.x. PubMed: 18331337.

61. Mertens B, Boon N, Verstraete W (2005) Stereospecific effect of hexachlorocyclohexane on activity and structure of soil methanotrophic communities. Environ Microbiol 7: 660-669. doi:10.1111/j.1462-2920.2005.00735.x. PubMed: 15819848.

62. Cole JR, Wang Q, Cardenas E, Fish J, Chai B et al. (2009) The Ribosomal Database Project: improved alignments and new tools for rRNA analysis. Nucleic Acids Res 37: D141-D145. doi:10.1093/nar/gkp353. PubMed: 19004872.

63. Hall TA (1999) BioEdit: a user-friendly biological sequence alignment editor and analysis program for Windows 95/98/NT. Nucleic Acids Symp Ser 41: 95-98.

64. Manz W, Amann R, Ludwig W, Wagner M, Schleifer KH (1992) Phylogenetic oligodeoxynucleotide probes for the major subclasses of *proteobacteria*: problems and solutions. Syst Appl Microbiol 15: 593-600. doi:10.1016/S0723-2020(11)80121-9.

65. Amann RI, Ludwig W, Schleifer KH (1995) Phylogenetic identification and in situ detection of individual microbial cells without cultivation. Microbiol Rev 59: 143-169. PubMed: 7535888.

66. Amann RI, Binder BJ, Olson RJ, Chisholm SW, Devereux R et al. (1990) Combination of 16s ribosomal-RNA-targeted oligonucleotide probes with flow-cytometry for analyzing mixed microbial-populations. Appl Environ Microbiol 56: 1919-1925. PubMed: 2200342.

67. Daims H, Brühl A, Amann R, Schleifer KH, Wagner M (1999) The domain-specific probe EUB338 is insufficient for the detection of all Bacteria: Development and evaluation of a more comprehensive probe set. Syst Appl Microbiol 22: 434-444. doi:10.1016/S0723-2020(99)80053-8. PubMed: 10553296.

68. Heydorn A, Nielsen AT, Hentzer M, Sternberg C, Givskov M et al. (2000) Quantification of biofilm structures by the novel computer program COMSTAT. Microbiology 146: 2395-2407. PubMed: 11021916.

There are several supplemental files that are not available in this version of the article. To view this additional information, please use the citation on the first page of this chapter.

Abundance and Diversity of Bacterial Nitrifiers and Denitrifiers and Their Functional Genes in Tannery Wastewater Treatment Plants Revealed by High-Throughput Sequencing

ZHU WANG, XU-XIANG ZHANG, XIN LU, BO LIU, YAN LI, CHAO LONG, AND AIMIN LI

4.1 INTRODUCTION

Tannery wastewater is characterized by high contents of nitrogen, organic compounds and salt ions [1],[2], and the complexity of the wastewater components can affect microbial nitrification/denitrification processes that have been frequently used for nitrogen removal in wastewater treatment plants (WWTPs). Nitrification, the biological oxidation of ammonium to nitrite and nitrate, is carried out in two sequential steps via several

Abundance and Diversity of Bacterial Nitrifiers and Denitrifiers and Their Functional Genes in Tannery Wastewater Treatment Plants Revealed by High-Throughput Sequencing. © Wang Z, Zhang X-X, Lu X, Liu B, Li Y, Long C, and Li A. PLoS ONE 9,1 (2014), doi:10.1371/journal.pone.0113603. Licensed under a Creative Commons Attribution 4.0 International License, http://creativecommons.org/licenses/by/4.0/.

phylogenetically distinct groups of microorganisms: ammonia-oxidizing archaea (AOA), ammonia-oxidizing bacteria (AOB) and nitrite-oxidizing bacteria (NOB) [3]. Denitrification consists of consecutive reactions including transformation of nitrate or nitrite into gaseous forms (N_2 or N_2O) by different groups of bacteria (nitrifiers and denitrifiers) [4].

Due to the technological difficulty in isolation and culture of bacterial nitrifiers and denitrifiers, molecular methods have been widely used to analyze diversity and abundance of microorganisms involved in nitrification/denitrification in various environments, such as soil [5] and activated sludge [4]. Previous studies have investigated the microbial community of activated sludge in tannery WWTPs through 16S rRNA gene amplification and DNA sequencing [1],[6]. However, information about abundance and diversity of nitrifying and denitrifying bacteria in tannery WWTPs is limited due to the low throughput of those techniques.

Recently, high-throughput sequencing techniques, such as 454 pyrosequencing and Illumina sequencing methods, have shown great advantages on microbial community analysis due to their unprecedented sequencing depth, which have been used to deeply explore microbial communities in drinking water [7], soil [8] and WWTPs [9]–[11]. However, the metagenomic methods have not been used for comprehensive insights into the communities of nitrifies and denitrifies in tannery WWTPs.

In this study, 454 pyrosequencing of 16S rRNA gene was conducted to determine the bacterial community structure of activated sludge sampled from two full-scale tannery WWTPs, and Illumina high-throughput sequencing was used to comprehensively investigate functional genes involved in the bacterial nitrification and denitrification. To confirm the metagenomic results, quantitative real time PCR (qPCR) was used to quantify the functional genes encoding subunits of the ammonia monooxygenase (*amoA*), nitrite reductases (*nirK* and *nirS*) and nitrous oxide reductase (*nosZ*). Since ammonia oxidation is a rate-limiting step in nitrification, the *amoA* gene was chosen for genetic diversity analysis using DNA cloning and sequencing technology. The results of this study may help to extend our knowledge about the microbial nitrification and denitrification processes and their underlying biological mechanisms for industrial wastewater treatment.

4.2 MATERIALS AND METHODS

4.2.1 SLUDGE SAMPLING AND DNA EXTRACTION

In this study, activated sludge samples were separately collected from two full-scale WWTPs treating tannery wastewater of Boao Leather Industry Co., Ltd. (WWTPA geographically located in Xiangcheng City, Henan Province, China) and Yige Leather Industry Co., Ltd. (WWTPB geographically located in Zhecheng County, Henan Province, China). We would like to state that the two companies have approved this study which did not involve endangered or protected species. The two WWTPs basically involve a biological treatment system preceded by preliminary treatment (homogenization, chemical coagulation and primary settling). As described in our previous study [12], the biological treatment system of WWTPA was composed of an up-flow anaerobic sludge reactor (UASB) and an integrated anoxic/oxic (A/O) reactor, while WWTPB was composed of an oxidation ditch and an integrated anoxic/oxic (A/O) reactor. Relevant operational parameters about the two biological systems are shown in Table S1. Although the two WWTPs differed in the influent water quality and operational conditions, they both achieved high ammonia removal efficiencies (>97%). Thus, we chose the two WWTPs to explore the functional nitrifiers and denitrifiers responsible for the high ammonia removal. Four sludge samples were collected from the two WWTPs, including one anaerobic sludge sample (A-A) from the UASB and one aerobic sludge sample (A-O) from the last aerobic tank of the A/O reactor in WWTPA, and two aerobic sludge samples separately collected from the oxidation ditch (B-D) and the last aerobic tank of the A/O reactor (B-O) in WWTPB. The sludge samples were fixed with 50% ethanol (v/v) on site before transporting to laboratory for DNA extraction. The fixed sludge was centrifuged at 4,000 rpm for 10 min to collect approximately 200 mg of the pellet for total genomic DNA extraction with the FastDNA Soil Kit (MP Biomedicals, CA, USA). The concentration and quality of the extracted DNA were determined with microspectrophotometry (NanoDrop ND-1000, NanoDrop Technologies, Willmington, DE, USA).

4.2.2 PCR AMPLIFICATION, CLONING, SEQUENCING AND PHYLOGENETIC ANALYSIS OF AMOA GENE

PCR of AOA and AOB *amoA* genes was conducted by using the primers listed in Table S2. The 30-μL reaction mixture contained 3 μL of 10×buffer, 1.8 μL of 25 mM $MgCl_2$, 1.2 μL of 10 mM dNTP mixture, 0.2 μL of Ex Taq polymerase (TaKaRa, Japan), 0.3 μL of each primer (10 μM) and 20–50 ng of genomic DNA. The PCR conditions were initial denaturation at 95°C for 3 min, followed by 35 cycles of 94°C for 60 s, 56°C (for AOB *amoA*) or 53°C (for AOA *amoA*) for 45 s, and 72°C for 60 s; with a final extension at 72°C for 10 min. PCR products were analyzed by gel electrophoresis using 1% (w/v) agarose in 1×TAE buffer.

PCR products were purified by using DNA Fragment Purification Kit (TaKaRa, Japan). The purified PCR products were cloned using pMD19-T Vector (TaKaRa, Japan). An AOB *amoA* gene clone library was constructed for each of the four sludge samples; however, AOA *amoA* gene library was constructed for only sample A-A since the other samples failed to generate PCR products. About 20 clones in each clone library were randomly selected for DNA sequencing. The obtained *amoA* gene sequences were aligned and the Jukes-Cantor distances between subsequent pairs of sequences were calculated with DNADIST from the PHYLIP package (http://www.phylip.com/), and operational taxonomic units (OTUs) were defined with a distance cut-off of 3% using the DOTUR program [13]. In order to construct the phylogenetic tree, one representative sequence in each OTU was selected and aligned with the reference sequences from National Center for Biotechnology Information (NCBI) (http://www.ncbi. nlm.nih.gov/). The neighbor-joining phylogenetic trees of AOB and AOA *amoA* genes sequences were separately created by MEGA 5.1 software (http://www.megasoftware.net/).

4.2.3 QPCR

AOB and AOA *amoA* genes were used as molecular markers to determine the abundance of the nitrifiers. Meanwhile, the abundance of denitrifying bacteria was investigated by quantifying the genes *nirK*, *nirS* and *nosZ*.

qPCR was performed in Corbett Real-Time PCR Machine with the Rotor-Gene 6000 Series Software 1.7 (QIAGEN, the Netherlands) using SYBR Green method. The 20-µL PCR mixture contained 10 µL of SYBR Premix Ex Taq Super Mix (TaKaRa Japan), 0.2 µL of each primer (10 µM) (Table S2), 8 µL of template DNA corresponding to 40 ng of total DNA and 1.6 µL of ddH$_2$O. The reaction was initially denatured at 95°C for 3 min, followed by 40 cycles of denaturation at 95°C for 20 s, annealing at the given temperatures (Table S2) for 20 s and extension at 72°C for 40 s.

To determine gene abundance in one ng of extracted DNA, all targeted genes were cloned to plasmids following the method recommended by Zhang and Zhang [14] to generate qPCR standard curves. In order to correct for potential variations in DNA extraction efficiencies, eubacterial 16S rRNA genes were also quantified using the method recommended by Lopez-Gutierrez et al. [15]. All samples and standards were analyzed in triplicate, and the specificity of qPCR products was checked by melt curves observation and agarose electrophoresis. PCR efficiency ranged from 85.9% to 105.5% with R^2 values over 0.991 for all calibration curves.

4.2.4 454 PYROSEQUENCING AND ILLUMINA HIGH-THROUGHPUT SEQUENCING

Pyrosequencing and Illumina sequencing were conducted at Beijing Genome Institute (Shenzhen, China). For pyrosequencing, the bacterial DNA was amplified with a set of primers targeting the hypervariable V3-V4 region (about 460 bp) of the 16S rRNA gene. The forward primer was V3F (5'-ACTCCTACGGGAG GCAGCAG-3') and the reverse primer was V4R (5'-TACNVGGGTATCTAATCC-3'). Equal amounts of purified amplicon products bearing individual 10 nucleotide barcode from different samples were mixed for pyrosequencing on the Roche 454 FLX Titanium platform (Roche, USA).

DNA samples (10 µg each) were subject to high-throughput sequencing using Illumina Hiseq 2000 (Illumina, USA). A library consisting of 180-bp DNA fragment sequences was constructed according to the manufacturer's instructions before DNA sequencing. The strategy "Index 101 PE" (Paired End sequencing, 101-bp reads and 8-bp index sequence) was used for the Illumina sequencing, generating nearly equal amount of data for each sample. Low quality reads were removed following the method recommended by Shi et al. [7]. The 100-bp clean reads (about 1.0 Gb per each sample) were used for subsequent metagenomic analysis.

4.2.5 BIOINFORMATIC ANALYSIS ON PYROSEQUENCING DATASETS

After pyrosequencing, downstream sequence analyses were performed using the Ribosomal Database Project (RDP) (http://rdp.cme.msu.edu/) [16] and Mothur (http://www.mothur.org/) [17] following the method recommended by Zhang et al. [18]. All the raw reads were assigned to the designated sample based on their nucleotide barcode with the Pyrosequencing Pipeline Initial Process of the RDP, which also trimmed off the adapters and barcodes and removed sequences containing ambiguous 'N' or shorter than 200 bps. All samples were denoised individually using Mothur's 'pre.cluster' command to remove the erroneous sequences due to pyrosequencing errors. PCR chimeras were filtered out using the 'chimera.slayer' command in Mothur platform. The reads flagged as chimeras was extracted out using a self-written Python script, and those being assigned to any known genus with 90% confidence using the RDP Classifier were merged with the non-chimera reads.

Taxonomic assignment of the sequences was conducted individually for each sample using the RDP Classifier with a bootstrap cutoff of 50% [9],[18]. Unexpected archaeal sequences were manually removed before biodiversity analysis. After denoising, filtering out chimeras and removing the archaeal sequences, 6,471~13,727 effective pyrosequencing reads were generated for each of the four sludge samples (Table S3). For fair comparison, the library size of each sample was normalized to the same sequencing depth (6,471 reads) by randomly removing the redundant reads. Richness and diversity indices including OTUs number and Shannon's diversity index, as well as rarefaction curves, were calculated using Mothur.Cluster analysis and principal component analysis (PCA) of the microbial community structures of the different samples were conducted with PAleontological STatistics software (PAST, version 3.01) [19] and R (version 3.0.1) [20], respectively.

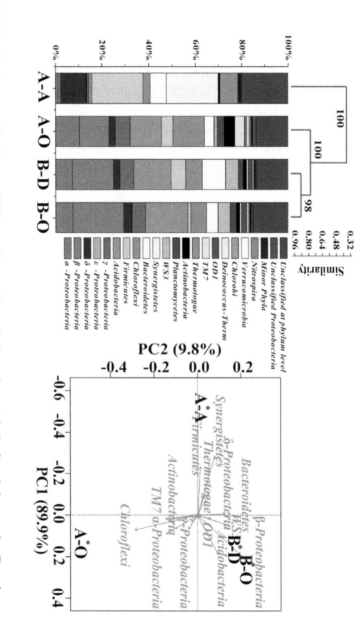

FIGURE 1: Abundance and distribution of different phyla and classes in *Proteobacteria* in the four sludge samples. Taxa shown occurred in at least one sample with abundance over 1%. Minor phyla refer to the taxa with their maximum abundance <1% in each sample. The effective pyrosequencing reads (6,471 sequences) were classified using RDP Classifier at a confidence threshold of 50%. Cluster analysis was conducted based on a distance matrix computed with Bray–Curtis similarity of four samples. Principal component analysis was conducted based on the phylum abundance using R (version 3.0.1).

4.2.6 BIOINFORMATIC ANALYSIS
ON ILLUMINA SEQUENCING DATASETS

The Illumina sequencing datasets were functionally annotated using the Clusters of Orthologous Groups (COG) [21] database in MG-RAST [22] with default parameters. In order to quantify the abundance of functional genes involved in nitrification and denitrification in the four samples, sequences of *amoA, nirK, nirS* and *nosZ* deposited in NCBI Nucleotide database (http://www.ncbi.nlm.nih.gov/nuccore/) were downloaded to construct a local database. After removing redundancies, the local database contained 17,483 sequences of AOA *amoA* gene, 12,211 sequences of AOB *amoA* gene, 5,716 sequences of *nirK* gene, 7,615 sequences of *nirS* gene and 5,436 sequences of *nosZ* gene. BLASTn was used to align all the sequencing reads against the local database, and a read was identified as *amoA, nirK, nirS* or *nosZ* if the BLAST hit (E-value cutoff at 10^{-5}) had a nucleotide sequence identity of above 90% over an alignment of at least 50 bp [10].

Additionally, the annotated reads were extracted by using a self-written Python script and then assigned to specific bacteria at the genus level by BLASTX against NCBI-nr database at an E-value of 1E-5. The BLASTX results were visualized with MEGAN (http://ab.inf.uni-tuebingen.de/software/megan/) at a threshold of bitscore>50 [10].

4.2.7 ACCESSION NUMBERS

The sequences of *amoA* for cloning library construction have been deposited in GenBank under accession numbers KF720406 to KF720513. The pyrosequencing datasets have been deposited into the NCBI Short Reads Archive Database (accession number: SRX396800). The Illumina metagenomic datasets are available at MG-RAST under accession numbers 4494863.3 (A-A), 4494888.3 (A-O), 4494854.3 (B-D) and 4494855.3 (B-O).

4.3 RESULTS AND DISCUSSION

4.3.1 BIODIVERSITY AND FUNCTIONS OF MICROBIAL COMMUNITIES OF THE WWTPS

OTUs and Shannon index analysis based on 16S rRNA gene pyrosequencing showed that the sample A-O from WWTPA had the richest biodiversity, followed by the anaerobic sludge sample A-A from the same WWTP (Table S3). This result is confirmed by the rarefaction curves (Figure S1) indicating that WWTPA samples had richer biodiversity than WWTPB. The biodiversity divergences between the two WWTPs may result from the differences in influent wastewater composition and biological treatment processes (Table S1). For example, WWTPB had higher levels of sodium salt than WWTPA in influent wastewater (Table S1), and it is known that that salinity is an important factor regulating and reducing bacterial diversity [23],[24]. In this study, the Good's coverage of each sample was above 0.93 and 0.96 at cutoff levels of 3% and 5%, respectively (Table S3), which indicated that the sequences generated at this sequencing depth could well represent the bacterial communities of the four sludge samples.

Over 30% of the pyrosequencing reads from all samples could not be assigned to any taxa at genus level (Figure S2), which is comparable to the proportions of unclassified sequences in 14 sewage sludge samples reported by Zhang et al. [18] who used V4 region for 454 pyrosequencing. Cluster analysis showed that bacterial community structure of anaerobic sludge was divergent from those of the three aerobic sludge samples, and PCA revealed that the different samples harbored characteristic bacterial communities at phylum level (Figure 1). Samples B-D and B-O, two aerobic sludge samples which collected from WWTPB, were grouped together. However, bacterial community structures of aerobic sludge (A-O) and anaerobic sludge (A-A) of WWTPA were clearly separated, both of which were dramatically different from those of WWTPB. The difference in wastewater quality may result in the divergence of the aerobic sludge communities of the two WWTPs, and available oxygen is considered a

crucial factor contributing to the difference between the aerobic and anaerobic sludge samples [25],[26].

Statistical analysis showed that 22.38% and 21.90% of the sequences
in anaerobic sludge sample A-A were assigned into *Synergistetes* and *Firmicutes*, respectively, while *Proteobacteria* were the dominant phylum in
the three aerobic sludge samples, accounting for about 33.64%, 34.31%,
and 39.42% in samples A-O, B-D, and B-O, respectively (Figure 1). *Synergistetes* is known to have low abundance in aerobic sludge, but usually
dominate in UASB reactors treating brewery wastewater [27] and tannery
wastewater [1].

Bacterial diversity and abundance were also analyzed more specifically at the genus level (Figure 2, Table S4). Cluster analysis also showed
that the bacterial community structure of the anaerobic sludge was dramatically different from those of the aerobic sludge samples (Figure 2).
Similar to the results obtained at phylum level, oxygen concentration is an
important factor shaping microbial community structures in WWTPs, and
may make huge contributions to the difference at genus level. It was found
that anaerobes such as *Aminobacterium* (11.47%), *Desulfobacter* (6.94%)
and *Kosmotoga* (5.75%) dominated in the anaerobic sludge sample A-A,
while had relative lower abundance in the three aerobic sludge samples
(<0.15%). Genus *Aminobacterium* belongs to the phylum *Synergistetes*,
which is known as amino acids degraders [28]. Other genera in *Synergistetes*, such as *Aminomonas* (2.21%), *Thermovirga* (3.00%) and *Clo-
acibacillus* (2.98%), also had higher abundance in the anaerobic sludge
sample A-A. Sulfate-reducing bacteria *Desulfobacter* (6.94%), *Desulfomi-
crobium* (1.70%) and *Desulfobulbus* (0.96%) had higher abundance in the
anaerobic sludge than in the aerobic samples, since sulfate is one common
pollutant in tannery wastewater [29] and dissolved oxygen (DO) is limited
in the UASB reactor.

In opposition, genera *Thauera*, Gp4 and *Caldilinea* dominated (>3%)
in all the three aerobic samples, but had lower richness in the anaerobic sample A-A (abundance <0.60%) (Figure 2, Table S4). *Thauera* had
extremely higher abundance in aerobic sludge samples B-D (12.95%),
B-O (15.81%) and A-O (4.64%) than in anaerobic sludge sample A-A
(0.11%). *Thauera* is known capable of denitrification and organic compounds biodegradation, which has been frequently detected in wastewater

treatment bioreactors [30],[31]. *Caldilinea* (3.08–5.95%) consists of some filamentous species capable of flocs stabilization of activated sludge [32]. The genus Gp4 (4.20–16.74%) has not been well-described to date, but is known to dominate in the activated sludge of industrial [9] and municipal [18] WWTPs.

The functions of the four tannery activated sludge metagenomes were predicted by alignment of the Illumina sequencing reads against COG database in MG-RAST (Figure S3). Similar to the structural patterns of the microbial communities, the three aerobic sludge samples A-O, B-D and B-O showed similar functional profiles at the highest level of the COG category system (Figure S3). However, the anaerobic sludge showed different COG functional categories distribution, especially on the categories of 'replication, recombination and repair', 'transcription' and 'translation, ribosomal structure and biogenesis' (Figure S3). Table S5 shows that a total of nine identified COGs were responsible for denitrification (COG1140, COG2223, COG3005, COG3256, COG3420, COG4263 and COG5013) and nitrogen fixation (COG1348 and COG2710) in the aerobic or anaerobic sludge. Each category of the denitrifying functional genes had lower abundance in anaerobic sludge than in aerobic sludge. A previous study has also indicated that the abundance of denitrifying bacteria in an aerobic reactor was higher than in a nitrate-free anaerobic digester [33].

4.3.2 DIVERSITY OF AOB AND AOA AMOA GENES

DNA cloning showed that a total of 8 OTUs were obtained at 3% distance cutoff from 81 AOB *amoA* gene sequences in 4 clone libraries (Figure 3A). The number of the OTUs recovered from individual libraries ranged from 3 (B-D) to 5 (B-O). OTU-1 and OTU-2 dominated in all samples, accounting for 86% of total sequences (Figure 3B). Phylogenetic analysis showed that all the bacterial *amoA* OTUs were affiliated to *Nitrosomonas europaea*, *Nitrosomonas communis* and *Nitrosomonas nitrosa* (Figure 3A), while no sequence in *Nitrosospira* lineage was observed. Previous studies have indicated that AOB *Nitrosomonas* instead of *Nitrosospira* dominates in WWTPs [3],[34]. Among the three species of *Nitrosomonas*, *N. europaea* dominated in all samples, and only a total of 3 sequences were affiliated to *N. communis* and *N. nitrosa* (Figure 3B). High concentration of ammonia (>170.0 mg/L) in the influent of the two WWTPs may contribute to the predominance of *N. europaea*, since previous studies have indicated that AOB of the *N. europaea* cluster are commonly found in the environments with high levels of ammonium [35]. In this study, similar AOB communities were observed in the UASB and the three aerobic bioreactors. Although DO is a limiting factor -of AOB growth in bioreactors, it has been indicated *N. europaea* lineage can enrich in both the high-DO and low-DO chemostats [36]. This is also supported by some previous studies revealing that *N. europaea* was the predominant AOB in various environments containing different levels of DO, such as laboratory-scale anaerobic ammonium-oxidizing reactors [37], full-scale modified Ludzack-Ettinger process WWTPs [38] and pilot-scale sequencing batch nitrifying reactors [39]. AOB in UASB may use the available oxygen to oxidize ammonia to nitrite, contributing to the anaerobic environments [40].

In the present study, we attempted to amplify AOA *amoA* gene from all the four samples by optimizing PCR conditions, but AOA *amoA* gene was detectable only in the anaerobic sludge sample A-A (Figure S4). Similarly, previous studies also revealed that AOA *amoA* is absent in aerobic sludge of different WWTPs [3],[41]. A total of 5 OTUs were obtained from 27 AOA *amoA* gene sequences in A-A clone library (Figure S5). It is known that most of environmental AOA are assigned to the clusters of Group I.1a (the marine and sediment lineage) and Group I.1b (the soil lineage) [42]. This study revealed that OTU-1 occupied 78% of the total AOA *amoA* gene sequences, and OTU-1, OTU-3 and OTU-4 had high similarity (>98%) to those AOA detected in sediment (clustered to the Group I.1a). Four sequences of OTU-2 and OTU-5 fell in the Group I.1b cluster. This result agrees with Mußmann et al. [41] indicating that all AOA *amoA* clones from tannery or refinery WWTPs were affiliated with Group I.1a.Ye and Zhang [43] also reported that AOA communities in a laboratory-scale bioreactor treating saline sewage were dominated by Group I.1a. However, the majority of AOA in many municipal WWTPs belonged to Group I.1b [3],[42]. The divergence may result from the difference in wastewater salinity, an important factor in shaping AOA communities [44].

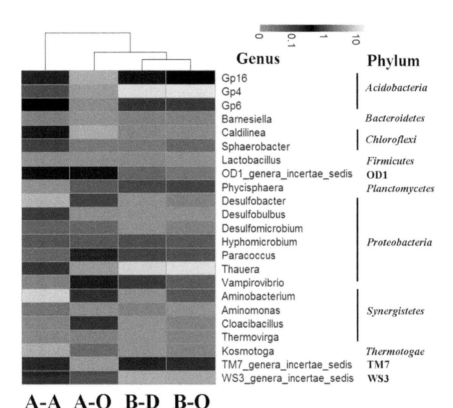

FIGURE 2: Heat map illustrating the abundance of all the major genera (with relative abundance over 1% in at least one sample). The color intensity (log scale) in each panel indicates the relative abundance of the genus in each sample.

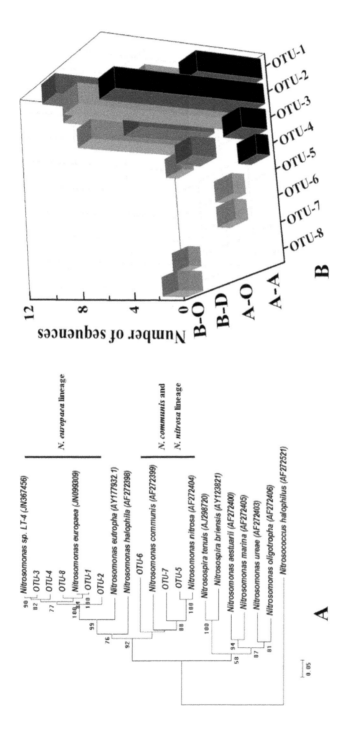

FIGURE 3: Neighbor-joining phylogenetic tree (A) and OTUs number (B) of AOB amoA gene in the four sludge samples. The evolutionary distances were computed using the Jukes–Cantor method. Bootstrap values (over 50) are indicated on branch nodes. Sequences obtained from this study are shown with "OTU-" in the names, and reference sequences were obtained from GenBank.

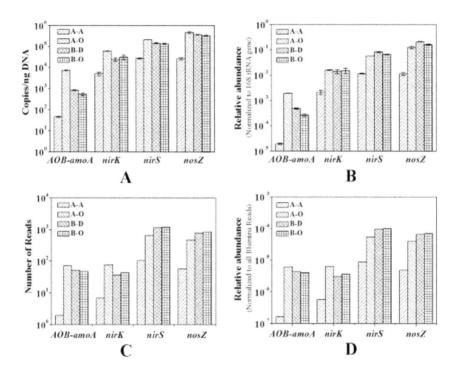

FIGURE 4: Abundance of *amoA, nirK, nirS* and *nosZ* genes in the four sludge samples revealed by qPCR (A and B) and metagenomic (alignment) approaches (C and D). A: relative abundance of the genes normalized to the mass of the extracted DNA; B: relative abundance of the genes normalized to the abundance of 16S rRNA gene; C: Number of the Illumina sequencing reads aligned to the genes sequences in the local database at identity cutoff>90% and alignment length>50 bp; D: Relative abundance of the genes in the Illumina dataset of each sludge sample.

4.3.3 ABUNDANCE OF AMOA, NIRK, NIRS AND NOSZ GENES

In this study, the abundance of *amoA*, *nirK*, *nirS* and *nosZ* genes was quantified using both qPCR and metagenomic (alignment) approaches, and the later method has also been frequently used to quantify functional genes in the environment, including *amoA* gene in soil [8] and activated sludge [10]. Correlation analysis showed that the two methods generated consistent results (Figure S6). qPCR showed that the copy numbers of bacterial *amoA* gene in the four samples ranged from 45.8 to 7.33×10^3 copies per ng DNA (Figure 4A). qPCR and metagenomic analysis consistently revealed that the relative abundance of bacterial *amoA* gene was higher in sample A-O than in the other sludge samples (Figure 4B and D). qPCR showed that the relative abundance of bacterial *amoA* gene to the overall bacterial population in the three aerobic sludge samples (0.03%–0.19%) was comparable to the previous reports of four membrane bioreactors (lower than 0.1%) capable of removing 70–95% of ammonium from different types of wastewater [45]. A previous study indicated that ammonia monooxygenase had high transcription activity in activated sludge [11]. Thus, the high ammonium removal efficiency in the two WWTPs may result from the high expression levels of ammonia monooxygenase and long hydraulic retention time (Table S1).

Nitrite reduction is usually catalyzed by two structurally different but functionally equivalent forms of nitrite reductase: copper and cytochrome cd_1-containing nitrite reductases encoded by the genes *nirK* and *nirS*, respectively. The copy numbers of *nirK* and *nirS* ranged from 5.05×10^3 to 6.02×10^4 and from 2.72×10^4 to 2.14×10^5 gene copies per ng DNA, respectively (Figure 4A). The average abundance of *nirS* and *nirK* relative to the overall bacterial population was 1.18% and 5.38%, respectively (Figure 4B). The relative abundance of *nirK* was found comparable to the results reported by Geets et al. (0.16–2.75%) [4]. qPCR and metagenomic analysis consistently indicated that A-O had higher abundance of *nirK* than the other samples. This study showed that *nirS* had higher abundance than *nirK* in each of the four samples (The ratio of *nirS/nirK* was over 3.0 in each sample). Similar results were observed in some WWTPs treating hospital wastewater, papermaking wastewater or pharmaceutical wastewater [4]. However, it cannot be indicated that *nirS* plays a greater role than *nirK* in the nitrogen removal, since expression level of the genes may affect their contribution to nitrogen removal [5]. Thus, further studies based on mRNA and protein expression have to be conducted.

N_2O is an important component of greenhouse gas, and reduction of N_2O to N_2 in the environment has been attributed exclusively to the denitrifiers carrying *NosZ* gene. qPCR showed that the copy number of *nosZ* gene in the four sludge samples ranged from 2.63×10^4 to 4.66×10^5 gene copies per ng DNA (Figure 4A) with relative abundance within the range from 1.1% to 21.0% (Figure 4B). qPCR and metagenomic analysis consistently indicated that *nosZ* gene preferred aerobic environment (Figure 4). In anaerobic sludge, the abundance of *nosZ* gene were lower than the total abundance of *nirK* and *nirS*, indicating that some denitrifiers in the anaerobic sludge might lack *nosZ*, which may lead to the accumulation of N_2O in anaerobic environments. Previous studies also indicated that low DO concentration in WWTPs favors N_2O production during nitrification/denitrification process [46],[47], although nitrous oxide reductase is the most oxygen sensitive enzyme in the denitrification pathway [48].

It was found the effluent nitrate concentration has a significantly positive correlation with the relative abundance of the denitrifying genes *nirK* (R = 0.977, P = 0.023), *nirS* (R = 0.851, P = 0.149) and *nosZ* (R = 0.811, P = 0.189). This is supported by a previous study revealing significant correlation between nitrate concentration and the abundance of *nirK* or *nirS* in soil [5].

4.3.4 METAGENOMIC ANALYSIS OF NITRIFYING AND DENITRIFYING BACTERIA POPULATIONS

16S rRNA gene pyroseqencing was conducted to investigate the abundance and diversity of AOB and NOB in the sludge samples. Results showed that *Nitrosomonas* and *Nitrosospira*, the two important genera of AOB in WWTPs [3], accounted for 0–0.34% and 0–0.20% of the total sequences of the samples, respectively (Table S6). Similarly, Zhang et al. [3] indicated that AOB occupied 0–0.64% of the total pyrosequencing sequences from six municipal WWTPs, and *Nitrosomonas* dominated in the WWTPs. Sample A-O had the highest abundance of AOB and NOB, followed by the other aerobic samples, and both AOB and NOB sequences were undetectable in the anaerobic sludge (A-A), which agrees with the qPCR results of *amoA* gene. A total of 76 AOB and 79 NOB sequences in all the samples were merged to construct phylogenetic tree to characterize the AOB and NOB communities, respectively. Similar to *amoA* gene diversity analysis, pyrosequencing showed that most of the AOB 16S rRNA sequences were grouped into *Nitrosomonas* sp., including *N. europaea* (5 OTUs), *N. ureae* (1 OTU) and *Nitrosomonas*-like species (3 OTUs) (Figure S7). Most of the *Nitrospira*-type NOB reads were grouped into *N. marina* (Figure S8), since tannery wastewater usually contains high concentration of salt (Table S1) [1].

Shapleigh [49] provided a list of 84 potential denitrifying bacterial genera, among which 17 genera were detectable in the four samples in this study (Table S7). The 11 types of potential denitrifying genera present in anaerobic sludge sample A-A accounted for 0.83% of the total pyrosequencing reads, which was lower than the corresponding proportions of the three aerobic sludge samples (8.22–20.52%). This agrees with the COG function alignment results, and lack of nitrate (Table S1) in UASB may be responsible for the low abundance of the denitrifiers.

Additionally, the Illumina reads annotated as *amoA*, *nirK*, *nirS* and *nosZ* genes were extracted and then assigned to specific host bacteria at genus level by BLASTX against NCBI-nr database and MEGAN. Unfortunately, most of the extracted reads had higher similarity to the corresponding genes of uncultured microorganisms. A total of 26 reads of *amoA*, 36 reads of *nirK*, 1,062 reads of *nirS* and 754 reads of *nosZ* in the samples were well assigned to specific host of 89 known genera (Figure S9). It should be noted that using the short Illumina sequencing reads in taxonomic assignment could produce unreliable results at low taxonomic levels [50], and the possible horizontal gene transfer of denitrifying genes among different taxa is likely to increase uncertainty of the results [51]. To maximize the reliability of the obtained results, only the highly abundant genera identified by both pyrosequencing and Illumina sequencing in this study were considered as the main nitrifiers or denitrifiers in the two tannery WWTPs.

All the *amoA* gene reads were assigned to *Nitrosomonas*, which agrees with *amoA* gene cloning results. Results showed that *Nitrosomonas* was the main host of nitrite reductase gene nirK in the samples A-A, B-D and

B-O. This result is supported by Beaumont et al. [52] indicating presence of *nirK* in *Nitrosomonas* europaea cells. Interestingly, *nirK* in sample A-O was mainly carried by the bacterial cells of *Rhizobium*, not *Nitrosomonas* (Figure S9). This is consistent with the pyrosequencing results demonstrating that *Rhizobium* had high abundance in sample A-O (Table S4). *NirS* and *nosZ* genes were found mainly located in the bacterial cells of *Thauera, Paracoccus, Hyphomicrobium, Comamonas* and *Azoarcus* (Figure S9) dominating in the aerobic sludge (Table S4), revealing the crucial roles of the microorganisms in denitrification in aerobic environments. It is known that *Thaurea mechernichensis* [53] and *Paracoccus denitrificans* [54] can reduce nitrate even in the presence of high concentrations of O_2.

In conclusion, combined use of DNA cloning, qPCR, 454 pyrosequencing and Illumina sequencing provided a comprehensive insight in microbial community structure of nitrifiers and denitrifiers in tannery WWTPs. *Proteobacteria* and *Synergistetes* phyla had highest abundance in aerobic and anaerobic sludge, respecitively. *Nitrosomonas europaea* dominated the ammonia-oxidizing community. *Thauera*, P*aracoccus, Hyphomicrobium, Comamonas* and *Azoarcus* were the predominant potential denitrifying genera, which may greatly contribute to the nitrogen removal in the WWTPs. qPCR and metagenomic approaches consistently revealed that functional genes *amoA, nirK, nirS* and *nosZ* had higher abundance in aerobic sludge than in anaerobic sludge. The results can extend our biological knowledge about nitrifiers and denitrifiers in tannery WWTPs, and might be practically useful in efficiently removing nitrogen from industrial wastewater.

REFERENCES

1. Lefebvre O, Vasudevan N, Thanasekaran K, Moletta R, Godon JJ (2006) Microbial diversity in hypersaline wastewater: the example of tanneries. Extremophiles 10:505–513. doi: 10.1007/s00792-006-0524-1
2. Leta S, Assefa F, Gumaelius L, Dalhammar G (2004) Biological nitrogen and organic matter removal from tannery wastewater in pilot plant operations in Ethiopia. Appl Microbiol Biotechnol 66:333–339. doi: 10.1007/s00253-004-1715-2
3. Zhang T, Ye L, Tong AHY, Shao MF, Lok S (2011) Ammonia-oxidizing archaea and ammonia-oxidizing bacteria in six full-scale wastewater treatment bioreactors. Appl. Microbiol. Biotechnol 91:1215–1225. doi: 10.1007/s00253-011-3408-y

4. Geets J, de Cooman M, Wittebolle L, Heylen K, Vanparys B, et al. (2007) Real-time PCR assay for the simultaneous quantification of nitrifying and denitrifying bacteria in activated sludge. Appl. Microbiol. Biotechnol 75:211–221. doi: 10.1007/s00253-006-0805-8

5. Guo GX, Deng H, Qiao M, Yao HY, Zhu YG (2013) Effect of long-term wastewater irrigation on potential denitrification and denitrifying communities in soils at the watershed scale. Environ Sci Technol 47:3105–3113. doi: 10.1021/es304714a

6. Chen J, Tang YQ, Wu XL (2012) Bacterial community shift in two sectors of a tannery plant and its Cr (VI) removing potential. Geomicrobiol J 29:226–235. doi: 10.1080/01490451.2011.558562

7. Shi P, Jia SY, Zhang XX, Zhang T, Cheng SP, et al. (2013) Metagenomic insights into chlorination effects on microbial antibiotic resistance in drinking water. Water Res 47:111–120. doi: 10.1016/j.watres.2012.09.046

8. Leininger S, Urich T, Schloter M, Schwark L, Qi J, et al. (2006) Archaea predominate among ammonia-oxidizing prokaryotes in soils. Nature 442:806–809. doi: 10.1038/nature04983

9. Ibarbalz FM, Figuerola ELM, Erijman L (2013) Industrial activated sludge exhibit unique bacterial community composition at high taxonomic ranks. Water Res 47:3854–3864. doi: 10.1016/j.watres.2013.04.010

10. Ye L, Zhang T, Wang TT, Fang ZW (2012) Microbial structures, functions, and metabolic pathways in wastewater treatment bioreactors revealed using high-throughput sequencing. Environ Sci Technol 46:13244–13252. doi: 10.1021/es303454k

11. Yu K, Zhang T (2012) Metagenomic and metatranscriptomic analysis of microbial community structure and gene expression of activated sludge. PloS ONE 7:e38183. doi: 10.1371/journal.pone.0038183

12. Wang Z, Zhang XX, Huang KL, Miao Y, Shi P, et al. (2013) Metagenomic profiling of antibiotic resistance genes and mobile genetic elements in a tannery wastewater treatment plant. PLoS ONE 8:e76079. doi: 10.1371/journal.pone.0076079

13. Schloss PD, Handelsman J (2005) Introducing DOTUR, a computer program for defining operational taxonomic units and estimating species richness. Appl Environ Microbiol 71:1501–1506. doi: 10.1128/aem.71.3.1501-1506.2005

14. Zhang XX, Zhang T (2011) Occurrence, abundance, and diversity of tetracycline resistance genes in 15 sewage treatment olants across China and other global locations. Environ Sci Technol 45:2598–2604. doi: 10.1021/es103672x

15. Lopez-Gutierrez JC, Henry S, Hallet S, Martin-Laurent F, Catroux G, et al. (2004) Quantification of a novel group of nitrate-reducing bacteria in the environment by real-time PCR. J Microbiol Methods 57:399–407. doi: 10.1016/j.mimet.2004.02.009

16. Cole JR, Wang Q, Cardenas E, Fish J, Chai B, et al. (2009) The Ribosomal Database Project: improved alignments and new tools for rRNA analysis. Nucleic Acids Res 37:D141–145. doi: 10.1093/nar/gkn879

17. Schloss PD, Westcott SL, Ryabin T, Hall JR, Hartmann M, et al. (2009) Introducing Mothur: open-source, platform-independent, community-supported software for describing and comparing microbial communities. Appl Environ Microbiol 75:7537–7541. doi: 10.1128/aem.01541-09

18. Zhang T, Shao MF, Ye L (2012) 454 Pyrosequencing reveals bacterial diversity of activated sludge from 14 sewage treatment plants. ISME J 6:1137–1147. doi: 10.1038/ismej.2011.188

19. Hammer Ø, Harper DAT, Ryan PD (2001) PAST: Paleontological statistics software package for education and data analysis. Palaeontologia Electronica 4(1): 9 pp.

20. R Core Team (2013) R: A language and environment for statistical computing. R Foundation for Statistical Computing. Vienna, Austria. ISBN 3-900051-07-0. Available: http://www.R-project.org/.

21. Tatusov RL, Galperin MY, Natale DA, Koonin EV (2000) The COG database: a tool for genomescale analysis of protein functions and evolution. Nucleic Acids Res 28:33–36. doi: 10.1093/nar/28.1.33

22. Meyer F, Paarmann D, D'Souza M, Olson R, Glass EM, et al. (2008) The metagenomics RAST server - a public resource for the automatic phylogenetic and functional analysis of metagenomes. BMC Bioinformatics 9:386. doi: 10.1186/1471-2105-9-386

23. Cortés-Lorenzo C, Sipkema D, Rodríguez-Díaz M, Fuentes S, Juárez-Jiménez B, et al. (2014) Microbial community dynamics in a submerged fixed bed bioreactor during biological treatment of saline urban wastewater. Ecol Eng 71:126–132. doi: 10.1016/j.ecoleng.2014.07.025

24. Benlloch S, Lopez-Lopez A, Casamayor EO, Ovreas L, Goddard V, et al. (2002) Prokaryotic genetic diversity throughout the salinity gradient of a coastal solar saltern. Environ Microbiol 4:349–360. doi: 10.1046/j.1462-2920.2002.00306.x

25. Wang X, Xia Y, Wen X, Yang Y, Zhou J (2014) Microbial community functional structures in wastewater treatment plants as characterized by GeoChip. PLoS ONE 9:e93422. doi: 10.1371/journal.pone.0093422

26. Wang X, Hu M, Xia Y, Wen X, Ding K (2012) Pyrosequencing analysis of bacterial diversity in 14 wastewater treatment systems in China. Appl Environ Microbiol 78:7042–7047. doi: 10.1128/aem.01617-12

27. Novak D, Franke-Whittle IH, Pirc ET, Jerman V, Insam H, et al. (2013) Biotic and abiotic processes contribute to successful anaerobic degradation of cyanide by UASB reactor biomass treating brewery waste water. Water Res 47:3644–3653. doi: 10.1016/j.watres.2013.04.027

28. Vartoukian SR, Palmer RM, Wade WG (2007) The division "Synergistes". Anaerobe 13:99–106. doi: 10.1016/j.anaerobe.2007.05.004

29. Murugananthan M, Raju GB, Prabhakar S (2004) Separation of pollutants from tannery effluents by electro flotation. Sep Purif Technol 40:69–75. doi: 10.1016/j.seppur.2004.01.005

30. Liu B, Zhang F, Feng X, Liu Y, Yan X, et al. (2006) *Thauera* and Azoarcus as functionally important genera in a denitrifying quinoline-removal bioreactor as revealed by microbial community structure comparison. FEMS Microbiol Ecol 55:274–286. doi: 10.1111/j.1574-6941.2005.00033.x

31. Zhao Y, Huang J, Zhao H, Yang H (2013) Microbial community and N removal of aerobic granular sludge at high COD and N loading rates. Bioresour Technol 143:439–446. doi: 10.1016/j.biortech.2013.06.020

32. Yoon DN, Park SJ, Kim SJ, Jeon CO, Chae JC, et al. (2010) Isolation, characterization, and abundance of filamentous members of *Caldilineae* in activated sludge. J Microbiol 48:275–283. doi: 10.1007/s12275-010-9366-8

33. Reyes M, Borrás L, Seco A, Ferrer J (2014) Identification and quantification of microbial populations in activated sludge and anaerobic digestion processes. Environ Techno DOI: 10.1080/09593330.2014.934745.

34. Figuerola ELM, Erijman L (2010) Diversity of nitrifying bacteria in a full-scale petroleum refinery wastewater treatment plant experiencing unstable nitrification. J Hazard Mater 181:281–288. doi: 10.1016/j.jhazmat.2010.05.009

35. Limpiyakorn T, Kurisu F, Sakamoto Y, Yagi O (2007) Effects of ammonium and nitrite on communities and populations of ammonia-oxidizing bacteria in laboratory-scale continuous-flow reactors. FEMS Microbiol Ecol 60:501–512. doi: 10.1111/j.1574-6941.2007.00307.x

36. Park HD, Noguera DR (2004) Evaluating the effect of dissolved oxygen on ammonia-oxidizing bacterial communities in activated sludge. Water Res 38:3275–3286. doi: 10.1016/j.watres.2004.04.047

37. Quan ZX, Rhee SK, Zuo JE, Yang Y, Bae JW, et al. (2008) Diversity of ammonium-oxidizing bacteria in a granular sludge anaerobic ammonium-oxidizing (anammox) reactor. Environ Microbiol 10:3130–3139. doi: 10.1111/j.1462-2920.2008.01642.x

38. Figuerola EL, Erijman L (2010) Diversity of nitrifying bacteria in a full-scale petroleum refinery wastewater treatment plant experiencing unstable nitrification. J Hazard Mater 181:281–288. doi: 10.1016/j.jhazmat.2010.05.009

39. Gieseke A, Bjerrum L, Wagner M, Amann R (2003) Structure and activity of multiple nitrifying bacterial populations co-existing in a biofilm. Environ Microbiol 5:355–369. doi: 10.1046/j.1462-2920.2003.00423.x

40. Ma B, Peng Y, Zhang S, Wang J, Gan Y, et al. (2013) Performance of anammox UASB reactor treating low strength wastewater under moderate and low temperatures. Bioresour Technol 129:606–611. doi: 10.1016/j.biortech.2012.11.025

41. Mußmann M, Brito I, Pitcher A, Damste JSS, Hatzenpichler R, et al. (2011) Thaumarchaeotes abundant in refinery nitrifying sludges express *amoA* but are not obligate autotrophic ammonia oxidizers. Proc Natl Acad Sci USA 108:16771–16776. doi: 10.1073/pnas.1106427108

42. Limpiyakorn T, Sonthiphand P, Rongsayamanont C, Polprasert C (2011) Abundance of *amoA* genes of ammonia-oxidizing archaea and bacteria in activated sludge of full-scale wastewater treatment plants. Bioresource Technol 102:3694–3701. doi: 10.1016/j.biortech.2010.11.085

43. Ye L, Zhang T (2011) Ammonia-oxidizing bacteria dominates over ammonia-oxidizing archaea in a saline nitrification reactor under low DO and high nitrogen loading. Biotechnol Bioeng 108:2544–2552. doi: 10.1002/bit.23211

44. Mosier AC, Francis CA (2008) Relative abundance and diversity of ammonia-oxidizing archaea and bacteria in the San Francisco Bay estuary. Environ Microbiol 10:3002–3016. doi: 10.1111/j.1462-2920.2008.01764.x

45. Zhang B, Sun BS, Ji M, Liu HN, Liu XH (2010) Quantification and comparison of ammonia-oxidizing bacterial communities in MBRs treating various types of wastewater. Bioresource Technol 101:3054–3059. doi: 10.1016/j.biortech.2009.12.048

46. Jia W, Liang S, Zhang J, Ngo HH, Guo W, et al. (2013) Nitrous oxide emission in low-oxygen simultaneous nitrification and denitrification process: sources and mechanisms. Bioresource Technol 136:444–451. doi: 10.1016/j.biortech.2013.02.117

47. Tallec G, Garnier J, Billen G, Gousailles M (2006) Nitrous oxide emissions from secondary activated sludge in nitrifying conditions of urban wastewater treatment plants: effect of oxygenation level. Water Res 40:2972–2980. doi: 10.1016/j.watres.2006.05.037

48. Bonin P, Gilewicz M, Bertrand J (1989) Effects of oxygen on each step of denitrification on Pseudomonas nautica. Can J Microbiol 35:1061–1064. doi: 10.1139/m89-177

49. Shapleigh JP (2006) The Denitrifying Prokaryotes. Springer, New York, NY: 769–792.

50. Wang Q, Garrity GM, Tiedje JM, Cole JR (2007) Naive Bayesian classifier for rapid assignment of rRNA sequences into the new bacterial taxonomy. Appl Environ Microbiol 73:5261–5267. doi: 10.1128/aem.00062-07

51. Heylen K, Gevers D, Vanparys B, Wittebolle L, Geets J, et al. (2006) The incidence of nirS and nirK and their genetic heterogeneity in cultivated denitrifiers. Environ Microbiol 8:2012–2021. doi: 10.1111/j.1462-2920.2006.01081.x

52. Beaumont HJ, Lens SI, Westerhoff HV, van Spanning RJ (2005) Novel nirK cluster genes in Nitrosomonas europaea are required for NirK-dependent tolerance to nitrite. J Bacteriol 187:6849–6851. doi: 10.1128/jb.187.19.6849-6851.2005

53. Scholten E, Lukow T, Auling G, Kroppenstedt RM, Rainey FA, et al. (1999) Thauera mechernichensis sp nov., an aerobic denitrifier from a leachate treatment plant. Int J Syst Bacteriol 49:1045–1051. doi: 10.1099/00207713-49-3-1045

54. Baumann B, Snozzi M, Zehnder AJ, Van Der MeerJR (1996) Dynamics of denitrification activity of Paracoccus denitrificans in continuous culture during aerobic-anaerobic changes. J Bacteriol 178:4367–4374.

There are several supplemental files that are not available in this version of the article. To view this additional information, please use the citation on the first page of this chapter.

CHAPTER 5

Assessing Bacterial Diversity in a Seawater-Processing Wastewater Treatment Plant by 454-Pyrosequencing of the 16S rRNA and amoA Genes

OLGA SÁNCHEZ, ISABEL FERRERA, JOSE M. GONZÁLEZ, AND JORDI MAS

5.1 INTRODUCTION

Activated sludge constitutes a crucial tool in the biodegradation of organic materials, transformation of toxic compounds into harmless products and nutrient removal in wastewater treatment plants (WWTPs). It contains a highly complex mixture of microbial populations whose composition has been intensively studied in the past decades. By applying culture-dependent methods many species have been isolated from activated sludge (Dias and Bhat, 1964; Prakasam and Dondero, 1967; Benedict and Carlson, 1971). However, a great majority cannot be obtained by conventional

Assessing Bacterial Diversity in a Seawater-Processing Wastewater Treatment Plant by 454-Pyrosequencing of the 16S rRNA and amoA Genes. © *Sánchez O, Ferrera I, González JM, and Mas J. Microbial Biotechnology 6,4 (2013). doi: 10.1111/1751-7915.12052. Reprinted with permission from the authors.*

techniques (Wagner et al., 1993) and, consequently, current molecular techniques such as sequence analysis of 16S rRNA gene clone libraries (Snaidr et al., 2011), fingerprinting methods such as denaturing gradient gel electrophoresis (DGGE; Boon et al., 2002), thermal gradient gel electrophoresis (TGGE; Eichner et al., 1964) and terminal restriction fragment length polymorphism (Saikaly et al., 2011) along with fluorescence in situ hybridization (FISH) have been employed in wastewater microbiology to analyse and compare the microbial structure of activated sludge. Recently, PCR-based 454 pyrosequencing has been applied to investigate the microbial populations of activated sludge in different WWTPs as well as in full-scale bioreactors (Sanapareddy et al., 2009; Kwon et al., 2010; Kim et al., 2011; Ye et al., 2002; Zhang et al., 2011a; b), greatly expanding our knowledge on activated sludge biodiversity.

An important process in WWTPs is nitrification, in which ammonium is removed by converting it first into nitrite and then to nitrate. Different bacterial species involved in this process have been characterized by means of clone library analysis in addition to FISH (Juretschko et al., 1998; Purkhold et al., 2000; Daims et al., 2001; Zhang et al., 2011b). Several ammonia-oxidizing and nitrite-oxidizing bacterial populations belonging to the phylum *Nitrospira* and to *Beta-* and *Gammaproteobacteria* have been identified as key members in this process, such as the genera *Nitrosomonas*, *Nitrobacter*, *Nitrospira* and *Nitrosococcus* (Wagner et al., 2002; Zhang et al., 2011b). Nevertheless, most studies of microbial diversity in WWTPs refer to freshwater plants, either domestic or industrial, and yet very little is known about plants that utilize seawater for their operation, mainly because there are still very few of these running in the world. Their utilization responds to the deficiency in hydric resources prevailing in their locations and their use will probably increase in the near future due to water shortage associated to global warming as many areas are experiencing today (Barnett et al., 2005). As a consequence, knowledge of the microbial diversity becomes crucial to identify the key players in these systems.

In a recent survey (Sánchez et al., 2011), the prokaryotic diversity of a seawater-utilizing WWTP from a pharmaceutical industry located in the south of Spain was characterized using a polyphasic approach by means of three molecular tools that targeted the 16S rRNA gene, i.e. DGGE, clone libraries and FISH. The results showed that the composition of the bacterial community differed substantially from other WWTP previously reported, since *Betaproteobacteria* did not seem to be the predominant group; in contrast, other classes of *Proteobacteria*, such as *Alpha-* and *Gammaproteobacteria*, as well as members of *Bacteroidetes* and *Deinoccocus-Thermus* contributed in higher proportions. Besides, utilization of specific primers for amplification of the *amoA* (ammonia monooxygenase subunit A) gene confirmed the presence of nitrifiers corresponding to the *Beta*-subclass of *Proteobacteria*, although they were not identified in this study.

In the present article, we further investigated the diversity of this system by applying 454-pyrosequencing, a much more powerful molecular technique, which provides thousands of sequence reads. We analysed the bacterial assemblage by targeting the 16S rRNA gene and increased our knowledge on its diversity by one order of magnitude. Additionally, we characterized the nitrifying members of this sludge by pyrosequencing the *amoA* gene. As far was we know, this is the first study that analyses the *amoA* gene diversity in an activated sludge of a WWTPs with the particularity to utilize seawater.

5.2 RESULTS AND DISCUSSION

We investigated the bacterial community structure and identified the nitrifying members from the activated sludge of a seawater-utilizing WWTP located in Almeria (South-east Spain). The plant treats wastewater from a pharmaceutical industry. The mean influent flow of the plant is 300 m^3 h^{-1} and has a treatment volume of 32 000 m^3. Nitrogen and chemical oxygen demand sludge loads were about 150–170 kg h^{-1} and 900–1000 kg h^{-1} respectively. DNA was extracted from samples of aerated mixed activated sludge collected in 2 consecutive years (2007 and 2008).

5.2.1 DIVERSITY OF BACTERIAL COMMUNITIES IN ACTIVATED SLUDGE

After a rigorous quality control (see Experimental procedures and Table S1) a total of 16 176 16S rRNA tag sequences of sufficient quality were analysed (8010 sequences corresponding to year 2007 and 8166 sequences to year 2008) and grouped into operational taxonomic units using uclust at 3% cut-off level. The clustering resulted in a total of 320 different OTUs from which 107 were shared between samples (33.3%) as shown in the Venn diagram (Fig. 1A). The number of OTUs in 2007 was 201 and in 2008 was 226. Although the proportion of shared OTUs is rather low, the unique diversity in each sample corresponded mainly to rare OTUs (relative abundance below 1%). In the case of year 2007, the unique clones to that sample represented a 19% of the total reads, from which only three OTUs were above 1%. For the 2008 sample, the unique clones represented a 9% of total sequences and only two OTUs presented an abundance above 1%. These results are in agreement with previous observations in which DGGE analysis from both samples showed virtually the same pattern for universal primers amplifying Bacteria, suggesting that activated sludge was at a steady state at least for the most abundant phylotypes (Sánchez et al., 2011).

Richness was computed by the Chao1 estimator and analysis by rarefaction showed that the diversity in the two samples was within the same range, although slightly higher in 2008. However, we found that this depth of sequencing was not sufficient to saturate the curve and therefore, the actual diversity is likely much higher (Fig. 2). Nevertheless, if compared with the rarefaction curve from a clone library performed from the 2007 activated sludge sample, we observe that, by applying pyrosequencing we increased our knowledge on the diversity present by one order of magnitude. Rank-abundance curves (Fig. S1A) show that there were only a few abundant phylotypes and a long tail of rare taxa, therefore, most of the unknown diversity probably corresponds to rare diversity (Pedrós-Alió, 2007).

RDP Classifier was used to assign the representative OTU sequences into different phylogenetic bacterial taxa. Figure S2 shows the relative abundances of the different groups at the phylum and class level for both years. *Deinococcus-Thermus*, *Proteobacteria*, *Chloroflexi* and *Bacteroidetes* were abundant in both samples. Comparison with a previous survey (Sánchez et al., 2011) indicates that most of these groups were also retrieved by different molecular methodologies. However, the contribution of each group varied depending on the technique used. The bacterial clone library over-represented the *Deinococcus-Thermus* group, while the rest of procedures showed similar results concerning this phylum. In contrast, the *Alphaproteobacteria* were over-represented by FISH (Fig. S3).

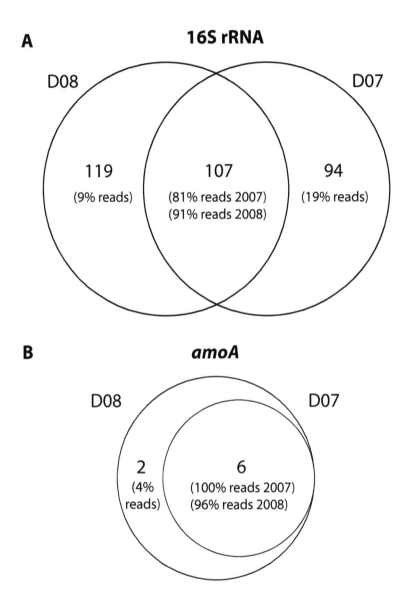

FIGURE 1: Venn diagrams of shared OTUs between the two samples (D07 and D08) for (A) 16S rRNA gene and (B) *amoA* gene. The number of reads that the OTUs represent is indicated in brackets.

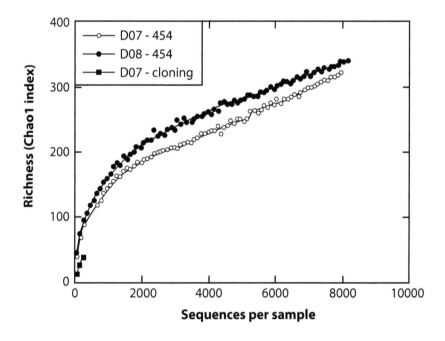

Figure 2: Rarefaction curves of 16S rRNA OTUs defined by 3% sequence variations in the activated sludge. Curves refer to the pyrosequence analyses of the 2 consecutive years (D07 and D08-454) compared with the clone library analysis of the 2007 sludge (D07 – cloning).

On the other hand, pyrosequencing allowed the detection of other groups that could not be recognized by other molecular techniques, such as the *Chloroflexi, Chlorobi, Deferribacteres, Verrumicrobia, Planctomycetes* and *Spirochaetes*, deepening our knowledge on the diversity of this activated sludge. Also, a certain percentage of sequences remained as unidentified bacteria (6.5% and 10.5% for years 2007 and 2008; Fig. S2). Except the *Chlorobi* and *Deferribacteres*, different pyrosequencing studies have reported the presence of these groups in conventional activated sludge samples (Sanapareddy et al., 2009; Kwon et al., 2010). However, it is remarkable that, in general, the proportions of the different groups in freshwater activated sludge were different from saline samples, and when going deeper into genus composition, the assemblage of our samples differs strongly from that previously reported. In general, prior pyrosequencing studies with different samples of activated sludge are in agreement with the predominance of the classes *Beta-* and *Gammaproteobacteria* and the phylum *Bacteroidetes* (Sanapareddy et al., 2009; Kwon et al., 2010), while in our saline activated sludge the groups that predominate are, within the phylum *Proteobacteria*, the *Alpha-* (8.0% and 7.3% in samples 2007 and 2008 respectively), *Gamma-* (19.0% and 21.4%) and *Deltaproteobacteria* (15.2% and 7.2%), as well as the *Deinococcus-Thermus* group (21.8% and 10.9%) and members of the phyla *Chloroflexi* (9.5% and 35.1%) and *Bacteroidetes* (18.3% and 2.8%). In contrast, Ye and colleagues (2002), who analysed by pyrosequencing the bacterial composition of a slightly saline activated sludge from a laboratory-scale nitrification reactor and a WWTP from Hong Kong, found that, in addition to *Proteobacteria* and *Bacteroidetes*, the phylum *Firmicutes* was also abundant in their samples; they also obtained similar groups as in the present study, such as the *Actinobacteria, Planctomycetes, Verrumicrobia, Deinococcus-Thermus, Chloroflexi* and *Spirochaetes*, although at different relative ratios, as well as different phyla not retrieved in the present work, for example the *Nitrospira, Chlamydiae* and TM7. Probably, differences are due to the feeding wastewater, since in our case the main influent corresponds to intermediate products of amoxicillin synthesis whereas in the other study the WWTP treated a slightly saline urban sewage from Hong Kong.

As in previous pyrosequencing studies (Keijser et al., 2008; Liu et al., 2008; Claesson et al., 2002), a part of sequences could only be assigned to the phylum/class level and the majority of taxa could not be classified at the genus level (74% for 2007 and 83.5% for 2008), demonstrating the extraordinary microbial diversity of activated sludge that cannot be classified using public 16S rRNA databases. Table S2 shows the taxa found in each sample in this study at the genus level, which are different from other reported genus of freshwater or either slightly saline activated sludge studies (Sanapareddy et al., 2009; Kwon et al., 2010; Ye et al., 2002). One of the most abundant genus in both samples is *Truepera*, a member of the phylum *Deinococcus-Thermus*, which includes radiation-resistant and thermophilic species. Although this phylum was also detected by pyrosequencing in a recent study with a slightly saline activated sludge (Ye et al., 2002), it accounted for no more than 0.6% of total community, while we found a significant percentage of sequences from this genus (21.8% and 10.9% for years 2007 and 2008 respectively).

5.2.2 DIVERSITY OF NITRIFYING COMMUNITY IN ACTIVATED SLUDGE

A total of 43 297 *amoA* gene sequences of good quality (11 236 reads for year 2007 and 32 061 reads for year 2008) were grouped into operational taxonomic units using uclust at 6% cut-off level. We selected a 6% cut-off to group closely related phylotypes of the *amoA* gene without losing potentially valuable information by the inclusion of phylogenetically distinct sequences. Interestingly, the diversity of the nitrifying bacterial community revealed by pyrosequencing of the *amoA* gene was very low and rarefaction analyses showed the depth of sequencing was sufficient to saturate the curve and recover the great majority. The clustering of 43 297 reads resulted in a total of only eight OTUs from which six were shared between samples as shown in the Venn diagram (Fig. 1B). The shared OTUs corresponded to 97% of total reads, which indicates that the nitrifying community was very similar both years.

All *amoA* sequences were highly related to previously described sequences in the GenBank database, both environmental and from isolates (Fig. 3). Phylogenetic analysis revealed that eight phylotypes formed two separate clusters. The first cluster, which contains three OTUs, was mostly retrieved in the 2008 library and represented 45.4% of sequences of that sample. The closest relatives in GenBank database (99% similarity) included sequences from organisms that have not been obtained from a WWTP, and *Nitrosomonas* sp. LT-2 and LT-5, isolated from a CANON reactor (98% identity). The second cluster, which contains five of the OTUs and represented the most abundant phylotypes in both samples, was most closely related (94% identity) to cultured representatives of strains of *Nitrosomonas* marina isolated from a biofilter of a recirculating shrimp aquaculture system (GenBank Accession No. HM345621, HM345612 and HM345618) and *Nitrosomonas* sp. NS20 isolated from coastal marine sediments. This cluster virtually represented all sequences (99.99%) in the sample of 2007 whereas in 2008 it comprised 54.6%.

These data are consistent with previous results found by Ye and colleagues (2002) in slightly saline activated sludge, which showed that *Nitrosomonas*, together with *Nitrospira*, was the dominant nitrifying genera, and also with the study by Park and colleagues (2009), who identified that a specific ammonia-oxidizing bacteria belonging to the *Nitrosomonas europaea* lineage was dominant in a full-scale bioreactor treating saline wastewater due to its adaptation to high-salt conditions. In general, nitrosomonads are also responsible for ammonia oxidation in conventional WWTPs (Purkhold et al., 2000; Zhang et al., 2011b). However, we did not retrieve *Nitrosomonas* in the pyrosequencing 16S rRNA libraries or previously in DGGE gels, clone libraries and FISH (Sánchez et al., 2011) probably due to their low abundance. Thus, pyrosequencing of functional genes such as *amoA* revealed the presence of particular groups which could not be retrieved when analysing the 16S rRNA, demonstrating its value to deepen into the functionality of microbial populations when targeting specific genes. The only nitrifier that could be retrieved in our samples by 16S rRNA pyrosequencing was *Nitrosococcus*, a *Gammaproteobacteria* which just represented 0.3% and 0.5% of the total reads for years 2007 and 2008 respectively, and has also been reported to be an important nitrifier in some activated sludges (Juretschko et al., 1998; Raszka et al., 2011). Thus, since it was actually detectable in the general bacterial 16S rRNA gene population, it could also participate in ammonia oxidization together with the *Betaproteobacteria*, despite previous efforts for amplifying the gammaproteobacterial *amoA* gene yielded negative results (Sánchez et al., 2011).

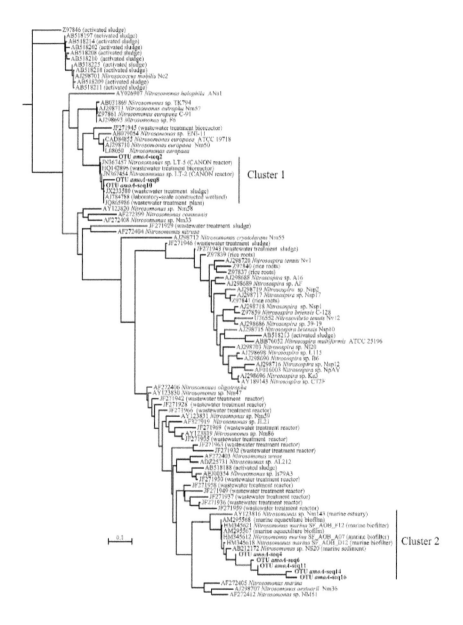

FIGURE 3: Maximum-likelihood tree of *amoA* gene. The tree was determined using approximately 491 unambiguously aligned positions of nucleic acid *amoA* sequences. Each sequence from this study (in bold) is representative of clustered *amoA* sequences in the WWTP activated sludge with an identity of 94%. Reference sequences from GenBank database are indicated by their accession number if they correspond to uncultured organisms or by the strain name if they belong to amoA sequences from bacterial strains. The tree was constructed with RAxML (http://bioinformatics.oxfordjournals.org/content/22/21/2688. full) using the GTRGAMMA model and an alignment made with MUSCLE (http://www. ncbi.nlm.nih.gov/pmc/articles/PMC390337/). The sequence of methane monooxygenase subunit A from *Methylococcus capsulatus* strain Bath served as outgroup (GenBank Accession No. YP_115248). The scale bar indicates substitutions per site.

On the other hand, we know that nitrification and denitrification are central processes in our system, since a nitrification fraction of 98% and a total nitrogen removal over 80% have been reported (M.I. Maldonado, pers. comm.). In fact, Yu and Zhang (2012), when applying both metagenomic and metatranscriptomic approaches to characterize microbial structure and gene expression of an activated sludge community from a municipal WWTP in Hong Kong found that nitrifiers such as *Nitrosomonas* and *Nitrospira* had a high transcription activity despite presenting a very low abundance (they accounted only 0.11% and 0.02% respectively in their DNA data set), and the results from Zhang and colleagues (2011b) indicated that the abundance of ammonia-oxidizing bacteria in the activated sludge from different WWTPs was very low. Similarly, our results suggest that *Nitrosomonas* could be responsible of nitrification although showing a low abundance. Also, it may be possible that other genera different from the well-known *Betaproteobacteria* could contribute to nitrification activity.

In fact, different studies have reported the autotrophic oxidation of ammonia by members of the domain *Archaea*. The crenarchaeon *Nitrosopumilus maritimus* is able to oxidize ammonia to nitrite under mesophilic conditions (Könneke et al., 2005), while ammonia-oxidizing *Archaea* occurred in activated sludge bioreactors used to remove ammonia from wastewater (Park et al., 2006). However, amplification of the *amoA* gene to detect the presence of archaeal nitrifiers yielded negative results in our samples. In fact, *Archaea* accounted only for 6% of DAPI counts, and all sequences retrieved previously in an archaeal clone library were related to methanogenic archaea (Sánchez et al., 2011). Other studies have also demonstrated the presence of methanogens in aerated activated sludge but, although active, they played a minor role in carbon and nitrogen turnover (Gray et al., 2002; Fredriksson et al., 2012).

Interestingly, different heterotrophic bacteria, such as *Bacillus* sp. (Kim et al., 2005), *Alcaligenes faecalis* (Liu et al., 2012), *Marinobacter* sp. (Hai-Yan et al., 2002), *Achromobacter xylosoxidans* (Kundu et al., 2012) and *Pseudomonas* sp. (Su et al., 2006) have been described as potential nitrifiers, and remarkably, some of these genera have been isolated from our activated sludge by culture-dependent techniques (data not shown); for instance, some strains were identified as *Bacillus* sp., *Alcaligenes* sp., *Marinobacter hydrocarbonoclasticus* and *Pseudomonas* sp.

In contrast, sequences of *Nitratireductor* sp., a denitrifying microorganism, have been retrieved with different molecular methods (pyrosequencing in this study and DGGE and clone library in Sánchez et al., 2011), while other sequences from potential denitrifiers have been recovered only by 454-pyrosequencing, such as *Leucobacter* sp., *Caldithrix* sp., *Castellaniella* sp. and *Halomonas* sp. Besides, other candidates for denitrifying bacteria have been isolated by culture-dependent techniques, such as *Alcaligenes* sp., *Bacillus* sp., *Paracoccus* sp., *Pseudomonas* sp. and *Marinobacter* sp. (data not shown). Other genera retrieved by pyrosequencing were related to the nitrogen fixation process, that is, *Microbacterium* sp., *Aminobacter* sp. and *Spirochaeta* sp., while *Sphingomonas* was detected by clone library and culture-dependent methods.

Summarizing, we can conclude that the bacterial diversity in the activated sludge of the seawater-processing plant was high as previously observed in conventional WWTPs. However, the composition of the bacterial community differed strongly from other plants, and was dominated by *Deinococcus-Thermus*, *Proteobacteria*, *Chloroflexi* and *Bacteroidetes*. Previous analyses by clone library, DGGE and FISH were not enough to reflect the profile of the bacterial community in wastewater sludge and although pyrosequencing was a powerful tool to define the microbial composition deeper sequencing is required. Despite nitrification rates were high in the system, known ammonia-oxidizing bacteria were not identified by means of 16S rRNA studies and analysis of the specific functional gene *amoA* was required to reveal the presence and identity of the bacteria responsible for this process. These results suggest that only a few populations of low abundant but specialized bacteria likely with high transcription activity are responsible for removal of ammonia in these systems. However, further studies to isolate the key microorganisms involved in ammonia-oxidation will be essential in order to understand this process in saline WWTPs.

REFERENCES

1. Barnett TP, Adam JC, Lettenmaier DP. Potential impacts of a warming climate on water availability in snow-dominated regions. Nature. 2005;438:303–309.
2. Benedict RG, Carlson DA. Aerobic heterotrophic bacteria in activated sludge. Water Res. 1971;5:1023–1030.
3. Boon N, De Windt W, Verstraete W, Top EM. Evaluation of nested PCR-DGGE (denaturing gradient gel electrophoresis) with group-specific 16S rRNA primers for the analysis of bacterial communities from different wastewater treatment plants. FEMS Microbiol Ecol. 39:101–112.
4. Claesson MJ, O'Sullivan O, Wang Q, Nikkilä J, Marchesi JR, Smidt H, et al. Comparative analysis of pyrosequencing and a phylogenetic microarray for exploring microbial community structures in the human distal intestine. PLoS ONE. 2002;4:e6669. 2009.
5. Daims H, Nielsen JL, Nielsen PH, Schleifer KH, Wagner M. In situ characterization of Nitrospira-like nitrite-oxidation bacteria active in wastewater treatment plants. Appl Environ Microbiol. 2001;67:5273–5284.
6. Dias FG, Bhat JV. Microbial ecology of activated sludge. Appl Environ Microbiol. 12:412–417.
7. Eichner CA, Erb RW, Timmis KN, Wagner-Döbler I. Thermal gradient gel electrophoresis analysis of bioprotection from pollutant shocks in the activated sludge microbial community. Appl Environ Microbiol. 1964;65:102–109. 1999.
8. Fredriksson NJ, Hermansson M, Wilén B-M. Diversity and dynamics of Archaea in an activated sludge wastewater treatment plant. BMC Microbiol. 2012;12:140.
9. Gray ND, Miskin EP, Kornilova O, Curtis TP, Head IM. Occurrence and activity of Archaea in aerated activated sludge wastewater treatment plants. Environ Microbiol. 4:158–168.
10. Hai-Yan Z, Ying L, Xi-Yan G, Guo-Min A, Li-Li M, Zhi-Pei L. Characterization of a marine origin aerobic nitrifying–denitrifying bacterium. J Biosci Bioeng. 2002;114:33–37. 2012.
11. Juretschko S, Timmerman G, Schmid M, Schleifer K-H, Pommerening-Roser A, Koops H, P., and Wagner M. Combined molecular and conventional analyses of nitrifying bacterium diversity in activated sludge: Nitrosococcus mobilis and Nitrospi-

ra-like bacteria as dominant populations. Appl Environ Microbiol. 1998;64:3042–3051.

12. Keijser BJ, Zaura E, Huse SM, van der Vossen JM, Schuren FH, Montijn RC, et al. Pyrosequencing analysis of the oral microflora of healthy adults. J Dent Res. 2008;87:1016–1020.

13. Kim JK, Park KP, Cho KS, Nam SW, Park TJ, Bajpai R. Aerobic nitrification-denitrification by heterotrophic Bacillus strains. Bioresour Technol. 2005;96:1897–1906.

14. Kim T-S, Kim H-S, Kwon S, Park H-D. Nitrifying bacterial community structure of a full-scale integrated fixed-film activated sludge process as investigated by pyrosequencing. J Microbiol Biotechnol. 2011;21:293–298.

15. Könneke M, Bernhard AE, de la Torre JR, Walker CB, Waterbury JB, Stahl DA. Isolation of an autotrophic ammonia-oxidizing marine archaeon. Nature. 2005;437:543–546.

16. Kundu P, Pramanik A, Mitra S, Choudhury JD, Mukherjee J, Mukherjee S. Heterotrophic nitrification by Achromobacter xylosoxidans S18 isolated from a small-scale slaughterhouse wastewater. Bioprocess Biosyst Eng. 2012;35:721–728.

17. Kwon S, Kim T-S, Yu GH, Jung J-H, Park H-D. Bacterial community composition and diversity of a full-scale integrated fixed-film activated sludge system as investigated by pyrosequencing. J Microbiol Biotechnol. 2010;20:1717–1723.

18. Liu Y, Li Y, Lv Y. Isolation and characterization of a heterotrophic nitrifier from coke plant wastewater. Water Sci Technol. 2012;65:2084–2090.

19. Liu Z, DeSantis TZ, Andersen GL, Knight R. Accurate taxonomy assignments from 16S rRNA sequences produced by highly parallel pyrosequencers. Nucleic Acids Res. 2008;36:e120.

20. Park H-D, Wells GF, Bae H, Criddle CS, Francis CA. Occurrence of ammonia-oxidizing archaea in wastewater treatment plant bioreactors. Appl Environ Microbiol. 2006;72:5643–5647.

21. Park H-D, Lee S-Y, Hwang S. Redundancy analysis demonstration of the relevance of temperature to ammonia-oxidizing bacterial community compositions in a full-scale nitrifying bioreactor treating saline wastewater. J Microbiol Biotechnol. 2009;19:346–350.

22. Pedrós-Alió C. Ecology. Dipping into the rare biosphere. Science. 2007;315:192–193.

23. Prakasam TBS, Dondero NC. Aerobic heterotrophic populations of sewage and activated sludge. I. Enumeration. Appl Environ Microbiol. 1967;15:461–467.

24. Purkhold U, Pommerening-Röser A, Juretschko S, Schmid MC, Koops HP, Wagner M. Phylogeny of all recognized species of ammonia oxidizers based on comparative 16S rRNA and amoA sequence analysis: implications for molecular diversity surveys. Appl Environ Microbiol. 2000;66:5368–5382.

25. Raszka A, Surmacz-Górska J, Zabczynski S, Miksch K. The population dynamics of nitrifiers in ammonium-rich systems. Water Environ Res. 83:2159–2169.

26. Saikaly PE, Stroot PG, Oerther DB. Use of 16S rRNA gene terminal restriction fragment analysis to assess the impact of solids retention time on the bacterial diversity of activated sludge. Appl Environ Microbiol. 2011;71:5814–5822. 2005.

27. Sanapareddy N, Hamp TJ, González LC, Hilger HA, Fodor AA, Clinton SM. Molecular diversity of a North Carolina wastewater treatment plant as revealed by pyrosequencing. Appl Environ Microbiol. 2009;75:1688–1696.

28. Sánchez O, Garrido L, Forn I, Massana R, Maldonado MI, Mas J. Molecular characterization of activated sludge from a seawater-processing wastewater treatment plant. Microb Biotechnol. 4:628–642.

29. Snaidr J, Amann R, Huber I, Ludwig W, Schleifer K-H. Phylogenetic analysis and in situ identification of bacteria in activated sludge. Appl Environ Microbiol. 2011;63:2884–2896. 1997.

30. Su JJ, Yeh KS, Tseng PW. A strain of Pseudomonas sp. Isolated from piggery wastewater treatment systems with heterotrophic nitrification capability in Taiwan. Curr Microbiol. 2006;53:77–81.

31. Wagner M, Amman R, Lemmer H, Schleifer K-H. Probing activated sludge with oligonucleotides specific for proteobacteria: inadequacy of culture-dependent methods for describing microbial community structure. Appl Environ Microbiol. 1993;59:1520–1525.

32. Wagner M, Loy A, Nogueira R, Purkhold U, Lee N, Daims H. Microbial community composition and function in wastewater treatment plants. Antonie Van Leeuwenhoek. 81:665–680.

33. Ye L, Shao M-S, Zhang T, Tong AHY, Lok S. Analysis of the bacterial community in a laboratory-scale nitrification reactor and a wastewater treatment plant by 454-pyrosequencing. Water Res. 2002;45:4390–4398. 2011.

34. Yu K, Zhang T. Metagenomic and metatranscriptomic analysis of microbial community structure and gene expression of activated sludge. PLoS ONE. 2012;7:e38183. doi: 10.1371/journal.pone.0038183.

35. Zhang T, Shao MF, Ye L. 454 pyrosequencing reveals bacterial diversity of activated sludge from 14 sewage treatment plants. ISME J. 2011a;6:1137–1147.

36. Zhang T, Ye L, Tong AHY, Shao M-F, Lok S. Ammonia-oxidizing archaea and ammonia-oxidizing bacteria in six full-scale wastewater treatment bioreactors. Appl Microbiol Biotechnol. 2011b;91:1215–1225.

There are several supplemental files that are not available in this version of the article. To view this additional information, please use the citation on the first page of this chapter.

CHAPTER 6

Assessment of Bacterial and Structural Dynamics in Aerobic Granular Biofilms

DAVID G. WEISSBRODT, THOMAS R. NEU, UTE KUHLICKE, YOAN RAPPAZ, AND CHRISTOF HOLLIGER

6.1 INTRODUCTION

Conventional activated sludge systems operated for biological nutrient removal (BNR) from wastewater require a high footprint for the integration of activated sludge tanks enabling microbial processes and of secondary clarifiers for the separation of biomass from the treated effluent. The aerobic granular sludge (AGS) process has recently deserved the attention of innovation analysts (Radauer et al., 2012). This technology has been developed for intensified BNR and secondary clarification in single sequencing batch reactors (SBR), and is related to definite savings in land area, construction, and operation costs according to reports on the performance of first full-scale plants (Giesen et al., 2012; Inocencio et al., 2013). AGS comprises suspended biofilm particles, called aerobic granules, formed by self-aggregation of microbial populations (Morgenroth et al., 1997). Although some full-scale plants are getting installed worldwide, the granulation mechanism at the microbial community level is not

Assessment of Bacterial and Structural Dynamics in Aerobic Granular Biofilms. © _Weissbrodt DG, Neu TR, Kuhlicke U, Rappaz Y, and Holliger C._ Frontiers in Microbiology _4,175 (2013). doi: 10.3389/ fmicb.2013.00175. Licensed under Creative Commons Attribution 3.0 Unported License, http://creativecommons.org/licenses/by/3.0/._

yet fully understood and improved knowledge of this phenomenon may enable further process optimization.

Nutrient removal deficiency and other process instabilities during granulation have been observed in several studies (Morgenroth et al., 1997; de Kreuk et al., 2005; Liu and Liu, 2006; Gonzalez-Gil and Holliger, 2011). Wash-out conditions that have been recommended for the selection of fast-settling biomass (Beun et al., 1999) have been shown to result in the deterioration of the settling properties and of the BNR performances caused by bacterial community imbalances with overgrowth of filamentous or zoogloeal populations, respectively (Weissbrodt et al., 2012a). Nevertheless, granule formation has been positively correlated with proliferation of *Zoogloea* spp. under wash-out conditions (Etterer, 2006; Adav et al., 2009; Ebrahimi et al., 2010; Weissbrodt et al., 2012a). Although *Zoogloea* can produce cohesive extracellular polymeric substances (EPS) (Seviour et al., 2012), it remains unclear whether these organisms are required to initiate granulation. Shifts in predominant populations in AGS systems have further been related to specific operation parameters. For instance, nutrient composition, temperature, carbon source, and selective excess sludge removal can impact the competition of polyphosphate-(PAO) and glycogen-accumulating organisms (GAO) related to "*Candidati Accumulibacter* and *Competibacter* phosphates" (hereafter referred to as *Accumulibacter* and *Competibacter*), respectively (de Kreuk and van Loosdrecht, 2004; Ebrahimi et al., 2010; Gonzalez-Gil and Holliger, 2011; Winkler et al., 2011; Bassin et al., 2012). Since some clades have been reported to denitrify, PAO and GAO can in addition be impacted by the type of terminal electron acceptor present in the medium (Kuba et al., 1997; Oehmen et al., 2010).

Granule structures comprise stratification of microbial niches oriented along radial substrate and microhabitat gradients (de Kreuk et al., 2005). Whereas bacterial community dynamics at reactor scale (Liu et al., 1998) can be assessed by terminal-restriction fragment length polymorphism (T-RFLP) analysis, biofilm architecture and microbial arrangements can be examined by confocal laser scanning microscopy (CLSM) combined with fluorescence in situ hybridization (FISH) (Wagner et al., 1994; Wuertz et al., 2004; Nielsen et al., 2009; Neu et al., 2010). Localization of bacterial populations along dissolved oxygen (DO) gradients in granules has been investigated with the latter methods (Tsuneda et al., 2003; Kishida et al., 2006; Lemaire et al., 2008a; Yilmaz et al., 2008; Gao et al., 2010; Filali et al., 2012).

In a lab-scale SBR operated for simultaneous nitrification, denitrification and phosphorus removal, Lemaire et al. (2008b) have for instance found that *Accumulibacter* and *Competibacter* have been predominant in the underlying bacterial community. According to FISH and oxygen microsensor measurements, *Accumulibacter* have dominated the oxygenated 200-μm outer layer of granules, while *Competibacter* have been more abundant in the anoxic granule core. However, among all studies having investigated microbial stratification within granules, it has been found that ammonium-oxidizing organisms (AOO) have formed clusters near the surface, but no consensus has been found on the position of PAO and GAO within the granules and their involvement in denitrification.

Since granules are biofilms, they are likely to exhibit complex spatial architecture depending on specific process conditions (Lawrence et al., 1996; Okabe et al., 1997; Alpkvist and Klapper, 2007). Granular biofilms have been composed of specific matrices of exopolymeric substances such as "granulan" (Seviour et al., 2011) or alginate (Lin et al., 2010). The combination of fluorescence lectin-binding analysis (FLBA) and CLSM (Neu and Lawrence, 1999; Staudt et al., 2003) could be relevant for mapping glycoconjugates in the granular biofilm matrix. Such analysis has previously been conducted for the staining of global exopolysaccharide matrices in different types of granules (Tay et al., 2003; McSwain et al., 2005; Adav et al., 2008). Since lectins display specific binding properties, a screening of different lectins can provide additional information on the type of polysaccharide residues present in granular biofilms.

The present research was conducted to elucidate the dynamics of the bacterial communities and of the structures of bioaggregates during transitions from activated sludge flocs to early-stage nuclei and to mature granular biofilms. Granulation was studied in one bubble-column (BC-SBR) and two stirred-tank (PAO-SBR, GAO-SBR) anaerobic-aerobic SBRs operated under conditions selecting for fast AGS cultivation and for enrichments of *Accumulibacter* and *Competibacter* in activated sludge, respectively. T-RFLP, pyrosequencing, CLSM, FLBA, and FISH methods were combined to investigate the mechanisms of bacterial selection, granule formation, and biofilm maturation in relation with the evolution of process variables.

6.2 MATERIALS AND METHODS

6.2.1 BUBBLE-COLUMN SBR OPERATION UNDER WASH-OUT DYNAMICS

The BC-SBR was operated at $23 \pm 2°C$ under wash-out conditions as reported in Weissbrodt et al. (2012a). This 2.5-L single-wall PVC reactor (height-to-diameter H/D ratio of 28) was inoculated with 3 g_{VSS} L^{-1} of activated sludge from a BNR wastewater treatment plant (WWTP Thunersee, Switzerland). The fixed 3-h SBR cycles comprised anaerobic feeding (60 min), aeration (110 min), settling (stepwise decrease from 15 to 3 min), and withdrawal (remaining cycle time; volume exchange ratio of 50%). Wash-out was generated by short settling and hydraulic retention times (HRT, 6 h). The sludge retention time (SRT) was not controlled. The synthetic wastewater composition comprised 4.8_{gP-PO4} and 12.5_{gN-NH4} per 100 g_{CODs} of acetate (Supplementary material 1). The system was fed over 220 days with a constant volumetric organic loading rate (OLR, 250 mg_{CODs} $cycle^{-1} L^{-1}_{R}$). The biomass specific OLR was initially 50 mg_{CODs} $cycle^{-1} g^{-1}_{CODx}$ and depended on the biomass amount remaining in the reactor. Aeration phases were run with up-flow superficial gas velocities of 0.025 m s^{-1}, free DO evolution up to saturation, and pH 7.0 ± 0.2. Temperature, DO, pH, and electrical conductivity were recorded on-line.

6.2.2 STIRRED-TANK PAO-SBR AND GAO-SBR OPERATION UNDER STEADY-STATE

The PAO- and GAO-SBRs were run to cultivate activated sludge enrichments over 1–2 years. The 2.5-L double-wall glass reactors (Applikon Biotechnology, The Netherlands, H/D = 1.3) were inoculated with 3 g_{VSS} L^{-1} of BNR activated sludge, and operated according to Lopez-Vazquez et al. (2009b). SBR cycles comprised N_2-flush (7 min), pulse feeding (7.3

min), N_2-flush (5 min), anaerobic, aerobic and settling phases (for timing see below), and withdrawal (5 min; 50%). During the anaerobic and aerobic (3.5 ± 0.5 mg_{O2} L^{-1}) phases, the reactor content was stirred at 300 rpm. Nitrification was inhibited in both reactors by addition of allyl-N-thiourea (Supplementary material 1). The SRTs were controlled by purge of excess sludge after aeration.

The PAO-SBR was operated at 17°C and pH 7.0–8.0, with 12-h HRT and 8-days SRT at steady-state, and with propionate, as well as with 9 g_{CODs} g^{-1}_{P-PO4} in the influent wastewater following Schuler and Jenkins (2003). Enhanced anaerobic propionate uptake and orthophosphate-cycling activities were ensured by stepwise adaptation of the volumetric OLR from 15 to 200 mg_{CODs} $cycle^{-1}$ L^{-1}_R in 12 days, and by proper control of the anaerobic and aerobic contact times (3–5 h) based on on-line conductivity profiles (Maurer and Gujer, 1995; Aguado et al., 2006). Since fast-settling biomass formed after 30 days, the settling time was decreased from 60 to 10 min to save cycle time, and to prevent prolonged endogenous respiration.

The GAO-SBR was operated at 30°C and pH 6.5 ± 0.2, with 12-h HRT and longer 16-days SRT at steady-state (Lopez-Vazquez et al., 2009a), acetate, and 200 gCOD g^{-1} P−PO_4. The volumetric OLR, anaerobic phase length, and settling time were fixed at 200 mgCODs $cycle^{-1}$ L^{-1} R, 3 h, and 60 min since start-up, respectively.

Schemes of experimental set-ups are available in Supplementary material 2. Each bubble-column and stirred-tank SBR experiment was operated on the long run. The definition of the operation conditions of the PAO- and GAO-SBRs nevertheless resulted from step-wise optimization based on multiple reactors experiments (Weissbrodt, 2012).

6.2.3 ANALYSES OF SOLUBLE COMPOUNDS AND BIOMASS

Concentrations of volatile fatty acids (VFA) and inorganic ions were measured by high performance liquid chromatography and ion chromatography, respectively. Sludge compositions were characterized as fractions of total (TSS), volatile (VSS), and inorganic suspended solids (ISS). Details on analytical methods are available in open access elsewhere (Weissbrodt et al., 2012a).

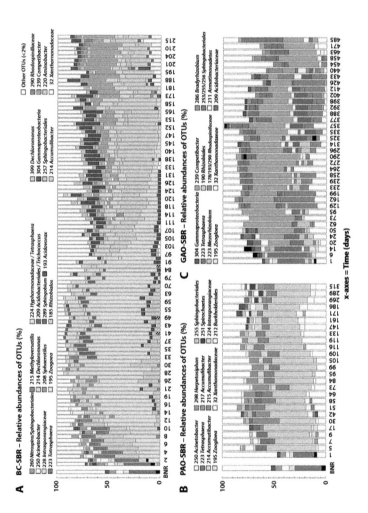

Figure 1: Bacterial community dynamics in the BC-SBR (A), PAO-SBR (B), and GAO-SBR (C). Phylotypes were identified with the PyroTRF-ID methodology (Weissbrodt et al., 2012b) and were compiled for each reactor in Supplementary material 5. In the BC-SBR, Dechloromonas was mainly contributing to OTU-214 during the first 60 days, whereas *Accumulibacter* was mainly contributing to this OTU in the mature granular sludge.

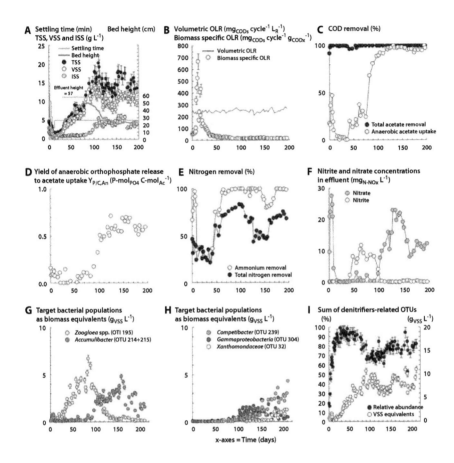

FIGURE 2: Dynamics in process variables during the operation of the BC-SBR under wash-out conditions. Evolution of the concentration and composition of biomass in terms of total (TSS), volatile (VSS), and inorganic suspended solids (ISS) during granule formation and maturation (A). Variations of the biomass specific organic loading rate (OLR) during operation under wash-out conditions with constant volumetric OLR (B). Dynamics in activities of COD removal (C), orthophosphate-cycling (D), nitrification and nitrogen removal (E,F). Shifts in predominant bacterial populations during granule formation and maturation (G,H). Rough estimation of the evolution of the sum of OTUs possibly affiliating with denitrifiers, according to the physiological properties reported in literature for the underlying phylotypes detected with PyroTRF-ID (I).

6.2.4 MOLECULAR ANALYSES
OF BACTERIAL COMMUNITY COMPOSITIONS

Bacterial community dynamics were investigated by T-RFLP analysis (Weissbrodt et al., 2012a). Biomass samples were homogenized by grinding, aliquoted in 1.5-mL Eppendorf tubes, and stored at −20°C. Eubacterial 16S rRNA gene pools were targeted and amplified by PCR with the labeled 8f (FAM-5′-AGAGTTTGATCMTGGCTCAG-3′) and unlabeled 518r (5′-ATTACCGCGGCTGCTGG-3′) primers. Amplicons were digested with *Hae*III. T-RFLP profiles were generated with predominant operational taxonomic units (OTU) >2%. Richness and Shannon's H' diversity indices were computed in R in order to assess the impact of operation conditions on the overall bacterial community structures (R Development Core Team, 2008). Three biomass samples were analyzed in triplicates, leading to a relative standard deviation of 6% on the T-RFLP method.

OTUs were affiliated to phylotypes with the pyrosequencing-based PyroTRF-ID bioinformatics methodology (Weissbrodt et al., 2012b) applied to biomass grab samples collected on days 2 and 59 (BC-SBR), 109 (PAO-SBR), and 398 (GAO-SBR). The DNA extracts of these samples were sent to Research and Testing Laboratory LLC (Lubbock, TX, USA) for pyrosequencing analysis according to the protocol developed by the company (Sun et al., 2011). The pyrosequencing datasets underwent denoising, mapping, and in silico restriction in the PyroTRF-ID pipeline. Bacterial affiliations of OTUs were obtained by cross-correlating digital and experimental T-RFLP profiles. Greengenes (DeSantis et al., 2006) was used as mapping database in the PyroTRF-ID pipeline. The pyrosequencing dataset of a mature AGS sample (BC-II) originating from a precedent reactor operated under similar conditions was used to affiliate OTUs at later stage in the BC-SBR. Technical and biological replicates previously measured in Weissbrodt et al. (2012b) demonstrated that the pyrosequencing-based PyroTRF-ID method is reliable.

In order to obtain estimates of the evolution of the mass of target OTUs, biomass equivalents were calculated by multiplying the relative abundances of these OTUs with the VSS concentration, according to Weissbrodt et al. (2012a). It should nevertheless be pointed out that such calculations are only very rough estimations since the number of

16S rRNA copies per cell can range from 1 to 15 depending on the genome and that cell biomass vary between bacterial species (Kembel et al., 2012). However, this simplified mass-based approach can provide from an engineering point of view a deeper insight in the dynamics of target populations by complementing relative abundance data. This is particularly meaningful under the high biomass dynamics generated by operation under wash-out conditions.

6.2.5 CONFOCAL LASER SCANNING MICROSCOPY ANALYSES OF FLOCS AND GRANULES

Structural transitions from flocs to granules were examined with CLSM. Collected biomass samples were washed twice in phosphate buffer saline (PBS) pH 7.4, and stored at 0–5°C in paraformaldehyde 4% (m/v in PBS). Fluorescent dyes were screened for mapping bioaggregates (Supplementary material 3). Rhodamine 6G was optimal for time series. Glycoconjugates were detected in selected AGS samples of the BC-SBR by FLBA, according to Staudt et al. (2003) and Zippel and Neu (2011). Spatial bacterial dynamics were followed by FISH-CLSM using rRNA oligonucleotide probes selected from probeBase (Loy et al., 2003) and targeting *Zoogloea*, *Accumulibacter* and *Competibacter* (BC-SBR), *Accumulibacter* and *Zoogloea* (PAO-SBR), as well as *Competibacter* (GAO-SBR) (Supplementary material 4). Samples were hybridized according to Nielsen et al. (2009). Granules smaller than 2 mm were cross-sectioned with a scalpel at ambient temperature in 0.5–1.0 mm deep CoverWell chambers mounted on microscopy slides (Life Technologies, Switzerland). Bigger granules were cryosectioned (80 µm) in a cryotome CM3050S (Leica, Germany) after freezing at −26°C in Tissue-Tek OCT compound (Sakura, The Netherlands).

The CLSM used was a TCS SP5X (Leica, Germany) equipped with upright microscope, an acusto optical beam splitter, and a supercontinuum light source. The system was controlled by the LAS AF software version 2.6.1. Samples were examined by the objective lenses 10 × 0.3 NA, 20 × 0.5 NA (overview), and 63 × 1.2 NA (high resolution). Excitation and emission wavelengths of fluorochromes are given in Supplementary materials 3 and 4. Each CLSM analysis was performed in at least two biological replicates. The CLSM reflection signal was collected as structural reference, and in order to distinguish voids from unstained regions. Multi-channel datasets were recorded in sequential mode to avoid cross-talk (Zippel and Neu, 2011). Images were collected by optical sample sectioning over thicknesses and stepsizes of 5–215 μm and 1–8 μm, respectively. Digital Leica Image Files (.lif) were processed in the Imaris software. Digital images were mainly represented as maximum intensity projections (MIP) with the Easy 3-D mode. Specific biofilm structures were examined in three dimensions either with the XYZ projection mode or the volume mode. The color intensity levels of the tagged image files generated in Imaris were slightly adjusted in the Photoshop software for improved resolution and contrast in the digital image datasets.

6.3 RESULTS

6.3.1 PROCESS AND BACTERIAL DYNAMICS IN THE BUBBLE-COLUMN SBR

The formation and maturation of granules in the BC-SBR operated under wash-out conditions were linked to dynamics in bacterial community compositions (Figure 1A and Supplementary material 5) and in process variables (Figure 2) over 220 days. The start-up period over the first 60 days was previously described in details in Weissbrodt et al. (2012a).

Wash-out conditions mainly impacted on the process before important accumulation of granular biomass. Under low residual biomass concentration, the operation with constant volumetric OLR and fixed anaerobic feeding phase length resulted in transiently high biomass specific OLR (Figures 2A–C). Leakage of VFA into aeration resulted in hampered BNR activities (Figures 2D–F) and selection of *Zoogloea* (OTU-195, 37–79%, 6.7 g_{VSS} L−1) over *Accumulibacter* (OTU-214) and *Competibacter* (OTU-239) affiliates (Figures 2G,H). After day 90, the shift from *Zoogloea* to *Accumulibacter* (33 ± 5%, 3.2 ± 0.6 g_{VSS} L^{-1}) correlated with full acetate uptake and high apparent yield of orthophosphate release to acetate uptake ($Y_{P/C,An}$ = 0.60 ± 0.07 P-mol C-mol$^{-1}_{Ac}$) under anaerobic conditions. Phosphorus was fully removed (data not shown) leading to 32 ± 1% of ISS in

AGS. After day 110, the bed dropped from 40–50 to <30 cm together with a decrease in biomass from 20 to <15 g_{TSS} L^{-1} and spontaneous stabilization of the SRT at 24 ± 5 days (data not shown). Proliferation of *Competibacter* (20–30%, 3.6 g_{VSS} L^{-1}) over *Accumulibacter* (10–15%, 0.6–1.9 g_{VSS} L^{-1}) after 150 days correlated with lower $Y_{P/C,An}$ (0.50 P-mol C-mol$^{-1}_{Ac}$) and instable dephosphatation (82 ± 25%).

Nitrification and nitrogen removal were inactive during wash-out, but recovered up to 90% and 70%, respectively, together with accumulation of AGS and lowering of the biomass specific OLR. Nitrite and then nitrate accumulated between days 50 and 150 (Figure 2E). Efficient nitrogen removal between days 50 and 100 (71–83%) and 150–200 (63–75%) correlated with high totals of denitrifiers-related OTUs (Figure 2I). Nitrifier-related OTUs were hardly detectable with T-RFLP. One OTU-260 affiliating with *Sphingobacteriales* and nitrite-oxidizing *Nitrospira* increased from 0.5 ± 0.4% (days 30–60) to 1.6 ± 0.7% (days 110–200). Further analysis of the pyrosequencing datasets in MG-RAST (Meyer et al., 2008) suggested shifts in low abundance ammonium- (AOO) and nitrite-oxidizing organisms (NOO) from flocculent sludge on day 2 (*Nitrosomonas* 0.12%, *Nitrosococcus* 0.24%, *Nitrosovibrio* 0.06%, *Nitrobacter* 0.06%, *Nitrospira* 1.02%) to early-stage AGS on day 59 (*Nitrosospira* 0.03%) and mature AGS in sample BC-II (*Nitrococcus* 6.2%, *Nitrosomonas* 0.1%, *Nitrospira* 0.2%).

6.3.2 PROCESS AND BACTERIAL DYNAMICS IN THE STIRRED-TANK PAO-SBR AND GAO-SBR

The operation of the PAO-SBR under full control of the anaerobic contact time (5–3 h), volumetric OLR from 15 to 200 mg_{CODs} cycle^{-1} L^{-1}_{R}, SRT (8–10 days), and biomass concentration (>1.5 g_{VSS} L^{-1}) resulted in the full uptake of propionate under anaerobic conditions, and in the rapid enrichment of *Accumulibacter* in the activated sludge (48% on day 5) (Figures 1B, 3). Enhanced orthophosphate-cycling activities were measured during the whole experimental period with $Y_{P/C,An}$ ranging between 0.56 and 0.64 P-mol C-mol$^{-1}_{Pr}$. Steady-state was reached after 15 days, with gradual stabilization of the SRT from 5 to 8–10 days by purging excess sludge. Interestingly, fast-settling biomass nuclei (<500 μm) formed after

20 days. By decreasing the settling time from 60 to 10 min to save cycle time, nuclei evolved toward 1–2 mm large granules over the next 40 days. *Tetrasphaera* declined from 22% to 3% within 30 days, but remained between 2–10% in the system (Figure 1B). The enrichment displayed constant predominance of *Accumulibacter* (41 ± 6%), and significant level of *Xanthomonadaceae* (OTU-32, 7–20%) right from start-up. *Herpetosiphon* (OTU-298) amounted to 7–17% on days 60–120. *Zoogloea* was only detected up to day 20 with relative abundances below 4%.

In the GAO-SBR, biomass remained flocculent over more than 450 days. With constant volumetric OLR of 200 mgCODs cycle−1 L−1R, full anaerobic acetate uptake was obtained from day 20 onward (data not shown). *Tetrasphaera* dominated over the first 2 weeks (31–47%), and remained at 5% abundance up to day 200 (Figure 1C). *Competibacter* prevailed in the enrichment (22–59%). Despite operation at steady-state, accompanying guilds displayed quite high dynamics. For instance abundances of *Alphaproteobacteria* related to *Rhodospirillaceae* (OTUs 178, 193, and 290, 2–40%), *Rhizobiales* (OTU-190, 2–12%), *Bradyrhizobium* (OTU-285, 2–8%), and *Sphingomonas* (OTU-287, 2–20%), as well as *Acidobacteriaceae* (OTU-209, 7–23%) and *Thiobacillus* relatives (OTU-216, 21% on day 433) fluctuated considerably during rector operation. After more than 450 days, fast-settling nuclei (<500 μm) were observed in the system, and evolved toward 1–2 mm granules from day 480 onward. Granulation correlated with transient over-aeration caused by biofilm growth on DO sensors. An increase in *Sphingobacteriales* relatives up to more than 40% (OTUs 253–256) was observed during granule formation and *Zoogloea*, *Accumulibacter*, and *Rhodocyclaceae* relatives were almost absent (<5%).

6.3.3 STRUCTURAL AND BACTERIAL TRANSITIONS FROM FLOCS TO GRANULES IN THE BUBBLE-COLUMN SBR

CLSM examinations of structural dynamics of bioaggregates in the BC-SBR with Rhodamine 6G staining are presented in Figure 4 and Supplementary material 6. Amorphous flocs (150–200 μm) initially present in the BC-SBR (day 1) underwent transformation by swelling of microbial colonies around flocs (day 9), followed by granulation of nuclei with dense rounded structures of 450–750 μm (day 23). Early-stage granules (850–1500 μm) displayed smooth and folded biofilm structures (day 30). Between days 50 and 140, round and compact microcolonies (10–100 μm) followed by larger ones (120–300 μm) proliferated from the inner core of granules outwards. Granular biofilm detachment was detected on day 60, and increased during maturation (day 112). Mature granules comprised internal voids, large biofilm clusters, and slimy interfacial matrices (days 209 and 218).

Different glycoconjugates were detected by fluorescent lectin-binding analysis (FLBA) of cross-sections of granules collected on days 85 and 105. According to CLSM data provided in Figure 5A and in Table SM3.2 in Supplementary material 3, FLBA showed (1) embedded continuous matrices revealed by STA, PHA-L, and IAA lectins, (2) matrices surrounding larger microbial clusters (HAA, LEA, SBA), (3) matrices of microcolony interfaces (WGA, LcH), and (4) direct binding to cell surfaces (ConA). HAA also revealed filamentous sheaths, and ConA the outwards palisade-like orientation of the biofilm continuum that surrounded dense colony clusters. Further interesting structures were detected in the architecture of mature granules with the use of other fluorescent probes (Figure 5B), e.g., spherical dense colony clusters of about 140 μm stained with SYPRO Red, aggregation of cells containing bright reflecting intracellular storage compounds after staining membranes with FM4-64, and extracellular DNA stained with DDAO.

FISH-CLSM confirmed T-RFLP analyses and provided information on spatial dynamics of active bacteria inside the structure of bioaggregates (Figure 6 and Supplementary material 7). *Zoogloea* proliferated during granulation as microcolonies of 20–45 μm swelling around flocs (day 6), and formed the early-stage granular biofilm continuum (day 50). It can be observed on the latter image that early-stage granules displayed loose core and dense surface aggregation. *Zoogloea* disappeared from the biofilm architecture of mature granules, as displayed on day 170. PAO and GAO were present as microcolonies (<10 μm) in the flocculent sludge. After 100 days, PAO and GAO established over *Zoogloea* from the granule cores outwards by forming large and dense clusters (>300 μm) exhibiting bright and low reflection, respectively. After initial presence in flocs as compact microcolonies of up to 30 μm (day 6), AOO proliferated as dense microcolonies (20–70 μm) across granules after 105 days and into wider biofilm matrices near the granule surface after 170 days.

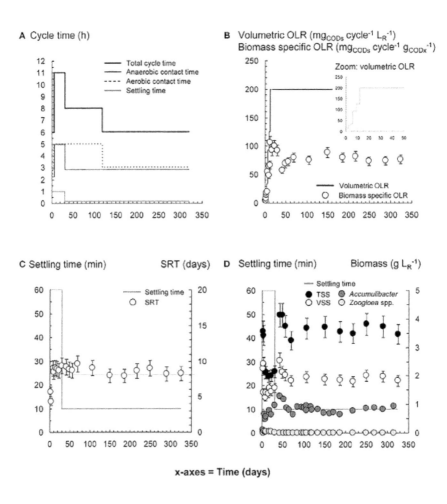

FIGURE 3: Process and bacterial dynamics in the PAO-SBR. The reactor was operated with proper control of anaerobic and aerobic contact times (A), volumetric organic loading rate (OLR) (B), and sludge retention time (SRT) (C) for rapid and preferential selection of *Accumulibacter* (D).

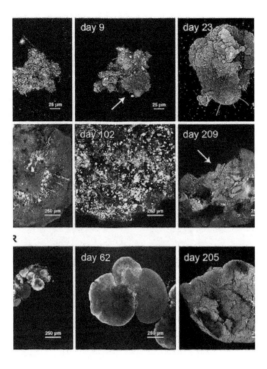

FIGURE 4: Temporal evolution of the architecture of bioaggregates from activated sludge flocs to early-stage granules and mature granular biofilms in the BC-SBR, PAO-SBR, and GAO-SBR. CLSM datasets were recorded on full bioaggregates from samples taken up to day 23, 62, and 484 in the BC-SBR, PAO-SBR, and GAO-SBR, respectively. Granules from later samples were analyzed on cross-sections. The sample taken on day 209 in the BC-SBR was analyzed as 80-μm cryosection. The green fluorescent dye Rhodamine 6G was used to map cells and biofilm matrices. In 8 bit data sets, 256 green levels were allocated to this dye. The reflection signal was used as reference with 256 gray/white color allocation. In the BC-SBR, swelling of microbial colonies around the floc structure can be observed on day 9. Early-stage granule nuclei on day 23 were 4–5 times bigger than flocs, and displayed compact biofilm aggregation. The evolution of the internal architecture of granules with proliferation of dense microcolonies from the granule core outwards can be observed from CLSM images taken on day 60 and 102. After more than 200 days, mature granules displayed a heterogeneous internal structure with internal voids and detachment. Granules formed in the PAO-SBR by proliferation of dense microbial clusters around the floc structure (day 49). Dense granule nuclei obtained after 62 days evolved toward mature granules exhibiting folded biofilm structures (day 205). In the GAO-SBR dense nuclei (day 484) evolved in 2 weeks toward 4–5 times larger granular biofilms (day 496). On day 510, granules were characterized as heterogeneous conglomerates of dense microbial clusters.

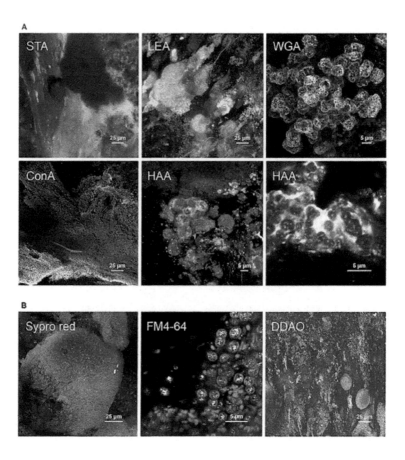

FIGURE 5: Examples of cellular and extracellular features detected in cross-sectioned granular biofilms collected after 85 days in the BC-SBR. (A) Selected glycoconjugate signals by means of FLBA. The STA lectin revealed the presence of a wide glycoconjugate matrix across the granule sections. Biofilm growth directions from the inner core to the outer sphere can be observed on the left part of the image with the growth lines displayed with the lectin staining. LEA showed dense microcolonies surrounded by a glycoconjugate matrix. WGA clearly showed glycoconjugate matrices surrounding specific types of microcolonies. ConA stained cell surface glycoconjugates also indicating biofilm growth lines and directions. HAA was used in combination with SYTO 60 and FM4-64 fluorescent probes. Color allocations: lectins binding to glycoconjugates (green), cell staining (red). (B) Further structures detected with additional fluorescent probes in the architecture of mature granules collected after 111 days in the BC-SBR. Dense spherical microbial clusters stained with SYPRO Red. Detection of microbial cells comprising bright reflecting intracellular storage compounds (cell membranes were stained with FM4-64). Presence of extracellular DNA stained with DDAO in the biofilm matrix and around cell clusters.

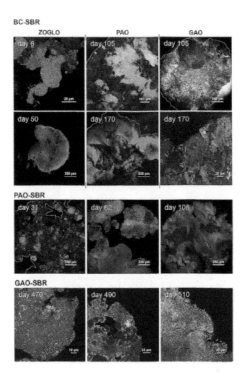

FIGURE 6: FISH-CLSM analyses of spatial dynamics of target bacterial populations during granule formation and maturation in the BC-SBR, PAO-SBR, and GAO-SBR. In the BC-SBR, colonies of *Zoogloea* (ZOGLO gene probe, red) swole around the floc structure on day 6 and formed the smooth biofilm continuum of early-stage granules on day 50. Biofilm growth lines from granule core outwards can be observed on this image. After more than 100 days, *Accumulibacter* (PAO, red) and *Competibacter* (GAO, green) proliferated inside mature granules from inner core outwards and were predominant in mature granules analyzed on day 170. On days 105 and 170, *Zoogloea* (shown in red in the GAO-related images) were only detected in low abundances in interstices between the different other microbial clusters. In the PAO-SBR, active *Accumulibacter* affiliates (red) predominated during granule formation. This population formed the dense bacterial clusters that proliferated in the floc structure (day 31), and that led to the formation of heterogeneous dense granules composed of bright reflecting clusters (day 62). The image of day 108 exhibits *Accumulibacter* coverage over the whole granule cross-section. In the GAO-SBR, *Competibacter* (GAO, green) was predominant during granulation. On day 470, the microbial cell membranes were stained with the fluorescent probe FM4-64 (red/yellow) after hybridization with the GAO gene probe. On day 490, the granules were composed of different GAO populations present in a continuous matrice and in dense microbial clusters, respectively. On day 510, granules displayed predominance of GAO populations with round cells.

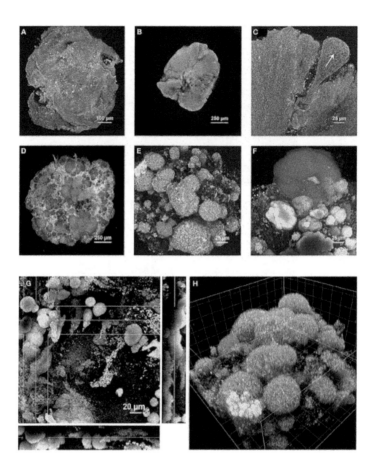

FIGURE 7: Main differences in the architectures of granules depending on the predominant bacterial populations involved, examined in two (A–F) and three dimensions (G,H). Under predominance of *Zoogloea* spp., smooth granular biofilm continua were obtained in early-stage granules collected after 20 days (A) and 60 days (B) in the BC-SBR. The formation and the outgrowth direction of biofilm petals can be observed in (C). The biofilm was composed of a homogeneous cell distribution in a gel matrix. Under conditions favoring PAO and GAO growth heterogeneous granular biofilm conglomerates were obtained as exemplified by day 503 (D) and day 484 (E) in the GAO-SBR, and by day 140 in the BC-SBR (F). During the transition on day 80 from early-stage to mature granules in the BC-SBR, the granular biofilm was composed of heterogeneous microbial clusters proliferating from the granule core outwards (G, XYZ projection). The cauliflower-like structure of heterogeneous granules present in the GAO-SBR is comparable to the mushroom-like structures of biofilm growth on a solid substratum under substrate-limiting conditions (H, 3-D volume projection, grid size of 20 μm).

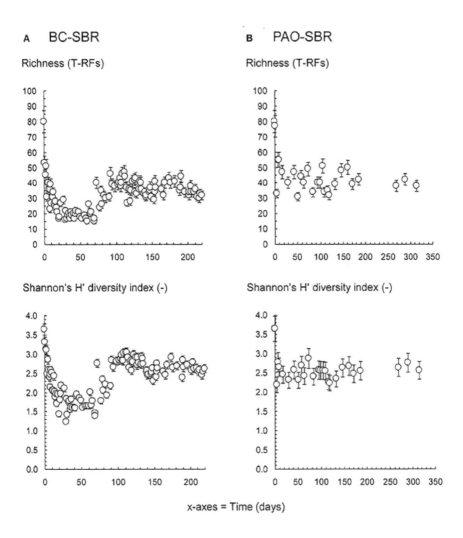

FIGURE 8: Evolutions of the richness and Shannon's H' diversity indices of the bacterial communities present in the BC-SBR (A) and PAO-SBR (B) computed based on the measured T-RFLP profiles. The application of wash-out conditions in the BC-SBR impacted on both indices over the first 70 days. A richer and more diverse community re-established after more than 80 days under the process conditions that impaired the *Zoogloea* proliferation. The operation of the PAO-SBR under steady-state conditions by full control of operation conditions resulted in stable richness and diversity of the underlying community. In both SBRs, initial decreases in richness from 80 to <50 T-RFs and in diversity from 3.7 to <3.0 were observed over the first 5 days after inoculation with the activated sludge from the full-scale plant and resulted from reactor feeding with VFA-based synthetic wastewaters.

6.3.4 GRANULATION IN THE STIRRED-TANK PAO-SBR AND GAO-SBR

According to Figure 4 and Supplementary material 8, granulation occurred in the PAO-SBR successively (1) by floc smothering (day 30), (2) by proliferation of round and dense clusters of 90–180 μm in flocs (day 49), and (3) by formation of smooth and dense nuclei that evolved up to 1.3-mm early-stage granules (day 62) and 1.5–2.0 mm mature granules (day 205). Mature granules displayed folded biofilm structures that contained aggregation of cells in dense biofilm clusters (day 205 and 215).

FISH-CLSM measurements (Figure 6) confirmed that PAO dominated during granulation by forming small dense clusters (10–120 μm) in smooth globular flocs (day 31), and larger dense clusters (<500 μm) in early-stage granules (day 62). PAO occupied the entire cross-sections of mature granules after 108 days. PAO cells (1.5 μm) had a high content of bright-reflecting intracellular storage compounds, and were more densely aggregated over the first 200–300 μm from the granule surface. *Zoogloea* were almost absent during granulation in the PAO-SBR, by forming only low abundant colonies (10–20 μm) in flocs (day 5) and patches at the surface of granule nuclei (day 31) (Supplementary material 7). *Zoogloea* were only present in biofilm interstices after 62 days.

In the GAO-SBR, flocs turned into compact nuclei of 200–600 μm after 484 days that evolved up to larger 1.3–1.5 mm granules (day 496) (Figure 4). Granules present in the GAO-SBR were heterogeneous conglomerates of large and dense clusters of 50–350 μm comprising 2.0–2.5 μm sized cells (day 510). FISH-CLSM analyses revealed that granulation occurred in this reactor under predominance of GAO across granules (Figure 6).

6.3.5 CORRELATION BETWEEN GRANULE STRUCTURES AND PREDOMINANT POPULATIONS

Based on the different structural time series described above for the three SBRs, the structure of granules correlated with the predominant organisms involved. A compilation of main structural differences is provided in Figure 7. Over the first 60 days in the BC-SBR, *Zoogloea*-dominated granules displayed smooth external and internal granular biofilm architectures (Figures 7A,B). In this case, granules displayed homogeneous embedment of cells in a biofilm matrix growing outwards like petals (Figure 7C). Granules dominated by slower-growing PAO and GAO displayed heterogeneous aggregation of population clusters (Figures 7D–F). In the GAO-SBR, cauliflower-like structures were observed from the surface of granules (Figure 7E).

Additional features of granules are presented in Supplementary material 9. Different cell distributions were observed in cross-sections of early-stage granules collected on day 50 (A) and of mature granules collected on day 111 in the BC-SBR (B). Mature granules dominated by PAO in the BC-SBR displayed heterogeneous internal architectures closer to the ones of granules cultivated in the PAO-SBR (C). Further specific structures are the initial microcolonies that swelled in flocs after 6 days in the BC-SBR (D), the filamentous populations that were detected after 11 days in the flocs of the PAO-SBR (E), and the slimy matrices present at the surface of mature granules after 209 days in the BC-SBR (F).

6.3.6 THREE-DIMENSIONAL ANALYSES OF GRANULE STRUCTURES

Specific biofilm and microbial structures examined by CLSM were finally analyzed in three dimensions (3-D) either as XYZ projections or volumetric presentations. During the transition from early-stage to mature granules after 80 days, the internal architecture of granules was composed of various types of microcolonies such as dense spherical microbial clusters proliferating from the granule core and populations aggregated in looser amorphous structures (Figure 7G). The volumetric presentation of the granular aggregates obtained in the GAO-SBR on day 484 exhibit the cauliflower-like or mushroom-like cluster evolving from the granule core outwards (Figure 7H).

Additional 3-D examinations are provided in Supplementary material 10. The early-stage granule obtained after 20 days in the BC-SBR displayed surface roughness with biofilm protuberances and valleys (A). After 111 days, a zoom on a microcolony revealed a spherical shape and denser cell aggregation at the edge of the microcolony (B). On the same day, dual staining with SYTOX Green (nucleic acids label) and Nile Red (label for hydrophobic components such as poly-β-hydroxyalcanoates) enabled to detect different types of cells based on their membrane and intracellular properties (C). Other volumetric presentations display the 3-D orientation of the glycoconjugate matrices detected with the WGA lectin around microcolony interfaces (D) and the finger-like population structures detected in early-stage nuclei obtained after 15 days in the BC-SBR that was dominated by *Zoogloea* (E).

6.4 DISCUSSION

6.4.1 GRANULATION CAN OCCUR UNDER WASH-OUT AND STEADY-STATE CONDITIONS

Granulation of activated sludge occurred under wash-out and steady-state conditions with different SBR designs. Wash-out conditions in the BC-SBR resulted in fast granulation over 2 weeks, which agrees with the granule formation periods reported by Beun et al. (1999). Mosquera-Corral et al. (2011) have succeeded to cultivate granules in a fully aerated stirred-tank SBR by imposing wash-out with short settling time (1 min) and HRT (6 h) right from start-up. In the present study, spontaneous granulation occurred in the stirred-tank SBRs run at steady-state to cultivate PAO and GAO enrichments. This phenomenon has also been observed in various stirred-tank SBRs operated for BNR (Dangcong et al., 1999; de Villiers and Pretorius, 2001; Dulekgurgen et al., 2003; You et al., 2008; Barr et al., 2010). Granulation in the GAO-SBR correlated with over-aeration that could have induced a physiological stress by increased endogenous respiration processes. Although three times slower than with wash-out, spontaneous granulation occurred in the PAO-SBR operated at steady-state under conditions selecting for *Accumulibacter*, with full control of the anaerobic uptake of VFA by step-wise adaptation of the OLR and anaerobic contact time, with stable SRT between 8 and 10 days, and with not very short settling time (60–10 min), and HRT (12 h). Fast-settling nuclei formed despite initially long 60-min settling periods. Such enhancement of biomass settling can be explained by the fact that dephosphatating sludge is denser than conventional activated sludge because of the presence of organic and inorganic intracellular polymers inside PAO cells (Schuler et al., 2001). The decrease of settling time from 60 to 10 min probably further selected for faster settling aggregates and full granulation. Hence, although wash-out conditions are efficient for fast granulation, granulation can also occur spontaneously. The granulation phenomenon in stirred-tank SBRs should further be investigated in a statistical approach in order to identify the trigger factors of granulation in this particular type of reactors. In addition, wash-out conditions can lead to prolonged deteriorated BNR since proper conditions to select for active PAO and nitrifiers are often not fulfilled (Weissbrodt et al., 2012a). Intermediate conditions between wash-out and steady-state should therefore be investigated for rapid formation of granules with good BNR activities.

6.4.2 BACTERIAL ECOLOGY CONSIDERATIONS

Since *Tetrasphaera* have been reported as important phosphorus-removing and glucose-fermenting organisms of full scale BNR-WWTP (Nielsen et al., 2012a,b), their regular presence in AGS-SBRs can be explained by the alternation of anaerobic-aerobic conditions and the high content of exopolysaccharides present in granules. The longest persistence of *Tetrasphaera* in the GAO-SBR at high abundances revealed that this population can cope with operation at higher mesophilic temperature and slight acidic pH, what could be relevant for dephosphatation of warm and acidic wastewaters. The GAO-enrichment culture revealed abundant *Rhodospirillaceae*-related *Alphaproteobacteria*. *Defluviicoccus vanus* which belongs to *Rhodospirillaceae* has been reported as a putative alphaproteobacterial GAO (Meyer et al., 2006), that is selected for with propionate as substrate (Lopez-Vazquez et al., 2009b). Here, *Rhodospirillaceae* relatives were abundant despite the presence of only acetate as substrate.

Nitrogen removal in the BC-SBR correlated with dynamics of denitrifiers. Although clades of *Accumulibacter* (and *Competibacter*) can denitrify and are desired for denitrifying dephosphatation (Yilmaz et al., 2008; Oehmen et al., 2010), denitrification in anaerobic-aerobic AGS-SBRs is not restricted to only PAO and GAO. Other denitrifiers that presumably do not take up VFA anaerobically were present in the full anaerobic-aerobic AGS ecosystems. Denitrifying metabolic activities and utilization of exopolysaccharides as electron donors in AGS systems should be investigated further based on previous knowledge gained from activated sludge systems (Finkmann et al., 2000; Thomsen et al., 2007; Ni et al., 2009; Nielsen et al., 2012b).

The richness and diversity patterns, which are commonly used to characterize biological and wastewater systems (Liu et al., 1997; Borcard et al., 2011; Gonzalez-Gil and Holliger, 2011; Winkler et al., 2012), were used to compare the evolution of the overall bacterial community structure under wash-out and steady-state conditions (Figure 8). In addition to the drop in richness and diversity during early-stage granulation reported in Weissbrodt et al. (2012a), wash-out conditions more intensively impacted on these indices. Under steady-state conditions in the PAO-SBR, granulation occurred with stable and relatively high community indices, despite first decrease due to synthetic laboratory conditions in the very beginning.

6.4.3 GRANULATION MECHANISMS DEPEND ON PROCESS CONDITIONS AND PREDOMINANT ORGANISMS

Granulation occurred under predominance of *Zoogloea* (BC-SBR), *Accumulibacter* (PAO-SBR), as well as *Competibacter* and *Sphingobacteriales* relatives (GAO-SBR). Thus, *Zoogloea* seem not to be essential for granule formation, answering an open question from earlier work where *Zoogloea* was predominant in dense fast-settling granules in different start-up experiments of AGS BC-SBRs (Weissbrodt et al., 2012a). According to Hesselsoe et al. (2009) and Nielsen et al. (2010), the predominant organisms that were observed in granules share physiological functions required for biofilm formation, namely production of exopolysaccharides and surface adhesins.

Whereas *Zoogloea*, *Xanthomonadaceae*, *Sphingomonadales*, and *Rhizobiales* relatives can contribute to the production of e.g., zooglan, xanthan, and sphingan exopolysaccharides (Lee et al., 1997; Pollock et al., 1998; Denner et al., 2001; Dow et al., 2003), *Sphingobacteriales* affiliates consume exopolysaccharides (Matsuyama et al., 2008). Granules can therefore be considered as exopolysaccharide-based ecosystems of bacterial producers and consumers. Seviour et al. (2011, 2012) have isolated a specific exopolysaccharide (granulan) from GAO-dominated granules, whereas Lin et al. (2008, 2010) have highlighted alginate as key granular exopolymer. The here applied in situ assessment of specific glycoconjugates by FLBA can inform on the type of abundant sugar residues (Zippel and Neu, 2011) present in the granules examined. N-acetyl-glucosamine (GlcNAc) multimers were observed in the biofilm continuum (STA lectin) and monomers in matrices surrounding microcolonies (LEA, WGA), N-acetyl-galactosamine (GalNAc) around bacterial clusters (HAA, SBA), and α -glucose or α -mannose at (*Zoogloea-*) cell surfaces dispersed in the continuum (ConA). Whereas the GAO-related granulan heteropolysaccharide isolated by Seviour et al. (2012) comprises GalNAc residues, the exopolysaccharides produced by *Zoogloea ramigera* contain glucose and galactose. The diversity of polymer matrices detected here indicated, however, that granules are composed of a complex mixture of glycoconjugates and that there is probably not only one key exopolymer present in all aerobic granules. Since exopolysaccharide presence depends on predominant microorganisms involved and on specific growth and operation conditions (Nielsen et al., 2004), functional screening should operate on a systems microbiology approach. Glycoconjugates and bacterial populations could be co-localized by combined lectin-binding and FISH analyses (Böckelmann et al., 2002). The microbial ecology data collected in this study indicated that only a low number of microorganisms would have to be screened in BNR AGS systems.

Different granulation mechanisms were observed depending on process conditions and predominant organisms involved. Predominance of fast-growing *Zoogloea* under non-limiting substrate conditions resulted in homogeneous granular biofilm matrices of cells dispersed in a gel. This biofilm continuum was formed by microcolony swelling around flocs, and embedded further proliferation of dense clusters of slower-growing nitrifiers, PAO, and GAO. This picture is similar to experimental and model-based descriptions of multispecies biofilms and underlying cooperative and non-cooperative microbial interactions (Picioreanu et al., 2000, 2004; Alpkvist et al., 2006; Alpkvist and Klapper, 2007; Xavier and Foster, 2007). Fast-growing heterotrophic competitors apparently formed the embedding biofilm continuum by production of exopolysaccharides and rapid proliferation outwards. Biofilm growth against substrate gradients explains the palisade-like orientation of cell lines, which has also been shown in methanogenic granules (Batstone et al., 2006). This converges to the first hypothesis of Barr et al. (2010) on the granulation mechanism by microcolony outgrowth, and on the early statement of Characklis (1973) that attached "microbial growth originates in a mixture of slime and zoogloeal bacteria (microorganisms that form gelatinous aggregates)."

Proliferation of nitrifiers, PAO, and GAO as dense clusters transported by the zoogloeal matrix may rely on the slower growth rates exhibited by these populations compared to fast-growing heterotrophic ones (de Kreuk and van Loosdrecht, 2004; Okabe et al., 2004). Accumulation of *Accumulibacter* resulted in denser granular biofilms, something which can explain, in addition to higher polyphosphate contents, the slight decrease in bed height in the BC-SBR despite increase of the biomass concentration between days 90 and 110. At mature stage in all reactors, the *Accumulibacter-* and *Competibacter*-dominated granules displayed heterogeneous aggregation of dense bacterial clusters in a cauliflower-like structure. Formation of heterogeneous biofilms has also been related to growth rate considerations under substrate limitations (Picioreanu et al., 2000; Alpkvist et al., 2006) that occurred as soon as full anaerobic feast and aerobic growth under starvation were achieved. Whereas heterogeneous architectures of mature granules can be explained by microcolony re-aggregation after detachment (Barr et al., 2010). Differences in bacterial physiologies can also lead to formation of different granular shapes by proliferation in single granules of either fast-growing organisms in a smooth continuum or slower-growing ones in heterogeneous dense clusters.

Biofilm growth is limited by detachment (Morgenroth, 2008). Detachment not only occurred at granule surfaces, but on entire biofilm whorls starting from the core of granules. Such detachment can result in dual access of substrates by the surface and the core of granules, what is comparable to biofilms growing on porous membranes with substrate penetrating from both sides (Downing and Nerenberg, 2008), and should be considered in mass transport phenomena across granules. The FISH-CLSM micrographs suggested that the proliferation of *Accumulibacter, Competibacter*, and nitrifiers clusters occurred from granule core outwards. This might be sustained by detachment phenomena and the access of substrates to central granule zones. Similarly to planar biofilms and to what has also been observed by different authors (Lemaire et al., 2008a; Lee et al., 2009; Barr et al., 2010), granules exhibit more complex structures than stratified architectures considered in conceptual and mathematical models. Granules heterogeneities can also explain the differences in the spatial organization of target organisms in the granular biofilm ecosystems depending on the process operation.

In conclusion, the knowledge gained on the complex bacterial and structural dynamics during granule formation and maturation led to the following findings. *Zoogloea, Accumulibacter*, and *Competibacter* can form granules and therefore *Zoogloea* are not essential for granulation. Granulation mechanisms depend on operation conditions and predominant organisms involved. *Zoogloea* form homogenous biofilms embedding the development of nitrifiers, PAO, and GAO colonies. *Accumulibacter* and *Competibacter* form heterogeneous aggregates of dense clusters. Mature granules display complex internal architectures exhibiting interspersing channels and biofilm detachment that can favor substrate access and growth of bacterial clusters from granule core outwards. Granulation of active *Accumulibacter* populations was possible under steady-state conditions under full control of the OLR and anaerobic contact time. Under such conditions, granulation of activated sludge was, however, three times slower than under wash-out conditions. Further studies targeting rapid formation of actively dephosphatating granules from flocculent sludge should therefore find a compromise between wash-out and steady-state conditions. The present fundamental study was performed with VFA-based synthetic wastewater. As additional research objective, one may investigate the impact of real wastewater compositions and particulate substrates on granulation mechanisms.

REFERENCES

1.   Adav, S. S., Lee, D. J., and Lai, J. Y. (2009). Functional consortium from aerobic granules under high organic loading rates. Bioresour. Technol. 100, 3465–3470. doi: 10.1016/j.biortech.2009.03.015

2. Adav, S. S., Lee, D. J., and Tay, J. H. (2008). Extracellular polymeric substances and structural stability of aerobic granule. Water Res. 42, 1644–1650. doi: 10.1016/j. watres.2007.10.013

3. Aguado, D., Montoya, T., Ferrer, J., and Seco, A. (2006). Relating ions concentration variations to conductivity variations in a sequencing batch reactor operated for enhanced biological phosphorus removal. Environ. Model. Softw. 21, 845–851. doi: 10.1016/j.envsoft.2005.03.004

4. Alpkvist, E., and Klapper, I. (2007). A multidimensional multispecies continuum model for heterogeneous biofilm development. Bull. Math. Biol. 69, 765–789. doi: 10.1007/s11538-006-9168-7

5. Alpkvist, E., Picioreanu, C., Van Loosdrecht, M. C. M., and Heyden, A. (2006). Three-dimensional biofilm model with individual cells and continuum EPS matrix. Biotechnol. Bioeng. 94, 961–979. doi: 10.1002/bit.20917

6. Barr, J. J., Cook, A. E., and Bond, P. L. (2010). Granule formation mechanisms within an aerobic wastewater system for phosphorus removal. Appl. Environ. Microbiol. 76, 7588–7597. doi: 10.1128/AEM.00864-10

7. Bassin, J. P., Winkler, M. K. H., Kleerebezem, R., Dezotti, M., and van Loosdrecht, M. C. M. (2012). Improved phosphate removal by selective sludge discharge in aerobic granular sludge reactors. Biotechnol. Bioeng. 109, 1919–1928. doi: 10.1002/ bit.24457

8. Batstone, D. J., Picioreanu, C., and van Loosdrecht, M. C. M. (2006). Multidimensional modelling to investigate interspecies hydrogen transfer in anaerobic biofilms. Water Res. 40, 3099–3108. doi: 10.1016/j.watres.2006.06.014

9. Beun, J. J., Hendriks, A., van Loosdrecht, M. C. M., Morgenroth, E., Wilderer, P. A., and Heijnen, J. J. (1999). Aerobic granulation in a sequencing batch reactor. Water Res. 33, 2283–2290. doi: 10.1016/S0043-1354(98)00463-1

10. Böckelmann, U., Manz, W., Neu, T. R., and Szewzyk, U. (2002). Investigation of lotic microbial aggregates by a combined technique of fluorescent in situ hybridisation and lectin-binding-analysis. J. Microbiol. Methods 49, 75–87. doi: 10.1016/ S0167-7012(01)00354-2

11. Borcard, D., Gillet, F., and Legendre, P. (2011). Numerical Ecology With R. Heidelberg: Springer-Verlag GmbH. doi: 10.1007/978-1-4419-7976-6

12. Characklis, W. G. (1973). Attached microbial growths: I. Attachment and growth. Water Res. 7, 1113–1127. doi: 10.1016/0043-1354(73)90066-3

13. Dangcong, P., Bernet, N., Delgenes, J. P., and Moletta, R. (1999). Aerobic granular sludge-a case report. Water Res. 33, 890–893. doi: 10.1016/S0043-1354(98)00443-6

14. de Kreuk, M. K., Heijnen, J. J., and van Loosdrecht, M. C. M. (2005). Simultaneous, COD, nitrogen, and phosphate removal by aerobic granular sludge. Biotechnol. Bioeng. 90, 761–769. doi: 10.1002/bit.20470

15. de Kreuk, M. K., and van Loosdrecht, M. C. M. (2004). Selection of slow growing organisms as a means for improving aerobic granular sludge stability. Water Sci. Technol. 49, 9–17.

16. de Villiers, G. H., and Pretorius, W. A. (2001). Abattoir effluent treatment and protein production. Water Sci. Technol. 43, 243–250.

17. Denner, E. B. M., Paukner, S., Kampfer, P., Moore, E. R. B., Abraham, W. R., Busse, H. J., et al. (2001). Sphingomonas pituitosa sp. nov., an exopolysaccharide-produc-

ing bacterium that secretes an unusual type of sphingan. Int. J. Syst. Evol. Microbiol. 51, 827–841. doi: 10.1099/00207713-51-3-827

18. DeSantis, T. Z., Hugenholtz, P., Larsen, N., Rojas, M., Brodie, E. L., Keller, K., et al. (2006). Greengenes, a chimera-checked 16S rRNA gene database and workbench compatible with ARB. Appl. Environ. Microbiol. 72, 5069–5072. doi: 10.1128/AEM.03006-05

19. Dow, J. M., Crossman, L., Findlay, K., He, Y. Q., Feng, J. X., and Tang, J. L. (2003). Biofilm dispersal in Xanthomonas campestris is controlled by cell-cell signaling and is required for full virulence to plants. Proc. Natl. Acad. Sci. U.S.A. 100, 10995–11000. doi: 10.1073/pnas.1833360100

20. Downing, L. S., and Nerenberg, R. (2008). Effect of oxygen gradients on the activity and microbial community structure of a nitrifying, membrane-aerated biofilm. Biotechnol. Bioeng. 101, 1193–1204. doi: 10.1002/bit.22018

21. Dulekgurgen, E., Ovez, S., Artan, N., and Orhon, D. (2003). Enhanced biological phosphate removal by granular sludge in a sequencing batch reactor. Biotechnol. Lett. 25, 687–693. doi: 10.1023/A:1023495710840

22. Ebrahimi, S., Gabus, S., Rohrbach-Brandt, E., Hosseini, M., Rossi, P., Maillard, J., et al. (2010). Performance and microbial community composition dynamics of aerobic granular sludge from sequencing batch bubble column reactors operated at 20°C, 30°C, and 35°C. Appl. Microbiol. Biotechnol. 87, 1555–1568. doi: 10.1007/s00253-010-2621-4

23. Etterer, T. J. (2006). Formation, Structure and Function of Aerobic Granular Sludge. Ph.D. thesis, Munich: Technische Universität München.

24. Filali, A., Bessiere, Y., and Sperandio, M. (2012). Effects of oxygen concentration on the nitrifying activity of an aerobic hybrid granular sludge reactor. Water Sci. Technol. 65, 289–295. doi: 10.2166/wst.2012.795

25. Finkmann, W., Altendorf, K., Stackebrandt, E., and Lipski, A. (2000). Characterization of N2O-producing Xanthomonas-like isolates from biofilters as Stenotrophomonas nitritireducens sp. nov., Luteimonas mephitis gen. nov., sp. nov. and Pseudoxanthomonas broegbernensis gen. nov., sp. nov. Int. J. Syst. Evol. Microbiol. 50, 273–282. doi: 10.1099/00207713-50-1-273

26. Gao, J. F., Chen, R. N., Su, K., Zhang, Q., and Peng, Y. Z. (2010). Formation and reaction mechanism of simultaneous nitrogen and phosphorus removal by aerobic granular sludge. Huan Jing Ke Xue 31, 1021–1029.

27. Giesen, A., Niermans, R., and van Loosdrecht, M. C. M. (2012). Aerobic granular biomass: the new standard for domestic and industrial wastewater treatment. Water 21, 28–30.

28. Gonzalez-Gil, G., and Holliger, C. (2011). Dynamics of microbial community structure and enhanced biological phosphorus removal of propionate- and acetate-cultivated aerobic granules. Appl. Environ. Microbiol. 77, 8041–8051. doi: 10.1128/AEM.05738-11

29. Hesselsoe, M., Fureder, S., Schloter, M., Bodrossy, L., Iversen, N., Roslev, P., et al. (2009). Isotope array analysis of Rhodocyclales uncovers functional redundancy and versatility in an activated sludge. ISME J. 3, 1349–1364. doi: 10.1038/ismej.2009.78

30. Inocencio, P., Coehlo, F., van Loosdrecht, M. C. M., and Giesen, A. (2013). The future of sewage treatment: Nereda technology exceeds high expectations. Water 21, 28–29.

31. Kembel, S. W., Wu, M., Eisen, J. A., and Green, J. L. (2012). Incorporating 16S gene copy number information improves estimates of microbial diversity and abundance. PLoS Comp. Biol. 8:e1002743. doi: 10.1371/journal.pcbi.1002743

32. Kishida, N., Kim, J., Tsuneda, S., and Sudo, R. (2006). Anaerobic/oxic/anoxic granular sludge process as an effective nutrient removal process utilizing denitrifying polyphosphate-accumulating organisms. Water Res. 40, 2303–2310. doi: 10.1016/j.watres.2006.04.037

33. Kuba, T., van Loosdrecht, M. C. M., and Heijnen, J. J. (1997). Biological dephosphatation by activated sludge under denitrifying conditions: pH influence and occurrence of denitrifying dephosphatation in a full-scale waste water treatment plant. Water Sci. Technol. 36, 75–82. doi: 10.1016/S0273-1223(97)00713-0

34. Lawrence, J. R., Korber, D. R., and Wolfaardt, G. M. (1996). Heterogeneity of natural biofilm communities. Cells Mater. 6, 175–191.

35. Lee, C. C., Lee, D. J., and Lai, J. Y. (2009). Labeling enzymes and extracellular polymeric substances in aerobic granules. J. Taiwan Inst. Chem. Eng. 40, 505–510. doi: 10.1016/j.jtice.2009.04.002

36. Lee, J. W., Yeomans, W. G., Allen, A. L., Gross, R. A., and Kaplan, D. L. (1997). Production of zoogloea gum by Zoogloea ramigera with glucose analogs. Biotechnol. Lett. 19, 799–802. doi: 10.1023/A:1018304729724

37. Lemaire, R., Webb, R. I., and Yuan, Z. (2008a). Micro-scale observations of the structure of aerobic microbial granules used for the treatment of nutrient-rich industrial wastewater. ISME J. 2, 528–541. doi: 10.1038/ismej.2008.12

38. Lemaire, R., Yuan, Z., Blackall, L. L., and Crocetti, G. R. (2008b). Microbial distribution of Accumulibacter spp. and Competibacter spp. in aerobic granules from a lab-scale biological nutrient removal system. Environ. Microbiol. 10, 354–363. doi: 10.1111/j.1462-2920.2007.01456.x

39. Lin, Y., de Kreuk, M., van Loosdrecht, M. C. M., and Adin, A. (2010). Characterization of alginate-like exopolysaccharides isolated from aerobic granular sludge in pilot-plant. Water Res. 44, 3355–3364. doi: 10.1016/j.watres.2010.03.019

40. Lin, Y. M., Wang, L., Chi, Z. M., and Liu, X. Y. (2008). Bacterial alginate role in aerobic granular bio-particles formation and settleability improvement. Sep. Sci. Technol. 43, 1642–1652. doi: 10.1080/01496390801973805

41. Liu, W. T., Marsh, T. L., Cheng, H., and Forney, L. J. (1997). Characterization of microbial diversity by determining terminal restriction fragment length polymorphisms of genes encoding 16S rRNA. Appl. Environ. Microbiol. 63, 4516–4522.

42. Liu, W. T., Marsh, T. L., and Forney, L. J. (1998). Determination of the microbial diversity of anaerobic-aerobic activated sludge by a novel molecular biological technique. Water Sci. Technol. 37, 417–422. doi: 10.1016/S0273-1223(98)00141-3

43. Liu, Y., and Liu, Q.-S. (2006). Causes and control of filamentous growth in aerobic granular sludge sequencing batch reactors. Biotechnol. Adv. 24, 115–127. doi: 10.1016/j.biotechadv.2005.08.001

44. Lopez-Vazquez, C. M., Hooijmans, C. M., Brdjanovic, D., Gijzen, H. J., and van Loosdrecht, M. C. M. (2009a). Temperature effects on glycogen accumulating organisms. Water Res. 43, 2852–2864. doi: 10.1016/j.watres.2009.03.038

45. Lopez-Vazquez, C. M., Oehmen, A., Hooijmans, C. M., Brdjanovic, D., Gijzen, H. J., Yuan, Z., et al. (2009b). Modeling the PAO-GAO competition: effects of carbon source, pH and temperature. Water Res. 43, 450–462. doi: 10.1016/j.watres.2008.10.032

46. Loy, A., Horn, M., and Wagner, M. (2003). ProbeBase: an online resource for rRNA-targeted oligonucleotide probes. Nucleic Acids Res. 31, 514–516. doi: 10.1093/nar/gkg016

47. Matsuyama, H., Katoh, H., Ohkushi, T., Satoh, A., Kawahara, K., and Yumoto, I. (2008). Sphingobacterium kitahiroshimense sp. nov., isolated from soil. Int. J. Syst. Evol. Microbiol. 58, 1576–1579. doi: 10.1099/ijs.0.65791-0

48. Maurer, M., and Gujer, W. (1995). Monitoring of microbial phosphorus release in batch experiments using electric conductivity. Water Res. 29, 2613–2617. doi: 10.1016/0043-1354(95)00146-C

49. McSwain, B. S., Irvine, R. L., Hausner, M., and Wilderer, P. A. (2005). Composition and distribution of extracellular polymeric substances in aerobic flocs and granular sludge. Appl. Environ. Microbiol. 71, 1051–1057. doi: 10.1128/AEM.71.2.1051-1057.2005

50. Meyer, F., Paarmann, D., D'Souza, M., Olson, R., Glass, E., Kubal, M., et al. (2008). The metagenomics RAST server - a public resource for the automatic phylogenetic and functional analysis of metagenomes. BMC Bioinformatics 9:386. doi: 10.1186/1471-2105-9-386

51. Meyer, R. L., Saunders, A. M., and Blackall, L. L. (2006). Putative glycogen-accumulating organisms belonging to the Alphaproteobacteria identified through rRNA-based stable isotope probing. Microbiology 152, 419–429. doi: 10.1099/mic.0.28445-0

52. Morgenroth, E. (2008). "Modelling biofilms," in Biological Wastewater Treatment: Principles, Modelling and Design, eds M. Henze, M. C. M. Van Loosdrecht, G. A. Ekama, and D. Brdjanovic (London: IWA Publishing), 457–492.

53. Morgenroth, E., Sherden, T., van Loosdrecht, M. C. M., Heijnen, J. J., and Wilderer, P. A. (1997). Aerobic granular sludge in a sequencing batch reactor. Water Res. 31, 3191–3194. doi: 10.1016/S0043-1354(97)00216-9

54. Mosquera-Corral, A., Arrojo, B., Figueroa, M., Campos, J. L., and Mendez, R. (2011). Aerobic granulation in a mechanical stirred SBR: treatment of low organic loads. Water Sci. Technol. 64, 155–161. doi: 10.2166/wst.2011.483

55. Neu, T. R., and Lawrence, J. R. (1999). Lectin-binding analysis in biofilm systems. Methods Enzymol. 310, 145–152. doi: 10.1016/S0076-6879(99)10012-0

56. Neu, T. R., Manz, B., Volke, F., Dynes, J. J., Hitchcock, A. P., and Lawrence, J. R. (2010). Advanced imaging techniques for assessment of structure, composition and function in biofilm systems. FEMS Microbiol. Ecol. 72, 1–21. doi: 10.1111/j.1574-6941.2010.00837.x

57. Ni, B. J., Fang, F., Rittmann, B. E., and Yu, H. Q. (2009). Modeling microbial products in activated sludge under feast-famine conditions. Environ. Sci. Technol. 43, 2489–2497. doi: 10.1021/es8026693

58. Nielsen, J. L., Nguyen, H., Meyer, R. L., and Nielsen, P. H. (2012a). Identification of glucose-fermenting bacteria in a full-scale enhanced biological phosphorus removal plant by stable isotope probing. Microbiology 158, 1818–1825. doi: 10.1099/mic.0.058818-0

59. Nielsen, P. H., Saunders, A. M., Hansen, A. A., Larsen, P., and Nielsen, J. L. (2012b). Microbial communities involved in enhanced biological phosphorus removal from wastewater - a model system in environmental biotechnology. Curr. Opin. Biotechnol. 23, 452–459. doi: 10.1016/j.copbio.2011.11.027

60. Nielsen, P. H., Daims, H., and Lemmer, H. (2009). FISH Handbook for Biological Wastewater Treatment - Identification and Quantification of Microorganisms in Activated sludge and Biofilms by FISH. London: IWA Publishing.

61. Nielsen, P. H., Mielczarek, A. T., Kragelund, C., Nielsen, J. L., Saunders, A. M., Kong, Y., et al. (2010). A conceptual ecosystem model of microbial communities in enhanced biological phosphorus removal plants. Water Res. 44, 5070–5088. doi: 10.1016/j.watres.2010.07.036

62. Nielsen, P. H., Thomsen, T. R., and Nielsen, J. L. (2004). Bacterial composition of activated sludge - importance for floc and sludge properties. Water Sci. Technol. 49, 51–58.

63. Oehmen, A., Carvalho, G., Lopez-Vazquez, C. M., van Loosdrecht, M. C. M., and Reis, M. A. M. (2010). Incorporating microbial ecology into the metabolic modelling of polyphosphate accumulating organisms and glycogen accumulating organisms. Water Res. 44, 4992–5004. doi: 10.1016/j.watres.2010.06.071

64. Okabe, S., Kindaichi, T., Ito, T., and Satoh, H. (2004). Analysis of size distribution and areal cell density of ammonia-oxidizing bacterial microcolonies in relation to substrate microprofiles in biofilms. Biotechnol. Bioeng. 85, 86–95. doi: 10.1002/bit.10864

65. Okabe, S., Yasuda, T., and Watanabe, Y. (1997). Uptake and release of inert fluorescence particles by mixed population biofilms. Biotechnol. Bioeng. 53, 459–469.

66. Picioreanu, C., Kreft, J. U., and van Loosdrecht, M. C. M. (2004). Particle-based multidimensional multispecies biofilm model. Appl. Environ. Microbiol. 70, 3024–3040. doi: 10.1128/AEM.70.5.3024-3040.2004

67. Picioreanu, C., van Loosdrecht, M. C. M., and Heijnen, J. J. (2000). A theoretical study on the effect of surface roughness on mass transport and transformation in biofilms. Biotechnol. Bioeng. 68, 355–369.

68. Pollock, T. J., Van Workum, W. A. T., Thorne, L., Mikolajczak, M. J., Yamazaki, M., Kijne, J. W., et al. (1998). Assignment of biochemical functions to glycosyl transferase genes which are essential for biosynthesis of exopolysaccharides in Sphingomonas strain S88 and Rhizobium leguminosarum. J. Bacteriol. 180, 586–593.

69. R Development Core Team. (2008). R: A Language and Environment for Statistical Computing, Vienna: R Foundation for Statistical Computing. Available online at: http://cran.r-project.org/

70. Radauer, A., Rivoire, L., Eparvier, P., and Schwanbeck, H. (2012). Presenting the (Economic) Value of Patents Nominated for the European Inventor Award 2012 - Inventor file Mark van Loosdrecht. (Vienna: Technopolis Group Austria).

71. Schuler, A. J., and Jenkins, D. (2003). Enhanced biological phosphorus removal from wastewater by biomass with different phosphorus contents, part I: experimen-

tal results and comparison with metabolic models. Water Environ. Res. 75, 485–498. doi: 10.2175/106143003X141286

72. Schuler, A. J., Jenkins, D., and Ronen, P. (2001). Microbial storage products, biomass density, and setting properties of enhanced biological phosphorus removal activated sludge. Water Sci. Technol. 43, 173–180.

73. Seviour, T., Yuan, Z., van Loosdrecht, M. C. M., and Lin, Y. (2012). Aerobic sludge granulation: a tale of two polysaccharides. Water Res. 46, 4803–4813. doi: 10.1016/j. watres.2012.06.018

74. Seviour, T. W., Lambert, L. K., Pijuan, M., and Yuan, Z. (2011). Selectively inducing the synthesis of a key structural exopolysaccharide in aerobic granules by enriching for "Candidatus Competibacter phosphatis". Appl. Microbiol. Biotechnol. 92, 1297–1305. doi: 10.1007/s00253-011-3385-1

75. Staudt, C., Horn, H., Hempel, D. C., and Neu, T. R. (2003). "Screening of lectins for staining lectin-specific glycoconjugates in the EPS of biofilms," in Biofilms in Medicine, Industry and Environmental Biotechnology, eds P. Lens, A. P. Moran, T. Mahony, P. Stoodley, and V. O'Flaherty (London: IWA Publishing), 308–326.

76. Sun, Y., Wolcott, R. D., and Dowd, S. E. (2011). Tag-encoded FLX amplicon pyrosequencing for the elucidation of microbial and functional gene diversity in any environment. Methods Mol. Biol. 733, 129–141. doi: 10.1007/978-1-61779-089-8_9

77. Tay, J. H., Tay, S. T. L., Ivanov, V., Pan, S., Jiang, H. L., and Liu, Q. S. (2003). Biomass and porosity profiles in microbial granules used for aerobic wastewater treatment. Lett. Appl. Microbiol. 36, 297–301. doi: 10.1046/j.1472-765X.2003.01312.x

78. Thomsen, T. R., Kong, Y., and Nielsen, P. H. (2007). Ecophysiology of abundant denitrifying bacteria in activated sludge. FEMS Microbiol. Ecol. 60, 370–382. doi: 10.1111/j.1574-6941.2007.00309.x

79. Tsuneda, S., Nagano, T., Hoshino, T., Ejiri, Y., Noda, N., and Hirata, A. (2003). Characterization of nitrifying granules produced in an aerobic upflow fluidized bed reactor. Water Res. 37, 4965–4973. doi: 10.1016/j.watres.2003.08.017

80. Wagner, M., Amann, R., Lemmer, H., Manz, W., and Schleifer, K. H. (1994). Probing activated sludge with fluorescently labeled rRNA targeted oligonucleotides. Water Sci. Technol. 29, 15–23.

81. Weissbrodt, D. G. (2012). Bacterial Resource Management for Nutrient Removal in Aerobic Granular Sludge Wastewater Treatment Systems. Ph.D. thesis No. 5641. Lausanne: Ecole Polytechnique Fédérale de Lausanne. doi: 10.5075/epfl-thesis-5641

82. Weissbrodt, D. G., Lochmatter, S., Ebrahimi, S., Rossi, P., Maillard, J., and Holliger, C. (2012a). Bacterial selection during the formation of early-stage aerobic granules in wastewater treatment systems operated under wash-out dynamics. Front. Microbiol. 3:332. doi: 10.3389/fmicb.2012.00332

83. Weissbrodt, D. G., Shani, N., Sinclair, L., Lefebvre, G., Rossi, P., Maillard, J., et al. (2012b). PyroTRF-ID: a novel bioinformatics methodology for the affiliation of terminal-restriction fragments using 16S rRNA gene pyrosequencing. BMC Microbiol. 12:306. doi: 10.1186/1471-2180-12-306

84. Winkler, M. K. H., Bassin, J. P., Kleerebezem, R., de Bruin, L. M. M., van den Brand, T. P. H., and van Loosdrecht, M. C. M. (2011). Selective sludge removal in a segregated aerobic granular biomass system as a strategy to control PAO-GAO

competition at high temperatures. Water Res. 45, 3291–3299. doi: 10.1016/j.watres.2011.03.024

85. Winkler, M. K. H., Kleerebezem, R., de Bruin, L. M. M., Verheijen, P. J. T., Abbas, B., Habermacher, J., et al. (2012). Microbial diversity differences within aerobic granular sludge and activated sludge flocs. Appl. Microbiol. Biotechnol. doi: 10.1007/S00253-012-4472-7. [Epub ahead of print].

86. Wuertz, S., Okabe, S., and Hausner, M. (2004). Microbial communities and their interactions in biofilm systems: an overview. Water Sci. Technol. 49, 327–336.

87. Xavier, J. B., and Foster, K. R. (2007). Cooperation and conflict in microbial biofilms. Proc. Natl. Acad. Sci. U.S.A. 104, 876–881. doi: 10.1073/pnas.0607651104

88. Yilmaz, G., Lemaire, R., Keller, J., and Yuan, Z. (2008). Simultaneous nitrification, denitrification, and phosphorus removal from nutrient-rich industrial wastewater using granular sludge. Biotechnol. Bioeng. 100, 529–541. doi: 10.1002/bit.21774

89. You, Y., Peng, Y., Yuan, Z. G., Li, X. Y., and Peng, Y. Z. (2008). Cultivation and characteristic of aerobic granular sludge enriched by phosphorus accumulating organisms. Huan Jing Ke Xue 29, 2242–2248.

90. Zippel, B., and Neu, T. R. (2011). Characterization of glycoconjugates of extracellular polymeric substances in tufa-associated biofilms by using fluorescence lectin-binding analysis. Appl. Environ. Microbiol. 77, 505–516. doi: 10.1128/AEM.01660-10.

There are several supplemental files that are not available in this version of the article. To view this additional information, please use the citation on the first page of this chapter.

PART II

ENVIRONMENTAL FACTORS

CHAPTER 7

Abundance, Diversity and Seasonal Dynamics of Predatory Bacteria in Aquaculture Zero Discharge Systems

PREM P. KANDEL, ZOHAR PASTERNAK, JAAP VAN RIJN, ORTAL NAHUM, AND EDOUARD JURKEVITCH

7.1 INTRODUCTION

Aquaculture has become a major source of food, generating more than 45% of global fish production (FAO, 2011). This success comes with a heavy environmental price, as aquaculture requires huge quantities of water for operational uses, discharges heavy loads of organic and inorganic contaminants to water bodies, and consumes natural coastlines and their ecosystems (Subasinghe et al., 2009). Inland water recycling systems help reduce this environmental burden, shifting the operations away from sensitive areas, and diminishing the demand for water. Yet, pollutants are still discharged into the environment, as organic and inorganic nitrogen compounds are mostly untreated (van Rijn, 2013). In zero discharge systems (ZDS) where fish basins (FBs) and treatment compartments are connected

Printed with permission from Kandel PP, Pasternak Z, van Rijn J, Nahum O, and Jurkevitch E. Abundance, Diversity and Seasonal Dynamics of Predatory Bacteria in Aquaculture Zero Discharge Systems. FEMS Microbiology Ecology **89**,1 (2014), DOI: 10.1111/1574-6941.12342.

to form continuous water recycling units, nitrogen, sulfate and organic matter are removed by microbial processes (Neori et al., 2007). ZDS are thus based on structured but connected environments that form mesocosms reflecting rather varied ecological conditions. This structure brings about continuous and efficient water purification leading to adequate water quality conditions for aquacultural needs over time, with water requirements limited to compensation for evaporation losses. Significantly, high fish yields are obtained under relatively disease-free conditions (Shnel et al., 2002; Gelfand et al., 2003).

The major processes responsible for water purification occurring in ZDS are nitrification, denitrification, sulfur oxido/reduction and degradation of organic matter. They are performed by numerous groups of bacteria, some of which have been identified and partially characterized (Cytryn et al., 2003, 2005a, b). Other ecological processes, such as microbial predatory interactions may affect the functioning of the microbial communities sustaining water purification, for example predation by viruses and protists in microbial food webs cause large-scale bacterial mortality (Fuhrman & Noble, 1995; Pernthaler, 2005). In aquatic environments, microbial biomass can be consumed at approximately the same rate as it is produced (Fuhrman & Noble, 1995; Pernthaler, 2005) and as such, microbial predation has a significant ecological effect. It has also been shown that microbial predation affects the evolution and the adaptation of both predators and prey, and modifies the structure and function of microbial communities (Gallet et al., 2007, 2009).

The understanding of the ecological relevance of microbial predatory interactions reposes with research with protists and with phages. Much less is known on the effects of bacterial predators of bacteria in natural or man-made systems, although these organisms are commonly found in the environment and are widely distributed (Jurkevitch & Davidov, 2007). Recent studies have shown that *Bdellovibrio* and like organisms (BALO), a major group of predatory bacteria, rapidly respond to changes in the structure of bacterial communities (Chauhan et al., 2009; Chen et al., 2011).

BALOs mainly belong to the *Deltaproteobacteria*, where they form two distinct and very diverse families. The *Bdellovibrionaceae*, with two defined species, *Bdellovibrio bacteriovorus* and *Bdellovibrio exovorus*, are in freshwater (including wastewater) and soils. The *Bacteriovoracaceae* comprise the halophilic species *Bacteriovorax marinus* sp. and *Bacteriovorax litoralis* sp. (Baer et al., 2000; Davidov & Jurkevitch, 2004), the freshwater/soil *Bacteriolyticum stolpii* (Pineiro et al., 2008), and the *Peredibacter* genus (Davidov & Jurkevitch, 2004). An additional BALO, the obligate predator *Micavibrio* sp., forms a distinct group in the *Alphaproteobacteria* (Davidov et al., 2006a). As to their effect on microbial communities, BALOs differ in prey range, from generalists to being rather restricted in range (Jurkevitch et al., 2000; Shemesh et al., 2003; Chen et al., 2011). BALOs may exert significant yet unexplored effects on the marine 'microbial loop' (Pomeroy et al., 2007), altering the structure of bacterial communities and affecting population dynamics. More specifically, BALOs have the potential to alter particular ecological functions by affecting the abundance of prey populations expressing them. As an example, BALOs isolated from fish ponds were shown to reduce disease incidence caused by the fish pathogens *Aeromonas hydrophila* and *Vibrio alginolyticus* (Chu & Zhu, 2009; Wen et al., 2009) and contributed to the improved growth performance of fish, shrimp, crab and sea cucumber (Qi et al., 2009). Thus, BALOs may affect the functioning of water recycling systems, and in aquaculture may help keep fish stocks healthy.

To assess the impact of BALOs on microbial processes it is imperative to identify the predators and to quantify their absolute and relative abundance. In this study, we explored the dynamics of BALO populations, including the *Bdellovibrionaceae* and the *Bacteriovorax* and *Bacteriolyticum* genera of the *Bacteriovoracaceae*, in two ZDS with different salinities over a period of 7 months. In the freshwater system, we examined the community composition of the total bacterial community in detail and provided evidence for the hypothesis that environmental parameters may affect BALO predatory dynamics and population structure. Although the predators do not dominate numerically, they form fairly abundant populations that fluctuate over time and differ according to salinity levels. Remarkably, freshwater *Bdellovibrionaceae* were actually found to populate the saline ZDS also.

7.2 MATERIALS AND METHODS

7.2.1 THE ZERO DISCHARGE SYSTEMS

A pilot plant comprising two parallel ZDS operates at the Rehovot campus (Supporting Information, Fig. S1). One ZDS operates with freshwater, the other with water maintained at a salinity level of 20 p.p.t. Each system consists of a 5-m³ fish culture basin (FB) from which water is withdrawn through two parallel biofiltration loops: (1) Water from the upper layers of the FB is treated in a 7-m³ trickling filter (TF), filled with PVC substratum, for oxidation of ammonia to nitrate and for degassing of CO_2. The retention time of the water flowing through this loop is 30 min. (2) Water rich in particulate organic carbon from the bottom of the FB is diverted into a 2-m³ digestion/sedimentation basin (DB) where oxidation of organic material is mediated by a variety of electron acceptors including oxygen, nitrate and sulfate. Effluent from the DB is then pumped into a 13-L fluidized bed reactor (FBR) for additional nitrate and sulfide removal. The flow rate through this loop is 0.8 m³ h⁻¹. Following treatment, purified water from both loops is returned to the FB. Fish were stocked on 31 October 2010. The fresh water basin was stocked with 650 tilapia hybrids (*Oreochromis niloticus* × *Oreochromis aureus*) with an average weight of around 50 g. The salt water FB was stocked on 31 October 2010 with 738 gilt-head sea bream (*Sparus aurata*) with an average weight of 1.5 g. The fish were fed a high protein and lipid diet (45% and 19%, respectively). The diet is assimilated to up to 60%, but about 80% of the nitrogen is excreted in the form of ammonium. The tilapia and gilt-head sea bream reached an average of about 250 and 240 g at harvest, respectively. Harvesting was performed in July 2011.

7.2.2 SAMPLING

Water samples were collected in triplicate from each of the compartments of the fresh and salt water systems on a monthly basis, totalling 21 samples per time point, from December 2010 to June 2011, except for February

2011. Samples of 500 mL from the fish and DBs were sterilely collected from the upper 25 cm of the water phase using a self-constructed sampler. For the TF, water trickling down at the bottom of the filter was collected, whereas for the FBR water and some sand were collected from the base of the reactor. Collected samples were immediately stored at 4 °C before processing. Biofilm samples were collected by inserting a long plastic strip in the TF and allowing a biofilm to develop on it. To relate biofilm measurements with planktonic phase measurements, the biofilm was assumed to be 20 μm thick. Water quality parameters in the FBs were measured daily (temperature and dissolved oxygen) or twice monthly (pH, nitrite, nitrate, total ammonia, orthophosphate and sulfide levels).

7.2.3 DNA EXTRACTION

A 50-mL sample of those collected was filtered through a 0.2-μm filter (Schleicher and Schuell) to capture the bacteria. Half of the filter was cut into pieces and used for DNA extraction using a Powersoil™ DNA Isolation Kit (Mo Bio Laboratories, Inc.) according to the manufacturer's protocol. DNA was isolated from biofilms by cutting 10-cm^2 strips into sterile water followed by vigorous vortexing until the attached material was cleaned off the surface, and then proceeding as above. Finally, the DNA bound to the silica gel of the extraction column was eluted with 50 μL PCR grade double-distilled water (DDW). DNA concentration was measured with a Nanodrop 2000 spectrophotometer (Olympus) and stored at 4 °C until use.

7.2.4 BALO-SPECIFIC PCR

PCR was performed with the DNA samples extracted from the ZDS using 16S rRNA gene primers designed by Davidov et al. (2006b). These primers specifically target the *Bdellovibrionaceae* (Bd) family and the following clades within the *Bacteriovoracaceae* (Bc) family: *Bacteriovorax* (the halophilic BALOs) and *Bacteriolyticum* (freshwater and soil BALOs) (Table S2) but not *Peredibacter*, as confirmed by PCR testing using two strains from this genus (data not shown). The PCR reaction included 12.5 μL of PCR mastermix (Cat no. D123P; Lambda Biotech), 1 μL of each primer (10 μm), 2 μL of DNA and 8.5 μL of PCR grade DDW. Positive (BALO DNA) and negative (no DNA) controls were always included. The PCR was performed with a Mastercycler Gradient (Eppendorf AG, Germany). Thermal profiles for *Bdellovibrionaceae* were: denaturation and enzyme activation at 95 °C, 2 min, followed by 36 cycles of 95 °C for 30 s, annealing at 51 °C for 30 s, extension at 72 °C for 30 s, and final extension at 72 °C for 5 min. For the *Bacteriovoracaceae* the thermal profiles were similar except that annealing was at 55 °C for 30 s followed by an extension step at 72 °C for 45 s. PCR products (6 μL) were verified by gel electrophoresis [1% agarose gel in 0.5% Tris acetate-EDTA (TAE) buffer stained with 1 μL of ethidium bromide; 100 V for 30 min] and the gel was digitized with a B.I.S Bioimaging System (Pharmacia Biotech, Sweden).

7.2.5 STANDARD CURVES FOR BALO-SPECIFIC QPCR

The predatory strains used as reference strains in the PCR and QPCR assays (Table S1) were grown as previously described (Jurkevitch, 2012). Bd and Bc abundances were calculated using standard curves generated by serially diluting plasmid preparations containing a fragment of the 16S rRNA gene from *B. bacteriovorus* 109J or from *B. stolpii* UKi2, respectively. The plasmid containing the *B. bacteriovorus* gene fragment was constructed by inserting a 492-bp fragment of the gene, amplified with primers BbsF216 and BbsR707 (Table S2), and using the following thermal cycling profile (modified from Van Essche et al., 2009): 95 °C, 2 min, 29 cycles of 95 °C for 30 s, 56 °C for 30 s, 72 °C for 30 s, with a final extension at 72 °C for 5 min. The plasmid containing the *B. stolpii* gene fragment was constructed using a 981-bp fragment from the 16S rRNA gene amplified with the primers Bac69F and Bac1049R (Table S2) designed using the NCBI primer-blast program (http://www.ncbi.nlm.nih.gov/tools/primer-blast/). The PCR reaction parameters used to amplify this fragment were as above, except that annealing was performed at 56 °C and extension was run for 60 s.

PCR products were cleaned with High Yield Gel/PCR DNA Fragment Extraction Kit (RBC Bioscience), and cloned into PGEM-T easy plasmid vector system (Promega). The recombinant plasmid vectors were transferred into competent *Escherichia coli* TG1 cells prepared in the lab and transformants were detected on Luria–Bertani (LB) plates by resistance to ampicillin and by the X-gal assay. Following growth of white colonies in LB/Amp, plasmids were extracted and purified with the Gene Elute Plasmid Miniprep Kit (Sigma Aldrich, Israel), and DNA was quantified using Nanodrop. The presence of inserts was confirmed by running uncut, single cut (SalI), and double cut (EcoRI) plasmids in gel electrophoresis. Ten-times serial dilutions from 9 to 9×10^6 and from 10 to 1×10^6 plasmid copies per reaction for Bd and Bc, respectively, were prepared. These dilutions were used to construct standard QPCR curves. Plasmid copy numbers were calculated as (Whelan et al., 2003):

$$\text{Copy number} = \frac{\text{DNA (ng)} \times 6.022 \times 10^{23}}{\text{length (bp)} \times 10^9 \times 650} \tag{1}$$

7.2.6 BALO-SPECIFIC QPCR

7.2.6.1 BDELLO*VIBRIO*NACEAE

QPCR for the Bd was performed based on the TaqMan technology, as in Van Essche et al. (2009), using primers Bd347F and Bd549R and probe Bd396P, labeled with FAM (6-carboxyfluorescein) at the 5' end, and TAMRA (6-carboxytetramethylrhodamine) at the 3' end (Sigma Aldrich; Table S2). Reaction tubes contained 12.5 µL of Jumpstart Taq Readymix for High Throughput Quantitative PCR (D6442; Sigma-Aldrich), 2 µL of each primer (10 µM), 1.25 µL of the Taqman probe (1 µM), 5 µL of DNA and 2.25 µL PCR grade DDW in 25 µL tubes. The thermal profile was as in Van Essche et al. (2009). The data were collected during the annealing and extension phase of the reaction. The reactions were performed in a MicroAMP 8-tube stripe (0.2 mL) with MicroAMP Optical 8-Cap Strip (Applied Biosystems). Reactions were run in a 7300 Real-Time PCR System and the data were collected and analyzed by the sds v1.4 software (Applied Biosystems).

7.2.6.2 BACTERIOVORACACEAE (BACTERIOVORAX AND BACTERIOLYTICUM)

This series of specific QPCR was performed based on SYBR green I chemistry. The reaction was carried out according to Zheng et al. (2008) using primers Bac 519F and Bac 677R, with some adjustments. Briefly, each 25-μL reaction consisted of 12.5 μL ABsolute Blue QPCR SYBR Green ROX Mix, 0.5 μL of each primer (10 μM), 1 μL of DNA and 10.5 μL of PCR grade DDW. Thermal profiles were 94 °C for 15 min (enzyme activation), and 45 cycles of 94 °C for 30 s, 62 °C for 30 s and 72 °C for 30 s followed by a dissociation stage. The data were collected during the extension phase of the reaction. The reactions were carried out in a 96-well plate (ABGene; Thermoscientific) with an adhesive seal applicator (AB-1391; Thermoscientific) on a Stratagene-MX 3000P Real-time PCR System (Stratagene). For both Bd and Bc reactions, seven standard plasmid samples (9 to 9×10^7 copies per reaction) were included in triplicates in every run, along with three controls without template (NTC).

7.2.7 QPCR FOR BACTERIA

This reaction was performed to measure the total abundance of bacteria in the sample, enabling the calculation of the relative abundance of each BALO group. Standards were prepared by inserting a 1467-bp fragment of the *E. coli* ML-35 16S rRNA gene amplified with primers 27F and 1494R into a Topo Cloning Vector (Invitrogen). The cloning, transformation and plasmid extraction and quantification were done as above. DNA samples were diluted five times before use in the reactions, which were performed with primers 519F and 907R (Muyzer et al., 1995; Table S2). Each 25-μL reaction consisted of 12.5 μL of ABsolute Blue QPCR SYBR Green ROX Mix (Thermoscientific) 1 μL of each primer, 1 μL of DNA and 9.5 μL of PCR grade DDW. Reactions were ran at 95 °C for 15 min followed by 40 cycles of 95 °C for 30 s, 58 °C for 30 s and 72 °C for 30 s. Data were collected during the extension phase of the reaction and included 24 standards (three replicates for building a standard curve with copy number of 102–109 per reaction) and six wells without template. The reactions were carried out in a 96-well plate (ABGene; Thermoscientific) with an adhesive seal applicator (AB-1391; Thermoscientific). QPCR was performed on a Stratagene-MX 3000P Real-time PCR System (Stratagene). Total QPCR counts (16S rRNA bacterial genes mL^{-1}) were estimated based on a standard curve with 100% efficiency factoring in the cycle threshold value of the particular sample obtained, taking into account that 25 mL of sample was used for DNA extraction, DNA was eluted in 50 μL of DDW and 1 μL of this DNA was used in the QPCR reaction.

7.2.8 HIGH THROUGHPUT SEQUENCING

Samples were collected and DNA extracted as above from the water phase of the TF, the DB, and the FBR, in duplicates, in December 2010 and April 2011. The April 2011 sampling set included one sample from the water phase and one from the biofilm phase of the TF. High throughput pyrosequencing (HTS) was performed at the Research and Testing Laboratory (Lubock, TX) on a 454 FLX Genome Sequencer System (Roche Applied Science, IN). All reads were cleaned using the mothur software (Schloss et al., 2009). First, Fasta and quality data were extracted from the raw SFF file. Sequences were grouped according to barcode and primer, allowing one mismatch to the barcode and two mismatches to the primer. Denoising was achieved with the AmpliconNoise algorithm (Quince et al., 2011), removing both 454 sequencing errors and PCR single base errors. Next, sequences were trimmed to remove barcode and primer sequences, all sequences with homopolymers (i.e. AAAA) longer than 8 bp, and all sequences < 100 bp long. All sequences were aligned to the silva reference alignment database (Schloss et al., 2009) and filtered so that they all overlapped perfectly (with no overhang or no-data base pairs). To further reduce sequencing errors, pre-clustering of the sequences was performed, based on the algorithm of Huse et al. (2010). Finally, chimeric reads were removed with the UChime method (Edgar et al., 2011), and all chloroplast, mitochondria and 'unknown' (i.e. unclassified at the Domain level) reads were deleted.

Sequences have been deposited at the MG-RAST metagenomic database, project #4588 (http://metagenomics.anl.gov/linkin.cgi?project= 4588).

7.2.9 STATISTICAL ANALYSIS

All 454 reads were clustered into genera based on current RDP-II taxonomy (Cole et al., 2009). The resulting genera were arranged in a data matrix where each row is a single sample and each column a genus; each data point in the matrix represents the abundance of the particular genus in the particular sample, relativized to the sampling effort (i.e. the number of 454 reads obtained from that sample). Multivariate analysis was performed in pc-ord v6.12 (mjm Software) with Sorensen distances. Ordinations were performed with non-metric multidimensional scaling (NMDS; Mather, 1976) at 500 iterations, and cluster analyses were performed with flexible beta linkages ($\beta = -0.25$). Differences between sample groups were calculated with multi-response permutation procedure (MRPP; Mielke, 1984), a test based on the assumption that if two groups are different from each other, the average within-group difference is smaller than the average between-group distance. The size of the difference between groups is represented by the A-statistic of the MRPP test, and its significance is identified by the MRPP P-value.

To detect which genera were mainly responsible for differences between groups, we use the method of Dufrêne & Legendre (1997). The basis for this procedure is the computation of Indicator Values (IVs), which are a combination of the frequency of occurrence and of the abundance of each genus in each group; IV spans between 0 and 100, and is larger if a genus is more frequent and/or more abundant in a given group than in the other group.

7.3 RESULTS

7.3.1 QPCR SENSITIVITY

Standard QPCR curves for Bd (Fig. S2a) and Bc (Fig. S2b) were generated by serial dilution of plasmids containing cloned fragments of the 16S rRNA gene amplified from *B. bacteriovorus* and *B. stolpii* reference strains for the Bd and Bc reactions, respectively. Reaction efficiency varied between 96% and 100% with correlation coefficients always > 0.99. The minimal detection limit for both Bd and Bc was 10 copies per 5-μL reaction, corresponding to 2 predatory cells mL^{-1} in the sample, considering that 25 mL of sample was used, DNA was eluted in 50 μL of DDW, 5 μL of this DNA was used per reaction, and there are two 16S rRNA genes in the *Bdellovibrio* and *Bacteriovorax* genomes (Rendulic et al., 2004; Crossman et al., 2013). However, highly reproducible results were obtained from 90 copies per reaction. Hence, the quantification limit was set at 100 copies per reaction, or 20 predatory cells mL^{-1} in the sample. Counts of *B. bacteriovorus* 109J cells growing on an *E. coli* prey were about 2.5 times higher when measured by QPCR than by counting plaques in a double-layered agar (Fig. S3). This difference may be due to decreased predation in soft agar, to the ability of QPCR to detect predators not forming plaques or to limited detection of small plaques by visual inspection (Van Essche et al., 2009). The correlation coefficients between copy numbers and cycles of standard QPCR curves for *Bacteria* were very high (r^2 > 0.99).

7.3.2 QUANTIFICATION OF THE BACTERIAL COMMUNITY USING QPCR

Total bacterial abundance was measured at three time points in the FB, TF and DB in both the FW and SW ZDS. In the former, 7×10^6 to 1×10^8 16S rRNA gene copies mL^{-1} and in the latter, 4×10^6 to 1.2×10^8 16S rRNA gene copy mL^{-1} were measured (Table 1). The 16S rRNA gene copy number per genome varies greatly between bacteria, from 1 to 15 (Lee et al.,

2009). Nonetheless, it has been estimated that cells in wastewater plants bear an average of 3.6 copies of the 16S rRNA gene per cell (Dionisi et al., 2002). Accordingly, cell concentrations in the compartments varied between 1.9×10^6 to 2.8×10^7 cell mL^{-1}, and 1.1×10^6 to 3.3×10^7 cell mL^{-1} in the FW ZDS and in the SW ZDS, respectively.

TABLE 1: Bacterial 16S rRNA gene mL^{-1} counts in ZDS samples over time

Compartment	December	April	June
FB/FW	$8 \times 10^7 \pm 3.1 \times 10^{6A}{}_1$	$7.3 \times 10^7 \pm 1.7 \times 10^{6A}{}_1$	$1.3 \times 10^7 \pm 6.3 \times 10^{6B}{}_1$
TF/FW	$1 \times 10^8 \pm 3.1 \times 10^{6C}{}_2$	$5.7 \times 10^7 \pm 7.1 \times 10^{6CD}{}_1$	$6.7 \times 10^6 \pm 9 \times 105^{DE}{}_1$
DB/FW	$2.2 \times 10^7 \pm 5.3 \times 10^{5F}{}_3$	$3.6 \times 10^7 \pm 2.8 \times 10^{6F}{}_1$	$7.3 \times 10^6 \pm 4.9 \times 10^{5G}{}_1$
FBR/FW	$2.6 \times 10^7 \pm 5.4 \times 10^5{}_3$	NA	NA
FB/SW	$4 \times 10^6 \pm 1.4 \times 10^{5H}{}_4$	$4.2 \times 10^7 \pm 2.4 \times 10^{6I}{}_1$	NA
TF/SW	$6 \times 10^6 \pm 4.2 \times 10^{4J}{}_4$	$3.3 \times 10^7 \pm 3.4 \times 10^{6K}{}_1$	$5.5 \times 10^7 \pm 4.3 \times 10^{6K}{}_2$
DB/SW	$4.8 \times 10^6 \pm 7.2 \times 10^{4L}{}_4$	$1.2 \times 10^8 \pm 1.7 \times 10^{7L}{}_1$	$3 \times 10^7 \pm 1.5 \times 10^{6L}{}_3$
FBR/SW	NA	$7.8 \times 10^7 \pm 6.6 \times 10^6{}_1$	NA

Values shown are means and standard deviations of three biological replicates. Cells within a row connected with different capital letters in superscript are significantly different. Within a column, cells connected with different numbers in subscript are significantly different (P = 0.05). FW, fresh water; NA, not available for sampling; SW, salt water.

7.3.3 ABUNDANCE AND DYNAMICS OF SPECIFIC BALOS IN THE ZDS

Bd populations were quantified by QPCR in both the FW and the SW systems, over 7 months. Bd was found in all the compartments of both systems but the predator was more abundant in the FW ZDS than in the SW ZDS at most sampling times, ranging from 1.5×10^3 to 4.4×10^5 cell mL^{-1} and from 1.5×10^2 to 4.9×10^3 cell mL^{-1}, respectively (Table S3). The presence of Bd in the SW ZDS is remarkable, as to date this genus was thought to be confined to non-saline environments (Baer et al., 2000; Davidov & Jurkevitch, 2004). Bd abundance within compartments in each system fluctuated significantly with time. Noteworthy, Bd seasonal dynamics in the FB and TF were similar, differing from those of the DB (Fig. 1a, b and d). Additionally, whereas FB and TF populations in the FW ZDS fluctuated by less than six-fold during the sampling period, in the SW ZDS they increased with time by more than 30-fold (Fig. 1a and b; Table S3). A biofilm that developed on PVC clips introduced in the FW TF supported high Bd populations: Bd abundance calculated assuming a depth of 20 μm deep was 1.3×10^7 cell mL^{-1}, but reached 1.2×10^4 cell mL^{-1} (difference is significant at P = 0.001) in samples from the planktonic phase of the same compartment taken at the same time.

Populations of Bc predators were more alike and stable than the Bd populations in both the FW and SW systems, and between the compartments in each system, spanning from 1.2×10^4 to 1.8×10^5 cell mL^{-1} (Fig. 1c, Table S4). Overall, and in both systems, Bc populations were almost always more abundant than Bd populations. The restricted number of FBR samples precluded comparative analysis but suggested that Bd population levels were similar to those of the other basins, whereas those of the Bc were lower (Fig. 1).

7.3.4 CORRELATION BETWEEN SPECIFIC BALO ABUNDANCE, BIOTIC AND ABIOTIC PARAMETER

The relative abundance of the detected BALOs (Bd + Bc) in both ZDS was separately calculated as the 16S rRNA gene count ratio of Bd and of Bc over total Bacteria. The total of the targeted BALOs accounted for 0.07–0.78% of the total bacterial 16S rRNA gene count, or 0.13–1.4% of the total cell counts, using 2 and 3.6 gene copies genome−1 for BALOs and for total bacteria, respectively. A significant correlation between BALOs and the total bacterial counts was found in the FW ZDS (r^2 = 0.78, P = 0.012) but not in the SW ZDS (r^2 = 0.51, P = 0.19), hinting at direct coupling between predator and prey abundance in the former system. The relative abundance of BALOs (Bd + Bc) was highest in most of the DB samples in both the FW and SW systems, reflecting the high population levels of predators reached in this compartment where heterotrophic degradation of organic matter occurs (Figs 1 and 2). A significant correlation was found between Bd abundance and reactive phosphate in the SW system ($r^2 = 0.92$, P < 0.0024). Since no information on phosphate thresholds in Bd exists, the reason for this correlation is unclear; it is noteworthy that sediment-trapped phosphate was released into the water starting about March. No other correlation between the targeted BALO abundance and temperature, oxygen, pH, nitrates, nitrites, ammonia, or sulfides was detected.

7.3.5 SEQUENCE-BASED COMMUNITY ANALYSIS

The bacterial composition of the FW ZDS, which includes the populations preyed upon by BALOs, was explored by HTS of the 16S rRNA gene. In total, 74 003 curated sequences were obtained at two time points, and from 10 samples: c. 6000–9600, 7500–9300, 5500–7100, and 5400 sequences from four, three, two, and one samples from the DB, the TF, the FBR, and the TF biofilm, respectively. The diversity was essentially covered, reaching 95–98% (Table S5), and totalling 238 OTUs. The DB was richer (125 OTUs) and the TF had fewer of OTUs (89). The structure of the bacterial communities did not differ between the different compartments, as revealed by NMDS and MRPP tests (P > 0.1; Fig. 3). However, a seasonal effect was clearly seen, with April and December samples clustering separately (MRPP A = 0.22, P = 0.002; Fig. 3), a separation contributed by 7 and 8 OTUs, respectively (Table S6).

The most abundant four OTUs constituted over 60% of the total sequences and belonged to: *Flavobacteria* (29.7 ± 12.9%), which strongly dominated all samples; unclassified Bacteria (12.8 ± 7.1%), which were mainly found in the BF, the DB and the FBR, and significantly less in the TF; *Bacteroidetes* (10.4 ± 3.2%), which were more abundant in the BF, DB, and FBR, and appeared to decrease in the TF; and *Fusobacteriaceae* (8.6 ± 5.0%), which were less common in the BF than in the other samples (Table S7). The 20 most commonly retrieved OTUs, representing 85.4 ± 4.7% of all sequences, included taxa commonly found in freshwater [*Flavobacteriaceae, Cytophagaceae, Polynucleobacter (Burkholderiaceae)*], *Lacibacter* (freshwater fish) (*Cetobacterium – Fusobac-*

teriaceae; Tsuchiya et al., 2008), *Bacteroidetes* and *Flavobacteria* (Kim et al., 2007; Sullam et al., 2012); and wastewater and sludge [*Verrucomicrobia*, *Flavobacteriaceae*, *Novosphingobium* (*Sphingomonadaceae*), *Burkholderiales*, *Rhizobiales*; Zhang et al., 2012]. Thus, the composition of the ZDS microbiota appears to reflect its dual purpose. Confirming this assumption, a phylogenetic analysis of *Flavobacteria*, the system's dominant phylum, revealed that the large majority of these sequences (79%) are more closely related to those of FW fish *Flavobacteria* than to those from other habitats (Fig. S4).

To further characterize the ZDS, potentially important ecological functions, that is nitrification, sulfate reduction, and predation were assigned, based on the identification of the populations in the different compartments (Table S8). The ammonium oxidizers *Nitrosomonadaceae* were barely detectable in the FBR, and only sporadically in the DB and TF, whereas the nitrite oxidizers (*Nitrospira*) formed dominant populations (c. 1–2% of total) in the FBR, and were also abundant in the BF. Sulfate reducers were only detected in the DB and at a lower concentration in the FBR.

The estimated fraction of BALO sequences detected by HTS was between two-thirds to a fifth of that found by QPCR, the relatively low number of sequences inducing a potentially large error (Fig. 2, Table S8). BALOs constituted 20–40% of the potential *Deltaproteobacteria* predators (*Myxococcales*); other potential predators included *Herpetosiphon*, which were present in the BF and DB, and *Ensifer* (Jurkevitch & Davidov, 2007), present in part of the DB and FBR samples. Some *Flavobacteria* strains have been shown to be predatory (Banning et al., 2010) but no 16S rRNA gene sequences closely related to these strains were found in our dataset.

A phylogenetic analysis of all the BALO sequences was performed (Fig. 4). The restricted length of the sequenced fragments precluded a precise analysis at the species level (Aharon et al., 2013) in both the *Bdellovibrionaceae* and the *Bacteriovoracaceae*. Yet, it could still be discerned that all *Bdellovibrio* sequences were similar to strain *B. bacteriovorus* HD100 but that no sequences related to the epibiotic predators B. *exovorus* or *Micavibrio* were found.

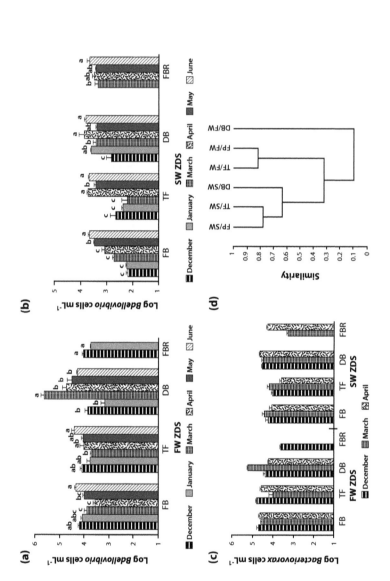

FIGURE 1: *Bdellovibrio* (a, b) and *Bacteriovorax* (c) abundance (cell mL−1) in the fresh water (FW) and salt water (SW) ZDS, as measured by clade-specific QPCR targeting the 16S rRNA gene, over time. (d) Cluster analysis relating the Bray–Curtis similarity between samples, using unweighted pair-group average (upgma) clustering. SE, standard error of mean.

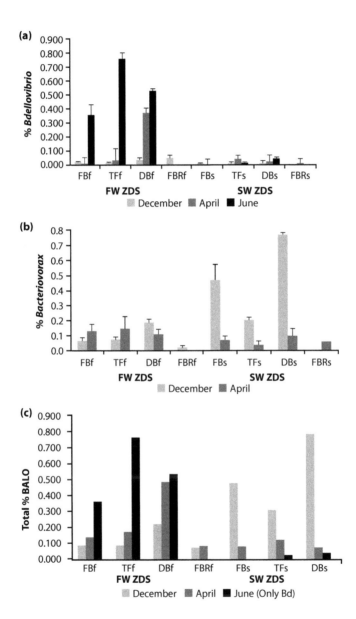

FIGURE 2: Percent Bd (a) and Bc (b) and total BALO (Bd + Bc) (c) from the total Bacterial community as determined by clade-specific (*Bdellovibrio*-Bd, *Bacteriovorax* + *Bacteriolyticum*-Bc) and general Bacteria QPCR targeting the 16S rRNA gene, over time. FW, fresh water; SE, standard error of mean; SW, salt water.

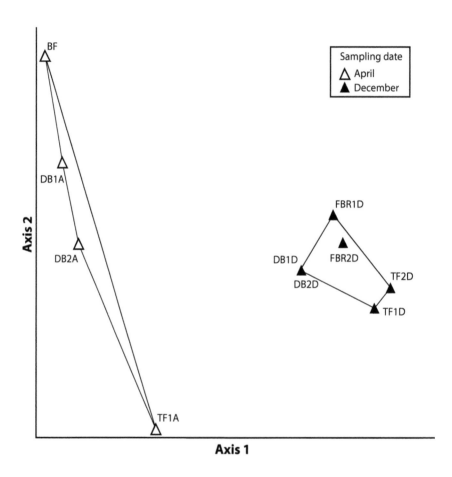

FIGURE 3: NMDS ordination of 16S rRNA gene sequences from December and April samples. Convex hulls connect samples from each sampling time. Ordination axes do not possess biological meaning; stress = 8.4%, confirming a good correlation with the data.

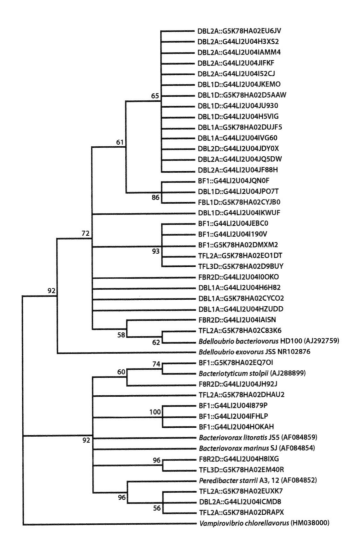

FIGURE 4: Phylogenetic tree of the BALO in the ZDS. All 16S rRNA gene sequences of BALO in the ZDS were compared with the BALO type-strain sequences retrieved from GenBank. The tree was constructed by maximum likelihood based on the Tamura–Nei model. Bootstrap consensus was inferred from 100 replicates (bootstrap values are shown next to the branches). Branches corresponding to partitions reproduced in < 50% bootstrap replicates were collapsed. BF, biofilm; FBR, fluidized bed reactor.

7.4 DISCUSSION

Predation is a major ecological process, affecting many parameters of ecosystem structure and function (Huntley & Kowalewski, 2007; Rodriguez-Valera et al., 2009; Vasseur & Fox, 2009; Vos et al., 2009; Estes et al., 2011). At the microbial level, bacteriophages and protists have been shown to make major contributions to bacterial mortality and turnover in aquatic environments (Fuhrman, 1999; Samuel et al., 2007); yet, our knowledge of bacterial predation by bacterial predators is still very fragmentary. Here, we analyzed the dynamics of defined BALO populations and the bacterial communities within which the predators are embedded, providing the first quantitative data for the BALO community, of its constituent populations and of their relative abundance as affected by salinity and other environmental parameters.

The relative abundance of BALOs as measured by QPCR ranged between 0.1% and 1.4% of the total bacterial population, suggesting that BALOs may be considered a relatively 'abundant species'. The fraction of sequences obtained in the HTS analysis was at the lower end of this distribution, reaching 0.059% after curation, or about 0.1% as expressed on a cell basis (based on two copies of the 16S rRNA gene in the BALOs and 3.6 in the general population, see 'Results'), a bias possibly due to the relatively small number of sequences obtained (42 in total).

A major difference between *Bdellovibrionaceae* and *Bacteriovoracaceae* is the environments they colonize. The former has so far been found in freshwater (including wastewater), soil and sometimes in animals (Baer et al., 2000; Davidov & Jurkevitch, 2004) and it cannot withstand salinity above 5 p.p.t. (Varon & Shilo, 1978; Baer et al., 2004); the latter includes the halophilic *Bacteriovorax*, which populate brackish and saline to hypersaline water, and the freshwater- and soil-dwelling Peredicter and *Bacteriolyticum* (Baer et al., 2004; Pineiro et al., 2004). Remarkably, *Bdellovibrionaceae* (Bd) were detected in the SW ZDS (salinity 20 p.p.t., or about two-thirds seawater), indicating that the ecological range of this taxon may be more extended than previously thought. This is further supported by a recent screen of published 16S rRNA amplicon datasets of environmental HTS studies that showed the presence of *B. bacteriovo-*

rus-related sequences in marine habitats (Dolinšek et al., 2013) but a full demonstration that Bd indeed populates saline water requires isolation. *Bdellovibrionaceae* and *Bacteriovoracaceae* may thus co-exist, sharing many environments, including saline ones (Davidov et al., 2006b). Differential utilization of prey within and between clades may enable this co-existence: Chen et al. (2011, 2012) elegantly showed that when particular prey are spiked into natural samples, the abundance of specific *Bacteriovorax* clusters is consequently affected. Strikingly, specific *Bacteriovorax* clades were differently amplified by spiking with prey representing freshwater vs. prey representing saline water habitats. Altogether, most Gram-negative bacteria in the environment may be susceptible to local BALOs. Rice et al. (1998) found that about 80% of Gram-negative isolates from an estuarine environment were susceptible to BALOs, including some *Flavobacteria*, the dominant taxon in the FW ZDS.

BALOs seem to differ widely in the breadth of their prey ranges. Some appear to be very restricted, preying on few different bacterial types, whereas others appear to be rather generalist predators (Chen et al., 2011; Koval et al., 2013; Chanyi et al., 2013). Since these BALOs can share the same habitats (Jurkevitch et al., 2000; Chen et al., 2011), the dynamics of the populations may not follow a 'kill the winner' pattern as observed with phages in wastewater (Shapiro et al., 2009). It has been shown that closely related (within families and even within species) phylogenetic clusters can form ecologically distinct populations with seasonally or spatially segregated populations. As an example, specific marine *Vibrio* ecotypes are free-living or are associated with zooplankton-free particles, whereas others are found with zooplankton and large particles (Hunt et al., 2008). Co-existence of related populations within the same ecosystem may be mediated by resource partitioning, a phenomenon observed not only in the water column (Hunt et al., 2008) but also in soil and in the phyllosphere (Wilson & Lindow, 1994; Oger et al., 1997). In the case of BALOs, the main partitioning resource is certainly prey. Different prey ranges may lead to sympatric separation of the predators' populations that may be reinforced as the prey themselves distribute differentially in the environment. In soils, the microbiota of the plant rhizosphere and of the bulk soil differs, selecting for BALO subpopulations exhibiting different ribotypes; these populations may further differ in prey preference (Jurkevitch et al., 2000).

Thus, the physiological capacities to use particular resources (e.g. plant exudates or suspended organic matter) may result in resource partitioning of consumer populations, leading to spatially segregated (potential) prey populations, which in turn affect BALO distribution and dynamics in the environment.

Correlative data between BALO community structure and environmental parameters have so far yielded few additional insights beyond the effects of salinity and temperature (Kelley et al., 1997; Pineiro et al., 2004). Our data suggest that water temperature (20 ± 1 °C in winter to 24 ± 1 °C in spring) had no direct effect on ZDS BALO populations as it does in natural ecosystems subjected to larger fluctuations (Kelley et al., 1997) and, for reasons that are not clear, only phosphate was found to correlate with Bd in the SW ZDS. In contrast, the general bacterial community was affected by season (Fig. 3), possibly due to temperature variations or to changes in stock density (fish biomass increased with time), a variable known to affect the microbiota (Zhou et al., 2011). The environmental conditions observed in the ZDS compartments are brought about by engineering, physically designing the specific habitats that enable different microbial activities (Cytryn et al., 2003, 2005a; Gelfand et al., 2003) while connecting them with a constant water flow. In ZDS, mixing may then be sufficient to blur local ecological selection. Indeed, the overall microbiota community structure of the FW ZDS remained similar in all basins (Fig. 3); however, the dominant taxa reflected the dual use of the recirculating system as a fish culture and wastewater treatment plant (Table S7; and see 'Results'). No differential distribution of specific BALOs between pools was found based on the short pyrosequencing reads obtained. Deeper sequencing with longer reads, functional analysis of BALO activities in the different components of the system, and a detailed examination of spatial distribution (e.g. biofilm and flocs vs. planktonic) may yet reveal more complex patterns of abundance. A striking example of such specific predator –prey interactions in confined niches was discovered by Dolinšek et al. (2013): *Micavibrio*-like cells that were barely detected in the planktonic phase of a nitrifying bioreactor were specifically associated with sub-lineage I *Nitrospira* clusters but with no other bacteria. Although only about 4% these *Nitrospira* aggregates were surrounded by *Micavibrio*-like bacteria, it was posited that the predators may prevent any particular nitrifier strain from becoming dominant, thus driving diversification and functional redundancy, and increasing resistance to perturbations. In the FW ZDS, *Nitrospira* sequences were quite abundant in all samples (Table S7) but *Micavibrio*-related sequences were not found, and it is not known whether other BALOs prey upon *Nitrospira*. Therefore, while the impact of BALOs on their potential prey populations as a whole may be relatively minor (Data S1), the effects of BALO predation may be significant when spatial structuring is taken into account. The impact of bacterial predation may not be restricted to BALOs. HTS data suggest that a potential and diverse bacterial predatory community may be at least two to five times more abundant than that of BALOs; these other predators exert predation through various strategies (Jurkevitch & Davidov, 2007). Consequently, bacterial predation by bacteria may affect the microbiota structure and dynamics in many intricate and unknown ways. As recently described by Yang et al. (2013), when intraspecific variation in the predatory community is taken in account, numerical and functional responses in predator–prey systems do not maintain proportionality.

REFERENCES

1. Aharon Y, Pasternak Z, Ben Yosef M, Behar A, Lauzon C, Yuval B & Jurkevitch E (2013) Phylogenetic, metabolic, and taxonomic diversities shape Mediterranean Fruit Fly microbiotas during ontogeny. Appl Environ Microbiol 79: 303–313.
2. Baer ML, Ravel J, Chun J, Hill RT & Williams HN (2000) A proposal for the reclassification of Bdellovibrio stolpii and Bdellovibrio starrii into a new genus, Bacteriovorax gen. nov. as Bacteriovorax stolpii comb. nov. and Bacteriovorax starrii comb. nov., respectively. Int J Syst Evol Microbiol 50: 219–224.
3. Baer ML, Ravel J, Pineiro SA, Guether-Borg D & Williams HN (2004) Reclassification of salt-water Bdellovibrio sp. as Bacteriovorax marinus sp. nov. and Bacteriovorax litoralis sp. nov. Int J Syst Evol Microbiol 54: 1011–1016.
4. Banning EC, Casciotti KL & Kujawinski EB (2010) Novel strains isolated from a coastal aquifer suggest a predatory role for flavobacteria. FEMS Microbiol Ecol 73: 254–270.
5. Chanyi RM, Ward C, Pechey A & Koval SF (2013) To invade or not to invade: two approaches to a prokaryotic predatory life cycle. Can J Microbiol 59: 273–279.
6. Chauhan A, Cherrier J & Williams HN (2009) Impact of sideways and bottom-up control factors on bacterial community succession over a tidal cycle. P Natl Acad Sci USA 106: 4301–4306.

7. Chen H, Young S, Berhane T-K & Williams H-N (2012) Predatory Bacteriovorax communities ordered by various prey species. PloSOne 3: e34174.

8. Chen H, Athar R, Zheng G & Williams HN (2011) Prey bacteria shape the community structure of their predators. ISME J 5: 1314–1322.

9. Chu WH & Zhu W (2009) Isolation of Bdellovibrio as biological therapeutic agents used for the treatment of Aeromonas hydrophila infection in fish. Zoonoses Public Health 57: 258–264.

10. Cole JR, Wang Q, Cardenas E et al. (2009) The Ribosomal Database Project: improved alignments and new tools for rRNA analysis. Nucleic Acids Res 37: D141–D145.

11. Crossman LC, Chen H, Cerdeno-Tarraga A-M et al. (2013) A small predatory core genome in the divergent marine Bacteriovorax marinus SJ and the terrestrial Bdellovibrio bacteriovorus. ISME J 7: 148–160.

12. Cytryn E, Gelfand I, Barak Y, Van Rijn J & Minz D (2003) Diversity of microbial communities correlated to physiochemical parameters in a digestion basin of a zero-discharge mariculture system. Environ Microbiol 5: 55–63.

13. Cytryn E, van Rijn J, Schramm A, Gieseke A, de Beer D & Minz D (2005a) Identification of bacteria potentially responsible for oxic and anoxic sulfide oxidation in biofilters of a recirculating mariculture system. Appl Environ Microbiol 71: 6134–6141.

14. Cytryn E, Minz D, Gelfand I, Neori A, Gieseke A, de Beer D & van Rijn J (2005b) Sulfide-oxidizing activity and bacterial community structure in a fluidized bed reactor from a zero-discharge mariculture system. Environ Sci Technol 39: 1802–1810.

15. Davidov Y & Jurkevitch E (2004) Diversity and evolution of Bdellovibrio- and-like organisms (BALOs), reclassification of Bacteriovorax starrii as Peredibacter starrii gen. nov., comb. nov., and description of the Bacteriovorax-Peredibacter clade as Bacteriovoracaceae fam. nov. Int J Syst Evol Microbiol 54: 1439–1452.

16. Davidov Y, Huchon D, Koval SF & Jurkevitch E (2006a) A new α-proteobacterial clade of Bdellovibrio-like predators: implications for the mitochondrial endosymbiotic theory. Environ Microbiol 8: 2179–2188.

17. Davidov Y, Friedjung A & Jurkevitch E (2006b) Structure analysis of a soil community of predatory bacteria using culture-dependent and culture-independent methods reveals a hitherto undetected diversity of Bdellovibrio- and -like organisms. Environ Microbiol 8: 1667–1673.

18. Dionisi HM, Layton AC, Harms G, Gregory IR, Robinson KG & Sayler GS (2002) Quantification of Nitrosomonas oligotropha-like ammonia-oxidizing bacteria and Nitrospira spp. from full-scale wastewater treatment plants by competitive PCR. Appl Environ Microbiol 68: 245–253.

19. Dolinšek J, Lagkouvardos I, Wanek W, Wagner M & Daims H (2013) Interactions of nitrifying bacteria and heterotrophs: identification of a Micavibrio-like putative predator of Nitrospira spp. Appl Environ Microbiol 79: 2027–2037.

20. Dufrêne M & Legendre P (1997) Species assemblages and indicator species: the need for a flexible and asymmetric approach. Ecol Monogr 67: 345–366.

21. Edgar RC, Haas BJ, Clemente JC, Quince C & Knight R (2011) UCHIME improves sensitivity and speed of chimera detection. Bioinformatics 27: 2194–2200.

22. Estes JA, Terborgh J, Brashares JS et al. (2011) Trophic downgrading of Planet Earth. Science 333: 301–306.

23. FAO (2011) World Aquaculture 2010, Vol. 500/1. FAO, Rome.

24. Fenton AK, Kanna M, Woods RD, Aizawa SI & Sockett RE (2010) Shadowing the actions of a predator: Backlit fluorescent microscopy reveals synchronous nonbinary septation of predatory Bdellovibrio inside prey and exit through discrete bdelloplast pores. J Bacteriol 192: 6329–6335.

25. Fuhrman JA (1999) Marine viruses and their biogeochemical and ecological effects. Nature 399: 541–548.

26. Fuhrman J & Noble RT (1995) Viruses and protists cause similar bacterial mortality in coastal seawater. Limnol Oceanogr 40: 1236–1242.

27. Gallet R, Alizon S, Comte P-A, Gutierrez A, Depaulis F, van Baalen M, Michel E & Muller-Graf CDM (2007) Predation and disturbance Interact to shape prey species diversity. Am Nat 170: 143–154.

28. Gallet R, Tully T & Evans MKE (2009) Ecological conditions affect evolutionary trajectory in a predator-prey system. Evolution 63: 641–651.

29. Gelfand I, Barak Y, Even-Chen Z, Cytryn E, Van Rijn J, Krom MD & Neori A (2003) A Novel Zero Discharge Intensive Seawater Recirculating System for the Culture of Marine Fish. J World Aquaculture Soc 34: 344–358.

30. Hespell RB, Thomashow MF & Rittenberg SC (1974) Changes in cell composition and viability of Bdellovibrio bacteriovorus during starvation. Arch Microbiol 97: 313–327.

31. Hunt DE, David LA, Gevers D, Preheim SP, Alm EJ & Polz MF (2008) Resource partitioning and sympatric differentiation among closely related bacterioplankton. Science 320: 1081–1085.

32. Huntley J & Kowalewski M (2007) Strong coupling of predation intensity and diversity in the Phanerozoic fossil record. P Natl Acad Sci USA 38: 15006–15010.

33. Huse SM, Welch DM, Morrison HG & Sogin ML (2010) Ironing out the wrinkles in the rare biosphere through improved OTU clustering. Environ Microbiol 12: 1889–1898.

34. Jurkevitch E (2012) Isolation and classification of Bdellovibrio and like organisms. Current Protocols in Microbiology (Coico R, Kowalik T, Quarles J, Stevenson B & Taylor R, eds), pp. 7B.1.1–7B.1.20. John Wiley and Sons, New York.

35. Jurkevitch E & Davidov Y (2007) Phylogenetic diversity and evolution of predatory prokaryotes. Predatory Prokaryotes, Vol. 4 (Jurkevitch E, ed.), pp. 11–56. Springer, Berlin.

36. Jurkevitch E, Minz D, Ramati B & Barel G (2000) Prey range characterization, ribotyping, and diversity of soil and rhizosphere Bdellovibrio spp. isolated on phytopathogenic bacteria. Appl Environ Microbiol 66: 2365–2371.

37. Kelley J, Turng B, Williams H & Baer M (1997) Effects of temperature, salinity, and substrate on the colonization of surfaces in situ by aquatic bdellovibrios. Appl Environ Microbiol 63: 84–90.

38. Kim DH, Brunt J & Austin B (2007) Microbial diversity of intestinal contents and mucus in rainbow trout (Oncorhynchus mykiss). J Appl Microbiol 102: 1654–1664.

39. Koval SF, Hynes SH, Flannagan RS, Pasternak Z, Davidov Y & Jurkevitch E (2013) Bdellovibrio exovorus sp. nov., a novel predator of Caulobacter crescentus. Int J Syst Evol Microbiol 63: 146–151.

40. Lane DJ (1991) 16S/23S rRNA sequencing. Nucleic Acid Techniques in Bacterial Systematics (Stackebrandt E, & Goodfellow M, eds), pp. 115–175. John Wiley and Sons, New York.

41. Lee ZM-P, Bussema C & Schmidt TM (2009) rrnDB: documenting the number of rRNA and tRNA genes in bacteria and archaea. Nucleic Acids Res 37: D489–D493.

42. Mather P (1976) Computational Methods of Multivariate Analysis in Physical Geography. J Wiley and Sons, London.

43. Mielke PW Jr (1984) Meteorological applications of permutation techniques based on distance functions. Handbook of Statistics, Vol. 4 (Krishnaiah PR & Sen PK, ed.), pp. 813–830. Elsevier, Amsterdam.

44. Muyzer G, Teske A, Wirsen CO & Jannash HW (1995) Phylogenetic relationship of Thiomicrospira species and their identification in deep-sea hydrothermal vent samples by denaturing gradient gel electrophoresis of 16S rDNA fragments. Arch Microbiol 164: 165–172.

45. Neori A, Krom MD & van Rijn J (2007) Biogeochemical processes in intensive zero-effluent marine fish culture with recirculating aerobic and anaerobic biofilters. J Exp Mar Biol Ecol 349: 235–247.

46. Oger P, Petit A & Dessaux Y (1997) Genetically engineered plants producing opines alter their biological environment. Nat Biotechnol 15: 369–372.

47. Pernthaler J (2005) Predation on prokaryotes in the water column and its ecological implications. Nat Rev Microbiol 3: 537–546.

48. Pineiro SA, Sahaniuk GE, Romberg E & Williams H (2004) Predation pattern and phylogenetic analysis of Bdellovibrionaceae from the Great Salt Lake, Utah. Curr Microbiol 48: 113–117.

49. Pineiro SA, Williams HN & Stine OC (2008) Phylogenetic relationships amongst the saltwater members of the genus Bacteriovorax using rpoB sequences and reclassification of Bacteriovorax stolpii as Bacteriolyticum stolpii gen. nov., comb. nov. Int J Syst Evol Microbiol 58: 1203–1209.

50. Pomeroy LR, Williams PJ leB, Azam F & Hobbie JE (2007) The microbial loop. Oceanography 20: 28–33.

51. Qi Z, Zhang X-H, Boon N & Bossier P (2009) Probiotics in aquaculture of China – Current state, problems and prospect. Aquaculture 290: 15–21.

52. Quince C, Lanzen A, Davenport R & Turnbaugh P (2011) Removing noise from pyrosequenced amplicons. BMC Bioinformatics 12: 38.

53. Rendulic S, Jagtap P, Rosinus A et al. (2004) A predator unmasked: life cycle of Bdellovibrio bacteriovorus from a genomic perspective. Science 303: 689–692.

54. Rice TD, Williams HN & Turng BF (1998) Susceptibility of bacteria in estuarine environments to autochthonous bdellovibrios. Microb Ecol 35: 256–264.

55. Rittenberg SC & Shilo M (1978) Early host damage in the infection cycle of Bdellovibrio bacteriovorus. J Bacteriol 102: 149–160.

56. Rodriguez-Valera F, Martin-Cuadrado A-B, Rodriguez-Brito B, Pasic L, Thingstad TF, Rohwer F & Mira A (2009) Explaining microbial population genomics through phage predation. Nat Rev Microbiol 7: 828–836.

57. Samuel P, Fernando U, Jean-Pierre D & Pierre S (2007) Fate of heterotrophic bacteria in Lake Tanganyika (East Africa). FEMS Microbiol Ecol 62: 354–364.

58. Schloss PD, Westcott SL, Ryabin T et al. (2009) Introducing Mothur: open-source, platform-independent, community-supported software for describing and comparing microbial communities. Appl Environ Microbiol 75: 7537–7541.

59. Seidler RJ, Mendel M & Baptist JN (1972) Molecular heterogeneity of the bdellovibrios: evidence for two new species. J Bacteriol 109: 209–217.

60. Shapiro OH, Kushmaro A & Brenner A (2009) Bacteriophage predation regulates microbial abundance and diversity in a full-scale bioreactor treating industrial wastewater. ISME J 4: 327–336.

61. Shemesh Y, Yaacov D, Susan K & Jurkevitch E (2003) Small eats big: ecology and diversity of Bdellovibrio and like organisms, and their dynamics in predator-prey interactions. Agronomie 23: 433–439.

62. Shnel N, Barak Y, Ezer T, Dafni Z & van Rijn J (2002) Design and performance of a zero-discharge tilapia recirculating system. Aquacult Eng 26: 191–203.

63. Subasinghe R, Soto D & Jia J (2009) Global aquaculture and its role in sustainable development. Rev Aquac 1: 2–9.

64. Sullam KE, Essinger SD, Lozupone CA et al. (2012) Environmental and ecological factors that shape the gut bacterial communities of fish: a meta-analysis. Mol Ecol 21: 3363–3378.

65. Tsuchiya C, Sakata T & Sugita H (2008) Novel ecological niche of Cetobacterium somerae, an anaerobic bacterium in the intestinal tracts of freshwater fish. Lett Appl Microbiol 46: 43–48.

66. Van Essche M, Sliepen I, Loozen G et al. (2009) Development and performance of a quantitative PCR for the enumeration of Bdellovibrionaceae. Environ Microbiol Rep 1: 228–233.

67. van Rijn J (2013) Waste treatment in recirculating aquaculture systems. Aquacult Eng 53: 49–56.

68. Varon M & Shilo M (1978) Ecology of aquatic bdellovibrios. Advances in Microbial Ecology (Droop MR & Jannash HW, eds), pp. 48. Academic Press, London.

69. Varon M, Dickbusch S & Shilo M (1974) Isolation of host-dependent and non-parasitic mutants of the facultative parasitic Bdellovibrio UKi2. J Bacteriol 119: 635–637.

70. Vasseur DA & Fox JW (2009) Phase-locking and environmental fluctuations generate synchrony in a predator prey community. Nature 460: 1007–1010.

71. Vos M, Birkett PJ, Birch E, Griffiths RI & Buckling A (2009) Local adaptation of bacteriophages to their bacterial hosts in soil. Science 325: 833.

72. Wen CQ, Lai XT, Xue M, Huang YL, Li HX & Zhou SN (2009) Molecular typing and identification of Bdellovibrio-and-like organisms isolated from seawater shrimp ponds and adjacent coastal waters. J Appl Microbiol 106: 1154–1162.

73. Whelan JA, Russell NB & Whelan MA (2003) A method for the absolute quantification of cDNA using realtime PCR. J Immunol Methods 278: 261–269.

74. Wilson M & Lindow SE (1994) Coexistence among epiphytic bacterial populations mediated through nutritional resource partitioning. Appl Environ Microbiol 60: 4468–4477.

75. Yang Z, Lowe CD, Crowther W, Fenton A, Watts PC & Montagnes DJS (2013) Strain-specific functional and numerical responses are required to evaluate impacts on predator-prey dynamics. ISME J 7: 405–416.
76. Zhang T, Shao M-F & Ye L (2012) 454 Pyrosequencing reveals bacterial diversity of activated sludge from 14 sewage treatment plants. ISME J 6: 1137–1147.
77. Zheng G, Wang C, Williams HN & Pineiro SA (2008) Development and evaluation of a quantitative real-time PCR assay for the detection of saltwater *Bacteriovorax*. Environ Microbiol 10: 2515–2526.
78. Zhou Z, He S, Liu Y, Shi P, Yao B & Ringo E (2011) Do stocking densities affect the gut microbiota of gibel carp Carassius auratus gibelio cultured in ponds? Aquac Res Dev S1: 003. doi:10.4172/2155-9546.S1-003.

CHAPTER 8

Microbial Community Functional Structures in Wastewater Treatment Plants as Characterized by GeoChip

XIAOHUI WANG, YU XIA, XIANGHUA WEN, YUNFENG YANG, AND JIZHONG ZHOU

8.1 INTRODUCTION

Biological activated sludge process is the most widely used biological process for treating municipal and industrial wastewater. By enriching selected functional microorganisms in the activated sludge of wastewater treatment plant (WWTP), microbial activities in the sludge community are accelerated, enabling removal of oxygen-depleting organics, toxics, and nutrients. Biological WWTPs must be functionally stable to continuously and steadily remove contaminants which rely upon the activity of complex microbial communities [1], [2]. Therefore, a better understanding of the

Microbial Community Functional Structures in Wastewater Treatment Plants as Characterized by GeoChip. © *Wang X, Xia Y, Wen X, Yang Y, and Zhou J.* PLoS ONE **9**,*3 (2014). doi:10.1371/journal. pone.0093422. Licensed under a Creative Commons Attribution 4.0 International License, http://creativecommons.org/licenses/by/4.0/.*

microbial community structure and functional genes of activated sludge in WWTPs can help elucidate the mechanisms of biological pollutant removal and improve the treatment performance and operational stability [3], [4].

Although microbial communities of activated sludge in WWTPs have been intensively studied [5]–[11], most of such efforts have been focused only on microbial phylogenetic composition. Knowledge is still lacking in regard to microbial community functional structures and their linkages to environmental variables. To date, there have been only a limited number of studies that characterized the overall functional profiles and metabolic pathways in the activated sludge of WWTPs using high-throughput sequencing [12], [13].

Functional gene arrays, which contain genes encoding key enzymes involved in a variety of biogeochemical cycling processes [14], can be used to study the overall microbial functional potentials of activated sludge in WWTPs. GeoChip 4.2, a powerful functional gene array, contains 83,992 oligonucleotide (50-mer) probes targeting 152,414 genes in 410 gene categories involved in nitrogen, carbon, sulfur and phosphorus cycling, metal resistance, antibiotic resistance as well as organic contaminant degradation [15], [16]. GeoChip has been widely used to examine microbial community functional structures in various environmental samples, such as soil and sediments [17]–[19], groundwater [20], acid mine drainage[21], deep-sea water [15], ocean crust [15], and hydrothermal vent [22].

In this study, 12 activated sludge samples from four WWTPs in Beijing were analyzed by GeoChip 4.2 to address the following two questions: (i) What were the microbial community functional structures of activated sludge in WWTPs? (ii) How do environmental factors and operational parameters affect microbial community functional structures? Our results revealed high similarities of microbial functional communities among activated sludge samples from the four WWTPs, and microbial functional potentials were highly correlated with water temperature, dissolved oxygen (DO), ammonia concentrations and loading rate of chemical oxygen demand (COD).

8.2 MATERIALS AND METHODS

8.2.1 WWTPS AND SAMPLING

Anaerobic/anoxic/aerobic (A^2O) is the most commonly used process for wastewater treatment in China, thus we examined four A^2O WWTPs in Beijing. Activated sludge samples were taken from the aeration tank of each WWTP once a day for three consecutive days in the summer of 2011. Therefore, three replicate samples were available from each WWTP, resulting in a total of 12 activated sludge samples. No specific permits were required for the described field studies. We confirm that: i) the locations were not privately-owned or protected in any way; and ii) the field studies did not involve endangered or protected species.

For microbial community analysis, each sample was dispensed into a sterile Eppendorf tube and centrifuged at 14,000 g for 10 min. After decanting the supernatant, the pellet was stored at −80°C prior to further analysis. Meanwhile, various chemical parameters such as COD, total nitrogen (TN), ammonia, total phosphorus (TP), pH and conductivity were analyzed according to standard methods [23]. Concentrations of chromium, cobalt, nickel, copper, zinc, and cadmium were measured by inductively coupled plasma mass spectrometry (ICP-MS)(Thermo Fisher Scientific, Waltham, MA, USA).

8.2.2 DNA PURIFICATION AND GEOCHIP HYBRIDIZATION

Microbial genomic DNA was extracted from the activated sludge samples by combining freeze-thawing and sodium dodecyl sulfate (SDS) for cell lysis as previously described [24]. Crude DNA was purified using the Wizard SV Genomic DNA Purification Kit (Promega, Madison WI, USA). The quality of purified DNA was assessed based on the ratios of 260/280 nm and 260/230 nm absorption measured by an ND-1000 spectrophotometer (Nanodrop Inc., Wilmington, DE, USA) and agarose gel electrophoresis. The quantity of community DNA was determined with Quant-It PicoGreen kit (Invitrogen, Carlsbad, CA, USA).

Purified DNA (1 μg) was labeled with Cy5, purified and resuspended in 10 μl hybridization solution as previously described (Lu et al. 2012). Then it was hybridized with GeoChip 4.2 on a MAUI hybridization station (BioMicro, Salt Lake City, UT, USA) at 42°C with 40% formamide for 16 hours. After hybridization, microarrays were scanned (NimbleGen MS200, Madison, WI, USA) at a laser power of 100%. The ImaGene version 6.0 (Biodiscovery, El Segundo, CA) was then used to determine the intensity of each spot, and identify poor-quality spots. The GeoChip raw data has been uploaded to the NCBI GEO with the accession number GSE54055.

8.2.3 DATA ANALYSIS

Raw data from ImaGene were submitted to Microarray Data Manager (http://ieg.ou.edu/microarray/) and analyzed using the data analysis pipeline with the following major steps: (i) The spots flagged as 1 or 3 by Imagene and with a signal to noise ratio (SNR) less than 2.0 were removed as low quality spots; (ii) After removing the bad spots, normalized intensity of each spot was calculated by dividing the signal intensity of each spot by the mean intensity of the microarray; (iii) If any of replicates had more than two times the standard deviation, this replicate was moved as an outlier [18].

Hierarchical cluster analysis was performed using the unweighted pairwise average-linkage clustering algorithm in the CLUSTER software and visualized in TREEVIEW software [25]. Canonical correspondence analysis (CCA) was used to examine the relationships of microbial communities and environmental variables. The significant environmental variables were identified by a forward selection procedure using Monte Carlo permutations ((999 permutations with a p-value<0.05). In addition, Variance inflation factors (VIFs; a measure for cross-correlation of explanatory variables) were checked and eliminated if VIFs were more than 20 [26].Based on partial CCA, variance partitioning analysis (VPA) was performed to attribute the variation observed in the microbial communities to the environmental variables. CCA and VPA were performed by the vegan package in R 2.14.0 (R Development Core Team, 2011).

8.3 RESULTS

8.3.1 PERFORMANCE OF WWTPS

Details of the influent and effluent characteristics and operational parameters of the four WWTPs are listed in Table S1 and S2 in File S1. Chemical oxygen demand (COD) in the influents ranged from 257 to 452 mg/L, while total nitrogen (TN) concentrations were between 46 and 64 mg/L. The COD removal efficiencies were more than 90% in all systems, and TN removal efficiencies varied from 68% to 75%. The removal efficiencies of COD and TN indicated that the four WWTPs were functionally stable.

8.3.2 OVERALL MICROBIAL COMMUNITY FUNCTIONAL STRUCTURE

Microbial community functional structures of 12 activated sludge samples from four WWTPs were analyzed with GeoChip 4.2 A total of 29,124 functional genes were detected. For individual samples, the numbers of detected genes were between 18,847 and 24,268 (Table 1). Hierarchical cluster analysis showed that the three replicate samples from each WWTP were clustered together, and the distances among the three replicate samples was always <2 (Figure 1). The samples from different WWTPs were well separated, and the distances among the four WWTPs were generally>2.

To assess α-diversity of microbial communities, Shannon-Weaver index (H), Simpson's (1/D) and evenness were calculated (Table 1). The values of H were very similar across the 12 samples, ranging from 9.9 to 10.1. Consistently, the Simpson's (1/D) and evenness did not show significant differences among these four plants.

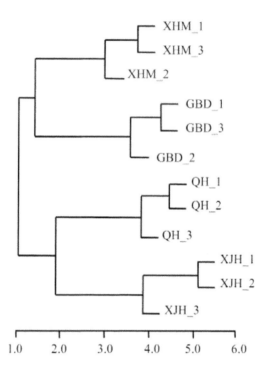

FIGURE 1: Hierarchical cluster analysis of functional genes in 12 activated sludge samples from four WWTPs named GBD, QH, XHM and XJH. Each sample is named after WWTP plant with "_1", "_2", or "_3" that indicates one of three replicate samples from each plant.

The overlapping genes were also determined. A large percentage (65–89%) of detected genes were shared among the samples (Table 1), indicative of high similarities of microbial community functional structures among activated sludge of four WWTPs. In consistency to the results of hierarchical clustering analysis, the percentages of the overlapping genes in samples from the same WWTP were higher than those from different WWTPs. For example, sample GBD_1 had 85% and 89% of the genes in common with sample GBD_2 and GBD_3 respectively, but 65% in common with QH_3.

TABLE 1: Percentage of overlapping genes and diversity indices of all functional genes in activated sludge samples.

Sample	GBD_1	GBD_2	GBD_3	QH_1	QH_2	QH_3	XHM_1	XHM_2	XHM_3	XJH_1	XJH_2	XJH_3
GBD_1												
GBD_2	84.67%											
GBD_3	89.16%	86.65%										
QH_1	69.41%	71.45%	71.96%									
QH_2	67.39%	69.49%	69.22%	85.78%								
QH_3	65.12%	68.03%	67.38%	81.68%	80.53%							
XHM_1	79.10%	80.08%	81.10%	72.82%	69.85%	67.72%						
XHM_2	75.56%	78.38%	77.87%	71.42%	68.90%	68.33%	85.82%					
XHM_3	77.18%	78.64%	79.68%	74.95%	72.89%	71.38%	88.14%	82.43%				
XJH_1	73.92%	77.03%	74.81%	71.96%	71.38%	71.08%	75.72%	76.30%	76.37%			
XJH_2	74.11%	76.22%	74.91%	72.62%	72.29%	72.01%	74.94%	75.19%	76.71%	89.15%		
XJH_3	67.12%	69.41%	67.89%	70.21%	71.17%	70.94%	67.74%	69.71%	69.96%	79.99%	80.27%	
Richness[a]	23064	22434	23554	20862	20610	19688	24268	22831	23413	20774	20838	18847
H[b]	10.1	10.0	10.1	10.0	9.9	9.9	10.1	10.0	10.1	10.0	10.0	9.9
1/D[c]	23301.2	22682.7	23805.7	21049.3	20777.9	19855.9	24452.9	23022.1	23624.7	20959.6	21033.5	18984.4
Evenness[d]	0.996	0.996	0.996	0.996	0.995	0.995	0.994	0.995	0.995	0.995	0.995	0.995

[a]Detected gene number. [b]Shannon-Weiner index, higher number represents higher diversity; [c]Reciprocal of Simpson's index, higher number represents higher diversity; [d]Evenness index.

At the phylogenetic level, 87.6–88.7% genes were derived from bacteria, 1.9–2.0% from archaea, and 8.7–9.7% from fungi. *Proteobacteria* was the predominant phylum of bacteria, constituting 50.8–52.2% of all detected microorganisms. *Actinobacteria, Firmicutes, Cyanobacteria* and *Bacteroidetes* were the subdominant bacterial groups, each containing 14.8–15.5%, 4.6–5.1%, 1.8–2.1% and 1.3–1.6% of detected microorganisms, respectively. *Ascomycota* was the most abundant phylum of fungi, constituting 6.3–7.4% of all detected microorganisms, followed by *Basidiomycota* (1.5–1.7%) as the second most abundant fungal group. In addition, *Euryarchaeota*, containing about 1.5% of detected microorganisms, was the most predominant phylum within archaea. (Table S3 in File S1). Taken together, these results indicated overall functional structures as well as phylogenetic diversities of microbial communities within WWTPs appeared to be quite high.

8.3.3 CARBON CYCLING GENES

Carbon cycling, especially carbon degradation, is a crucial biochemical process in WWTPs. The rate of carbon degradation depends on a number of factors, including the availability and types of carbon substrates as well as the microbial community [27]. Among detected carbon cycling genes, 3,375 genes were involved in the degradation of complex carbon substrates such as cellulose, starch, hemicelluloses, chitin and lignin. Generally, relative abundances and numbers of all detected carbon degradation genes were similar among activated sludge of four WWTPs (Figure 2 and Table S4 in File S1).

Endoglucanase are responsible for the initial steps of cellulose degradation. A total of 41 egl genes encoding endoglucanase were detected, and about half of them (20 genes) were commonly shared by all samples (Figure S1 in File S1). Nearly all of the egl genes were derived from isolated microorganisms, including the probes from the sequenced 149121623 (protein id number) of *Methylobacterium* sp. 4–46, 197934254 of *Streptomyces sviceus* ATCC 29083, and 171059025 of *Leptothrix cholodnii* SP-6.

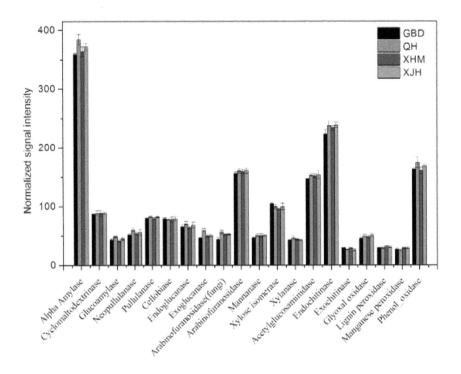

FIGURE 2: The normalized signal intensity of detected key genes involved in carbon cycling. The signal intensity for each functional gene was the average of the signal intensity from all the replicates. All data are presented as mean ± SE.

α-Amylase functions to hydrolyze starch to yield glucose and maltose, which plays an important role in starch degradation. A total of 366 *amyA* genes encoding α-amylase were detected, among which 156 genes were commonly shared across all samples. Almost all the detected *amyA* genes were derived from isolated organisms such as 113897923 of *Herpetosiphon aurantiacus* ATCC 23779, 50842594 of *Propionibacterium acnes* KPA171202, 240169491 of *Mycobacterium kansasii* ATCC 12478.

Xylanase is an enzyme that breaks down hemicellulose, which is a major component of plant cell walls. 33 xylanase genes were detected, and 13 of them were shared by all samples (Figure S2 in File S1). Most of xylanase genes were from isolates such as 113935746 of *Caulobacter* sp. K31, 116098389 of *Lactobacillus brevis* ATCC 367, 68230042 of *Frankia* sp. EAN1pec, etc. Also a total of 422 genes encoding chitin degradative enzymes, including endochitinase, acetylglucosaminidase and exochitinase, were found in all 12 samples. For genes encoding endochitinase, 99 out of 239 genes were shared by all samples.

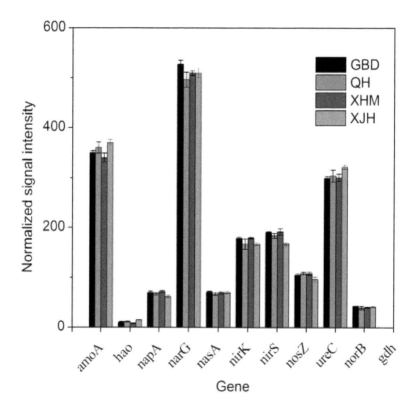

FIGURE 3: The normalized signal intensity of detected key genes involved in nitrogen cycling. The signal intensity for each functional gene was the average of the signal intensity from all the replicates. All data are presented as mean ± SE.

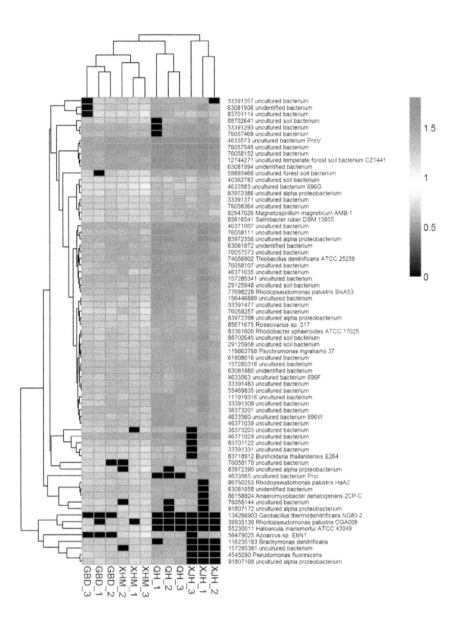

FIGURE 4: The hierarchical cluster analysis of *nosZ* genes encoding nitrous oxide reductases. The protein id number and its derived organism for each gene are indicated. The color intensity of each panel shows the normalized signal intensity of individual genes, referring to color key at the top right.

A total of 268 genes related to lignin degradation were detected, including genes encoding glyoxal oxidase (*glx*), ligninase (*lip*), manganese peroxidase (*mnp*) and phenol oxidase (*lcc*). Many of these genes were detected across all samples. For example, 13 out of 28 *lip* genes were shared by all samples and all of them were derived from isolates (Figure S3 in File S1). Similarly, almost all genes of *glx, mnp* and *lcc* were from isolated microorganisms rather than uncultured microorganisms.

8.3.4 NITROGEN CYCLING GENES

As nitrogen removal is one of the most important functions in WWTPs, we examined the nitrogen cycling genes and their relationship to nitrogen concentrations. A total of 2,055 genes involved in ammonification, nitrification, denitrification and assimilatory nitrate/nitrite reduction were detected by GeoChip 4.2. Similar to carbon degradation genes, the relative abundances of detected nitrogen cycling genes were similar among activated sludge of four WWTPs (Figure 3).

Of 197 detected *nirS* genes encoding nitrite reductases, 65 were present in all samples. 7 genes were derived from isolates, while the majority of genes were derived from environmental clones in various environments. The results of Mantel test showed that the abundance of *nirS* genes was positively correlated to the influent concentrations of TN ($r = 0.489$, $P<0.01$) and ammonia ($r = 0.498$, $P<0.01$) (Table S5 in File S1).

For *nirK* genes encoding nitrite reductases, 184 genes were detected. Among them, 62 genes were present in all samples. Most of them were from laboratory clones, while only 10 genes were derived from isolates. The results of Mantel test indicated that *nirK* genes were positively correlated to the concentrations of influent TN ($r = 0.421$, $P<0.01$) and ammonia ($r = 0.358$, $P<0.05$) (Table S5 in File S1).

For *nosZ* genes encoding nitrous oxide reductases, 110 genes were detected. Among them, 43 were shared by all samples. 94 *nosZ* genes were derived from uncultured microorganisms and 16 were from cultured organisms, including 39935130 of *Rhodopseudomonas palustris* CGA009, 83816541 of *Salinibacter ruber* DSM 13855, 82947026 of *Magnetospirillum magneticum* AMB-1, etc (Figure 4). The results of Mantel test analysis showed abundances of *nosZ* genes were positively correlated with the concentrations of influent TN (r = 0.348, P<0.05) and ammonia (r = 0.338, P<0.01) (Table S5 in File S1).

295 *ureC* genes encoding ureases were detected, and 128 genes were shared by all samples. Unlike other nitrogen genes, most of the *ureC* genes were form cultured microorganisms and only 9 from uncultured bacteria. The results of Mantel test showed a positive correlation between *ureC* genes and the concentrations of influent TN (r = 0.463, P<0.01) and ammonia (r = 0.393, P<0.01) (Table S5 in File S1).

Only four *hzo* genes encoding hydrazine oxidoreductases were detected, including two genes (208605292 and 224712632) from *Candidatus Brocadia* sp. and two genes (224712622 and 208605290) from uncultured *Planctomycete*. No significant correlation between hzo genes and the concentrations of influent TN (P>0.05) and ammonia (P>0.05) (Table S5 in File S1) was observed.

8.3.5 PHOSPHORUS CYCLING GENES

Phosphorus removal from WWTPs is a key factor in preventing eutrophication of surface waters. Genes encoding exopolyphosphatase (ppx) for inorganic polyphosphate degradation, polyphosphate kinase (ppk) for catalyzing the formation of polyphosphate from ATP and phytase for phytate degradation were detected among all samples (Figure 5).

A total of 90 of 246 detected ppx genes were shared by all samples, including genes derived from *Leptothrix cholodnii* SP-6 (171060286), *Streptomyces coelicolor* A3(2) (21221778), *Erythrobacter* sp. NAP1 (85709358). Nevertheless, there was no significant correlation between ppx genes and the concentrations of influent TP (Table S5 in File S1). For ppk, 57 of 165 detected genes were shared by all samples. Significantly positive correlation was observed between ppk gene abundances and the concentrations of influent TP (r = 0.242, P<0.05) (Table S5 in File S1). In addition, a total of 27 phytase genes were detected. Among them, 7 were shared by all samples (Figure S4 in File S1). The results of Mantel tests suggested that there was no significant correlation between phytase genes and influent TP (Table S5 in File S1).

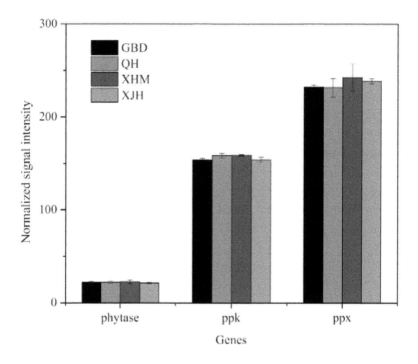

FIGURE 5: The normalized signal intensity of detected key genes involved in phosphorus cycling. The signal intensity for each functional gene was the average of the signal intensity from all the replicates. All data are presented as mean ± SE.

8.3.6 SULFUR CYCLING GENES

A total of 1,241 genes involved in sulfite reductase, adenylylsulfate re-ductase, sulphur oxidation, sulfide oxidation were detected. The relative abundances of sulfur cycling genes were presented in Figure 6. Among them, there were 46 *aprA* genes encoding dissimilatory adenosine-5'-phosposulfate reductase, a key enzyme involved in microbial sulfate re-duction and sulfur oxidation. 15 of 22 *aprA* genes present in all 12 samples were derived from isolates (Figure S5 in File S1). In addition, 327 *dsrA* genes encoding dissimilatory sulfite reductase were detected. Of the 100 genes shared by all samples, most genes were derived from uncultured microorganisms except for 18 genes from isolates such as 107784967 of *Desulfovibrio desulfuricans*, 13992710 of *Bilophila wadsworthia* and 83573380 of *Moorella thermoacetica* ATCC 39073. For *sox* genes encoding sulfite oxidase, 77 of 205 detected genes were shared by all samples, of which most were derived from isolated microorganisms such as 56677633 of *Silicibacter pomeroyi* DSS-3, 255258527 of *Sideroxydans lithotrophicus* ES-1, 214028622 of *Ruegeria* sp. R11.

8.3.7 DEGRADATION OF ORGANIC CONTAMINANTS

Organic contamination is a concern and a lot of research has been undertaken to examine the role of microorganisms in the degradation and remediation of organic contaminants [14]. In total, 7,012 genes from 165 gene categories involved in the degradation of organic contaminants related to aromatics, chlorinated solvents, herbicides and pesticides were detected. The relative abundances of *keg* genes involved in organic contaminant degradation were present in Figure 7. Among them, 24 *benAB* genes encoding benzoate 1, 2-dioxygenases were detected across all samples and 19 were present in all samples. The most abundant genes were derived from *Burkholderia xenovorans* LB400 (91779181), *Mycobacterium* sp. MCS (108768762), *Rhodococcus* sp. RHA1 (110818905).

8.3.8 METAL RESISTANCE GENES

Various heavy metals were often detected in WWTPs as a result of societal industrialization and modernization. Resistance genes of Ag, As, Cd, Co, Cr, Cu, Hg, Pb, Ni, Te and Zn were detected in this study. A substantial number (3,381) of genes involved in metal resistance were detected in all the samples, of which the vast majority was derived from isolated microorganisms. Notably, relative abundances of detected metal resistance genes were similar among activated sludges of four WWTPs (Figure S6 in File S1). In addition, the results of Mantel test showed that most of metal resistance genes, such as *chrA*, *nreB*, *copA*, *cueO*, *zntA*, *cadA* and *cadBD*, were positively correlated to the concentrations of Cr, Ni, Cu, Cu, Zn, Cd

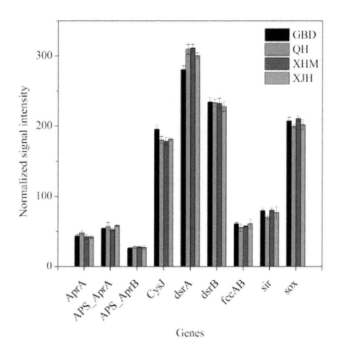

FIGURE 6: The normalized signal intensity of detected key genes involved in sulfur cycling. The signal intensity for each functional gene was the average of the signal intensity from all the replicates. All data are presented as mean ± SE.

and Cd in WWTPs, respectively. In contrast, other metal resistance genes (*corC, cusA, zitB*) did not show significant correlations to the respective metal concentrations (Table S5 in File S1).

8.3.9 ANTIBIOTIC RESISTANCE

Antibiotic resistance is a growing concern worldwide as more and more pathogens develop resistance to commonly used antibiotics. 9 gene categories families related to antibiotic resistance were detected in this study, including five transporters (ATP-binding cassette, multidrug toxic compound extrusion, major facilitator superfamily, Mex, and small multidrug

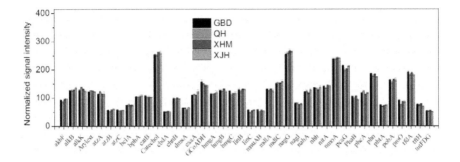

FIGURE 7: The normalized signal intensity of detected key genes involved in organic contaminant degradation. The signal intensity for each functional gene was the average of the signal intensity from all the replicates. All data are presented as mean ± SE.

resistance efflux pumps), three β-lactamase genes and several genes involved in tetracycline resistance (Figure S7 in File S1). A total of 1,072 antibiotic resistance genes were detected. Of 92 tetracycline resistance genes, 39 were shared by all samples and most of them were derived from isolated microorganisms, such as 214028433 of *Ruegeria* sp. R11, 259487609 of *Aspergillus nidulans* FGSC A4, 254407454 of *Streptomyces sviceus* ATCC 29083, etc. Since antibiotic resistance is often associated with metal resistance, Mantel test was performed to examine the correlation between heavy metal resistance genes and antibiotic resistance genes. The results indicated that the whole metal resistance genes were positively correlated to the whole antibiotic resistance genes (r = 0.325, P<0.01).

8.3.10 RELATIONSHIP OF ENVIRONMENTAL FACTORS TO THE MICROBIAL COMMUNITIES.

CCA was performed to correlate environmental variables with microbial community functional structure and determine major environmental variables in shaping the microbial community structure. On the basis of variance inflation factors, four variables were selected: DO, temperature, ammonia and COD (Figure 8). The results of CCA showed that the model was significant (P<0.05), suggesting that four variables were major environmental factor influencing microbial community functional structure. DO showed a strong positive correlation with the first axis and a negative correlation with the second axis, while ammonia and COD loading rate showed a strong positive correlation with the second axis and a negative correlation with the first axis. Temperature showed a strong negative correlation with both the first and second axes. Microbial community functional structures in WWTPs GBD and XHM were primarily linked to DO, while QH was linked to temperature. Ammonia and COD loading rate seemed to be major factors linking to microbial community functional structures in XJH.

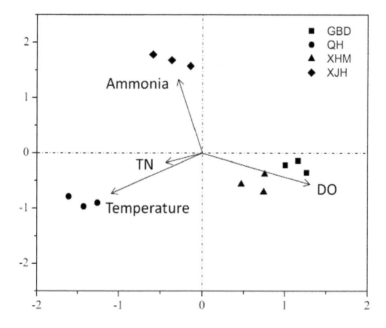

FIGURE 8: Canonical correspondence analysis (CCA) of GeoChip data and environmental variables. Environmental variables were chosen based on significance calculated from individual CCA data and variance inflation factors (VIFs).

Variance partitioning analyses (VPA) was further performed to assess contributions of wastewater characteristics and operational parameters to microbial community variances. Wastewater characteristics included influent concentrations of COD, TN, ammonia, TP, pH and conductivity. Operational parameters consisted of DO, temperature, hydraulic retention time (HRT) and mixed liquor suspended solids (MLSS). The results showed that 53% of total variances could be explained by these variables (Figure S8 in File S1). Wastewater characteristics and operational parameters can independently explain 25% and 23% of the variation of microbial communities, respectively. Interactions among the two major components (5%) appeared to be small.

8.4 DISCUSSION

In this study, microbial community functional structures in activated sludges of four WWTPs were analyzed by GeoChip 4.2. A total of 29,124 functional genes involved in carbon, nitrogen, phosphorous and sulfur cycles, metal resistance, antibiotic resistance and organic contaminant degradation were detected. It is reasonable to expect that the high diversity of functional genes can increase the functional redundancy, which is ensured by the presence of a pool of species able to perform the same ecological function. That is to say, many species share similar traits in activated sludge, therefore the loss of a few species has little impact on the system performance. Functional redundancy of microbial community in activated sludge is important to ensure the system stability when individual species are lost due to environmental changes, such as significant changes in the influent wastewater composition [4].

For individual gene category, such as *egl, amyA, lip, nirS, nirK, nosZ, ureC, ppx, ppk, aprA, dsrA, sox* and *benAB*, high diversity were detected. For example, 110 *nosZ* genes were detected in this study. *nosZ* gene encodes nitrous oxide reductase that is the only enzyme to mediate the conversion of N_2O to N_2 during denitrification process, which makes the *nosZ* gene an important molecular marker to trace complete denitrification [28]. This high diversity of *nosZ* genes ensures the presence of different species with similar function mediating the conversion of N_2O to N_2, which could lead to a more stable denitrification in WWTPs. Similarly, the high diversity of *egl, amyA*, etc. can increase the functional redundancy, and ensure the stable operation of plants.

GeoChip data revealed high similarities among microbial community functional structures within activated sludge of four WWTPs as measured by diversity indices and overlapped genes. This can be partly explained by the fact that all the WWTPs studied were located in the same city (Beijing), used to treat similar domestic wastewater, and operated with the same anaerobic/anoxic/aerobic (A^2O) process of similar operating parameters (COD and TN loading rate and MLSS concentration). For each gene category such as *egl, amyA, lip, nirS, nirK, nosZ, ureC, ppx, ppk, aprA, dsrA, sox, benAB*, there were a number of microorganisms commonly shared by all 12 samples, suggesting that there could be a core microbial community in activated sludge of four WWTPs. This agrees with the findings of zhang et al.[29], who suggested, based on pyrosequencing surveys of 14 WWTPs, that some core populations were shared by multiple samples. Similar results were also reported by other studies [30], [31].

Wastewater contains a broad spectrum of organic contaminants resulting from the mixing of wastewaters from different sources. Starch, cellulose, chitin and lignin could be the major carbon and energy sources for various microbial populations. Thus, we hypothesized that microbial communities would be abundant in functional genes involved in starch, cellulose, chitin and lignin. Our GeoChip data validated the hypothesis by evidence of detecting a wide variety of these genes across all samples such as *egl, amyA, lip*, xylanase genes, etc. Also, in WWTPs, biological nitrification coupled with denitrification is widely used to remove ammonia from wastewater. Nitrification is the main supply pathway of NO_3^- for denitrification, and thus it is reasonable to expect the genes involved in nitrogen cycle would be tightly linked with NH_4^+ and NO_3^- concentration. The results of Mantel test supported this expectation (Table S5 in File S1).

Understanding the factors that shape microbial community structure in WWTPs could potentially enhance treatment performance and control [11]. Of the 9 operational and environmental variables tested in this study, CCA ordination analysis indicated that DO was an important variable influencing microbial community structures. This agreed with the findings of Wells et al. [32], which suggested that DO was one of the most influential variables on microbial community structures of ammonia-oxidizing bacteria (AOB). Similar results have also been obtained in several lab-scale bioreactors by Park et al. [33].

Our results indicated that water temperature was also significantly linked to the microbial community structures in WWTPs, which was consistent with previous observations that water temperature was an important factor in shaping the microbial community structures. Ebrahimi et al. [34] found that, in two sequencing batch reactors (SBR) operated at 20°C, 30°C and 35°C, bacterial richness and diversity were clearly altered. In a membrane bioreactor treating domestic sewage, Ebie et al. [35] revealed shifts in AOB lineages in response to temperature changes by using denaturing gradient gel electrophoresis. Several other studies also demonstrated that temperature imposed a strong selective pressure on microbial communities of activated sludge [36]–[38].

In addition to DO and water temperature, ammonia concentrations within WWTPs were strongly and significantly linked to microbial community functional structures. Some previous studies have demonstrated the importance of ammonia concentration within bioreactor. In three gravel biofilters treating saline wastewater, Gregory et al. [39] observed that increasing ammonia concentration significantly influenced community structures of total bacteria and AOB. Miura et al. [40] found that ammonia concentration was an important structuring factor based on the DGGE analysis of AOB communities in four pilot-scale wastewater treatment systems with different ammonia concentrations.

The importance of COD loading rate for microbial community structures has been shown by previous studies. Pholchan et al. [41] reported that in laboratory-scale activated sludge reactors, high COD loading rate resulted in an increase and decrease of community diversity of total bacteria and AOB, respectively. In addition, Zhou et al. [42] showed that organic carbon (equivalent to COD loading rates) significantly affected microbial diversity in soil.

VPA was performed to attribute variations observed in microbial community functional structures to wastewater characteristics and operational parameters. VPA results showed that the deterministic wastewater characteristics and operational factors explained 53% of variances of microbial community functional structures, thus 47% of the variance was attributed to unknown factors. Some of them may result from unknown environmental variables or additional factors such as stochastic dispersal and immigration processes [43], [44], protozoan grazing [45], phage predation [46] and chaotic behavior [47], [48], which may play an influential role in shaping microbial communities in WWTPs.

Like other microarray technologies, GeoChip only detects genes that are represented on the array. Therefore, GeoChip 4.2 used in this study may not cover all of functional genes of microbial communities, though it contains as many as 83,992 probes targeting 152,414 genes involved in major microbial biogeochemical processes. Also, in our study, the DNA rather than cDNA from RNA was used for GeoChip hybridization. It is argued that DNA-based community analysis detects microbes irrespective of their viability or metabolic activity. Furthermore, DNA could persist extracellularly in environments after cell death, and it could result in biased population profiles [49].

In conclusion, GeoChip 4.2 was used to evaluate the microbial functional genes of activated sludge in four WWTPs. Our results revealed high similarities of microbial community functional structures among activated sludges of four WWTPs as measured by diversity indices and overlapped genes. CCA analysis indicated that microbial community functional structures were highly correlated to the water temperature, DO, ammonia concentrations and COD loading rate. Finally, a total of 53% of microbial community variations from GeoChip data were explainable by wastewater characteristics and operational parameters.

REFERENCES

1. Gentile ME, Jessup CM, Nyman JL, Criddle CS (2007) Correlation of functional instability and community dynamics in denitrifying dispersed-growth reactors. Appl Environ Microb 73: 680–690. doi: 10.1128/aem.01519-06
2. Wang XH, Wen XH, Yan HJ, Ding K, Hu M (2011) Community dynamics of ammonia oxidizing bacteria in a full-scale wastewater treatment system with nitrification stability. Front Environ Sci Engin China 5: 92–98. doi: 10.1007/s11783-010-0254-6
3. Rittmann BE, Hausner M, Loffler F, Love NG, Muyzer G, et al. (2006) A vista for microbial ecology and environmental biotechnology. Environ Sci Technol 40: 1096–1103. doi: 10.1021/es062631k

4. Briones A, Raskin L (2003) Diversity and dynamics of microbial communities in engineered environments and their implications for process stability. Curr Opin Biotech 14: 270–276. doi: 10.1016/s0958-1669(03)00065-x

5. Akarsubasi AT, Eyice O, Miskin I, Head IM, Curtis TP (2009) Effect of sludge age on the bacterial diversity of bench scale sequencing batch reactors. Environ Sci Technol 43: 2950–2956. doi: 10.1021/es8026488

6. Dytczak MA, Londry KL, Oleszkiewicz JA (2008) Activated sludge operational regime has significant impact on the type of nitrifying community and its nitrification rates. Water Res 42: 2320–2328. doi: 10.1016/j.watres.2007.12.018

7. Huang ZH, Gedalanga PB, Asvapathanagul P, Olson BH (2010) Influence of physicochemical and operational parameters on Nitrobacter and Nitrospira communities in an aerobic activated sludge bioreactor. Water Res 44: 4351–4358. doi: 10.1016/j. watres.2010.05.037

8. López-Vázquez CM, Hooijmans CM, Brdjanovic D, Gijzen HJ, van Loosdrecht MCM (2008) Factors affecting the microbial populations at full-scale enhanced biological phosphorus removal (EBPR) wastewater treatment plants in The Netherlands. Water Res 42: 2349–2360. doi: 10.1016/j.watres.2008.01.001

9. Miura Y, Hiraiwa MN, Ito T, Itonaga T, Watanabe Y, et al. (2007) Bacterial community structures in MBRs treating municipal wastewater: relationship between community stability and reactor performance. Water Res 41: 627–637. doi: 10.1016/j. watres.2006.11.005

10. Wang XH, Wen XH, Yan HJ, Ding K, Zhao F, et al. (2011) Bacterial community dynamics in a functionally stable pilot-scale wastewater treatment plant. Bioresource Technol 102: 2352–2357. doi: 10.1016/j.biortech.2010.10.095

11. Wang XH, Wen XH, Criddle C, Wells G, Zhang J, et al. (2010) Community analysis of ammonia-oxidizing bacteria in activated sludge of eight wastewater treatment systems. J Environ Sci-China 22: 627–634. doi: 10.1016/s1001-0742(09)60155-8

12. Sanapareddy N, Hamp TJ, Gonzalez LC, Hilger HA, Fodor AA, et al. (2009) Molecular Diversity of a North Carolina Wastewater Treatment Plant as Revealed by Pyrosequencing. Appl Environ Microb 75: 1688–1696. doi: 10.1128/aem.01210-08

13. Ye L, Zhang T, Wang T, Fang Z (2012) Microbial structures, functions, and metabolic pathways in wastewater treatment bioreactors revealed using high-throughput sequencing. Environ Sci Technol 46: 13244–13252. doi: 10.1021/es303454k

14. He Z, Deng Y, Van Nostrand JD, Tu Q, Xu M, et al. (2010) GeoChip 3.0 as a high-throughput tool for analyzing microbial community composition, structure and functional activity. ISME J 4: 1167–1179. doi: 10.1038/ismej.2010.46

15. Hazen TC, Dubinsky EA, DeSantis TZ, Andersen GL, Piceno YM, et al. (2010) Deep-Sea Oil Plume Enriches Indigenous Oil-Degrading Bacteria. Science 330: 204–208. doi: 10.1126/science.1195979

16. Lu Z, Deng Y, Van Nostrand JD, He Z, Voordeckers J, et al. (2012) Microbial gene functions enriched in the Deepwater Horizon deep-sea oil plume. ISME J 6: 451–460. doi: 10.1038/ismej.2011.91

17. Zhou JZ, Kang S, Schadt CW, Garten CT (2008) Spatial scaling of functional gene diversity across various microbial taxa. Proc Natl Acad Sci USA 105: 7768–7773. doi: 10.1073/pnas.0709016105

18. He ZL, Xu MY, Deng Y, Kang SH, Kellogg L, et al. (2010) Metagenomic analysis reveals a marked divergence in the structure of belowground microbial communities at elevated CO(2). Ecol Lett 13: 564–575. doi: 10.1111/j.1461-0248.2010.01453.x

19. Van Nostrand JD, Wu LY, Wu WM, Huang ZJ, Gentry TJ, et al. (2011) Dynamics of Microbial Community Composition and Function during In Situ Bioremediation of a Uranium-Contaminated Aquifer. Appl Environ Microb 77: 3860–3869. doi: 10.1128/aem.01981-10

20. Hemme CL, Deng Y, Gentry TJ, Fields MW, Wu LY, et al. (2010) Metagenomic insights into evolution of a heavy metal-contaminated groundwater microbial community. ISME J 4: 660–672. doi: 10.1038/ismej.2009.154

21. Xie JP, He ZL, Liu XX, Liu XD, Van Nostrand JD, et al. (2011) GeoChip-Based Analysis of the Functional Gene Diversity and Metabolic Potential of Microbial Communities in Acid Mine Drainage. Appl Environ Microb 77: 991–999. doi: 10.1128/aem.01798-10

22. Wang F, Zhou H, Meng J, Peng X, Jiang L, et al. (2009) GeoChip-based analysis of metabolic diversity of microbial communities at the Juan de Fuca Ridge hydrothermal vent. Proc Natl Acad Sci USA 106: 4840–4845. doi: 10.1073/pnas.0810418106

23. Association) AAPH (2012) Standard Methods for the Examination of Water and Waste Water. 22th ed. APHA, Washington DC, USA.

24. Zhou JZ, Bruns MA, Tiedje JM (1996) DNA recovery from soils of diverse composition. Appl Environ Microb 62: 316–322.

25. Eisen MB, Spellman PT, Brown PO, Botstein D (1998) Cluster analysis and display of genome-wide expression patterns. Proc Natl Acad Sci USA 95: 14863–14868. doi: 10.1073/pnas.95.25.14863

26. ter B, F CJ, Smilauer P (2002) CANOCO Reference Manual and CanoDraw for Windows User's Guide: Software for canonical community ordination (version 4.5). Microcomputer Power, Ithaca, New York, USA.

27. Wu L, Kellogg L, Devol AH, Tiedje JM, Zhou J (2008) Microarray-based characterization of microbial community functional structure and heterogeneity in marine sediments from the gulf of Mexico. Appl Environ Microb 74: 4516–4529. doi: 10.1128/aem.02751-07

28. Bergaust L, Bakken LR, Frostegard A (2011) Denitrification regulatory phenotype, a new term for the characterization of denitrifying bacteria. Biochem Soc Trans 39: 207–212. doi: 10.1042/bst0390207

29. Zhang T, Shao MF, Ye L (2012) 454 Pyrosequencing reveals bacterial diversity of activated sludge from 14 sewage treatment plants. ISME J 6: 1137–1147. doi: 10.1038/ismej.2011.188

30. Wang X, Hu M, Xia Y, Wen X, Ding K (2012) Pyrosequencing analysis of bacterial diversity in 14 wastewater treatment systems in china. Appl Environ Microb 78: 7042–7047. doi: 10.1128/aem.01617-12

31. Xia SQ, Duan LA, Song YH, Li JX, Piceno YM, et al. (2010) Bacterial Community Structure in Geographically Distributed Biological Wastewater Treatment Reactors. Environ Sci Technol 44: 7391–7396. doi: 10.1021/es101554m

32. Wells GF, Park HD, Yeung CH, Eggleston B, Francis CA, et al. (2009) Ammonia-oxidizing communities in a highly aerated full-scale activated sludge bioreactor:

betaproteobacterial dynamics and low relative abundance of Crenarchaea. Environ Microbiol 11: 2310–2328. doi: 10.1111/j.1462-2920.2009.01958.x

33. Park HD, Noguera DR (2004) Evaluating the effect of dissolved oxygen on ammonia-oxidizing bacterial communities in activated sludge. Water Res 38: 3275–3286. doi: 10.1016/j.watres.2004.04.047

34. Ebrahimi S, Gabus S, Rohrbach-Brandt E, Hosseini M, Rossi P, et al. (2010) Performance and microbial community composition dynamics of aerobic granular sludge from sequencing batch bubble column reactors operated at 20 A degrees C, 30 A degrees C, and 35 A degrees C. Appl Microbiol Biot 87: 1555–1568. doi: 10.1007/s00253-010-2621-4

35. Ebie Y, Matsumura M, Noda N, Tsuneda S, Hirata A, et al. (2002) Community analysis of nitrifying bacteria in. an advanced and compact Gappel-Johkasou by FISH and PCR-DGGE. Water Sci Technol 46: 105–111.

36. Li T, Bo L, Yang F, Zhang S, Wu Y, et al. (2012) Comparison of the removal of COD by a hybrid bioreactor at low and room temperature and the associated microbial characteristics. Bioresource Technol 108: 28–34. doi: 10.1016/j.biortech.2011.12.141

37. Szukics U, Abell GCJ, Hodl V, Mitter B, Sessitsch A, et al. (2010) Nitrifiers and denitrifiers respond rapidly to changed moisture and increasing temperature in a pristine forest soil. Fems Microbiol Ecol 72: 395–406. doi: 10.1111/j.1574-6941.2010.00853.x

38. Urakawa H, Tajima Y, Numata Y, Tsuneda S (2008) Low temperature decreases the phylogenetic diversity of ammonia-oxidizing archaea and bacteria in aquarium biofiltration systems. Appl Environ Microb 74: 894–900. doi: 10.1128/aem.01529-07

39. Gregory SP, Shields RJ, Fletcher DJ, Gatland P, Dyson PJ (2010) Bacterial community responses to increasing ammonia concentrations in model recirculating vertical flow saline biofilters. Ecol Eng 36: 1485–1491. doi: 10.1016/j.ecoleng.2010.06.031

40. Lydmark P, Almstrand R, Samuelsson K, Mattsson A, Sorensson F, et al. (2007) Effects of environmental conditions on the nitrifying population dynamics in a pilot wastewater treatment plant. Environ Microbiol 9: 2220–2233. doi: 10.1111/j.1462-2920.2007.01336.x

41. Pholchan MK, Baptista JD, Davenport RJ, Curtis TP (2010) Systematic study of the effect of operating variables on reactor performance and microbial diversity in laboratory-scale activated sludge reactors. Water Res 44: 1341–1352. doi: 10.1016/j.watres.2009.11.005

42. Zhou JZ, Xia BC, Treves DS, Wu LY, Marsh TL, et al. (2002) Spatial and resource factors influencing high microbial diversity in soil. Appl Environ Microb 68: 326–334. doi: 10.1128/aem.68.1.326-334.2002

43. Curtis TP, Sloan WT (2006) Towards the design of diversity: stochastic models for community assembly in wastewater treatment plants. Water Sci Technol 54: 227–236. doi: 10.2166/wst.2006.391

44. Sloan WT, Lunn M, Woodcock S, Head IM, Nee S, et al. (2006) Quantifying the roles of immigration and chance in shaping prokaryote community structure. Environ Microbiol 8: 732–740. doi: 10.1111/j.1462-2920.2005.00956.x

45. Petropoulos P, Gilbride KA (2005) Nitrification in activated sludge batch reactors is linked to protozoan grazing of the bacterial population. Can J Microbiol 51: 791–799. doi: 10.1139/w05-069

46. Kunin V, He S, Warnecke F, Peterson SB, Martin HG, et al. (2008) A bacterial meta-population adapts locally to phage predation despite global dispersal. Genome Res 18: 293–297. doi: 10.1101/gr.6835308

47. Graham DW, Knapp CW, Van Vleck ES, Bloor K, Lane TB, et al. (2007) Experimental demonstration of chaotic instability in biological nitrification. ISME J 1: 385–393. doi: 10.1038/ismej.2007.45

48. Ofiteru ID, Lunn M, Curtis TP, Wells GF, Criddle CS, et al. (2010) Combined niche and neutral effects in a microbial wastewater treatment community. Proc Natl Acad Sci USA 107: 15345–15350. doi: 10.1073/pnas.1000604107

49. Nogales B, Moore ER, Llobet-Brossa E, Rossello-Mora R, Amann R, et al. (2001) Combined use of 16 S ribosomal DNA and 16 S rRNA to study the bacterial community of polychlorinated biphenyl-polluted soil. Appl Environ Microbiol 67: 1874–1884. doi: 10.1128/aem.67.4.1874-1884.2001

There are several supplemental files that are not available in this version of the article. To view this additional information, please use the citation on the first page of this chapter.

CHAPTER 9

Key Design Factors Affecting Microbial Community Composition and Pathogenic Organism Removal in Horizontal Subsurface Flow Constructed Wetlands

JORDI MORATÓ, FRANCESC CODONY, OLGA SÁNCHEZ, LEONARDO MARTÍN PÉREZ, JOAN GARCÍA, AND JORDI MAS

9.1 INTRODUCTION

Water shortages in arid and semi-arid areas such as the Mediterranean have prompted a need for wastewater treatment and subsequent reuse. Reclamation can be achieved through conventional intensive systems or natural, ecologically engineered treatments such as horizontal subsurface flow (HSSF) constructed wetlands. Depending on wastewater type, some pathogenic microorganisms may be present and, therefore, wastewater reclamation processes with disinfection could be required (Asano and Levine, 1998). Thus, research into sewage treatment is needed in order to

Key Design Factors Affecting Microbial Community Composition and Pathogenic Organism Removal in Horizontal Subsurface Flow Constructed Wetlands. Morató J, Codony F, Sánchez O, Pérez LM, García J, and Mas J. Science of the Total Environment **481** (2014). doi:10.1016/j.scitotenv.2014.01.068. *Reprinted with permission from the authors. Reproduced with permission of ELSEVIER BV.*

reduce risks associated with improper sanitation, particularly in terms of wastewater reuse for crop irrigation.

Regardless of their location (water column, biofilms on surface material or sediment pore-water), pathogens must compete with the consortium of organisms surrounding them. As intestinal organisms, most may not survive and may also be destroyed by predation. Water temperature, organic matter concentration and hydraulic conditions such as flow, aspect ratio, and granular media type are some of the most important factors governing occurrence and growth of viable microbes in biofilms developed elsewhere (LeChevallier et al., 1988, LeChevallier et al., 1996, Reasoner et al., 1989, Block, 1992, Block et al., 1993 and Van der Kooij et al., 1995).

Molecular fingerprinting techniques (Friedrich et al., 2003 and Ibekwe et al., 2003; Vacca et al., 2005) have been used to study the dynamics and structure of microbial communities in constructed wetlands but few studies have been performed on the effect of wetlands on the removal of specific pathogens. Although the work with powerful new techniques, such as quantitative real time polymerase chain reaction (qPCR), will undoubtedly change the scene of health and environmental microbiology during the following years, until now routine examination for pathogenic microorganisms is not recommended because of the high cost of the analysis and the generally low number of a specific pathogen that is present in an environmental sample. Therefore, indicator organisms are routinely used to study microbial removal in constructed wetlands.

In general, most studies on fecal microorganism removal in constructed wetlands only describe total and fecal coliform removal (Kadlec and Knight, 1996, Wang et al., 2005, Tanaka et al., 2006 and Tunçsiper, 2007). Research using experimental, pilot and full-scale constructed wetlands has shown that fecal coliform bacteria inactivation usually ranges between 1.25 and 2.5 log units (Gersberg et al., 1989a, Gersberg et al., 1989b, Hiley, 1990, Rivera et al., 1995, Williams et al., 1995, Decamp et al., 1999, Arias et al., 2003, Vacca et al., 2005 and Vymazal, 2005). However, fecal coliform inactivation rates of 3.0 log units and higher have been recorded in tertiary HSSF constructed wetlands treating slaughterhouse wastewater with extremely high influent concentrations of fecal bacteria (from 6.0 to 11.0 log units/100 mL). Furthermore, removals of 2.4 to 5.3 orders of magnitude for cultivable *Salmonella* cells were found (Pundsack et al., 2001). The high degree of inactivation observed in these wetlands was related to the high influent microbial concentration (Rivera et al., 1995).

It must also be taken into account that the concentration of coliform bacteria in wastewaters is subject to significant daily fluctuations, so the highest concentration in the influent of a given constructed wetland system will not necessarily coincide with the highest concentration in the effluent (Cooper et al., 1996).

In any case, subsurface flow constructed wetlands offer a suitable combination of physical, chemical and biological mechanisms required to remove pathogenic organisms. Physical factors include filtration and sedimentation (Gersberg et al., 1989a and Pundsack et al., 2001), while chemical mechanisms combine oxidation and adsorption to organic matter (Gersberg et al., 1989a). The biological removal features include oxygen release and bacterial activity in the rhizosphere, aggregation and retention in biofilms (Hiley, 1995 and Brix, 1997), potential production of bactericidal compounds or antimicrobial activity of root exudates (Kickuth and Kaitzis, 1975, Seidel, 1976 and Axelrood et al., 1996), as well as predation by nematodes and protists (Decamp and Warren, 1998 and Decamp et al., 1999), attack by lytic bacteria and viruses (Axelrood et al., 1996), natural die-off (Gersberg et al., 1989a and Gersberg et al., 1989b) and competition for limiting nutrients or trace elements (Gersberg et al., 1987a and Gersberg et al., 1987b).

Fecal bacteria removal in constructed wetlands has been related to environmental factors such as granular medium or type of plant. Some studies appear to show that granular media and the presence or absence of plants (macrophytes) in constructed wetlands are important for fecal bacteria inactivation, while in others' works there is no evidence of this fact. On the one hand, wetland plants play several roles in the HSSF constructed wetland. Their root systems provide surfaces for the attachment of microorganisms, enhance filtration effects, and stabilize the bed surface. The roots contribute to the development of microorganisms by the release of oxygen and nutrients. Moreover, the plants give the treatment site an attractive appearance. The effect of macrophytes on the system efficiency seems to vary depending on the season, wastewater type and plant species (Stein and Hook, 2003). On the other hand, the use of a small size granular medium instead of a large size one seems to improve the microbial inactivation ratio between 1.0 and 2.0 log units for both fecal coliforms and somatic coliphage removal (Ottová et al., 1997 and García et al., 2003). Nevertheless, several researchers have obtained contradictory results, and links between microbial removal and environmental factors have still not been definitively understood.

Another important design parameter for constructed wetlands is water depth. From practical experience, water depth in subsurface flow constructed wetlands has been normally set at 0.60 m because this is the maximum depth at which the roots and rhizomes of the macrophytes grow, and it is therefore the maximum depth at which the macrophytes can have effects on the process (Cooper et al., 1996). The mass transfer theory dictates that water depth influences the oxygen transfer coefficient from the atmosphere to the water. Water depth also determines the fraction of water volume in contact with the underground biomass of the macrophytes. Therefore, despite the fact that only the surface area is in contact with atmosphere, it can be reasonably supposed that water depth influences the efficiency of the subsurface flow constructed wetlands. However, the information available on the effect of water depth is scarce and contradictory (US EPA, 2000 and Coleman et al., 2001).

Thus, the removal of pathogens in subsurface flow constructed wetlands used for wastewater treatment seems to be a very complex process and may vary in time and space, depending on many factors. Even though possible mechanisms of bacterial removal have been discussed in many papers (Burger and Weise, 1984, Armstrong et al., 1990, Morales et al., 1996 and Decamp and Warren, 2000), no systematic analyses on the removal processes and the fate of potential pathogenic bacteria in constructed wetlands are yet known. In terms of bacterial removal, constructed wetlands are generally considered to feature a combination of chemical and physical factors, including mechanical filtration and sedimentation (Pundsack et al., 2001). Also, the relative importance of the different biochemical reactions that can take place in these natural treatment systems plays a determinant role on the efficiency of the system and the microbial removal.

In view of all the above, the objective of the present work was to evaluate and clarify the role of different design key factors, such as granular media, water depth and season effect, that could affect the removal of microbial indicators in order to improve our understanding of the microbial reduction in constructed wetlands. We have also performed a molecular analysis using denaturing gradient gel electrophoresis (DGGE) with the aim to determine the effect of wetland design on the microbial diversity of gravel biofilms.

TABLE 1: Summary of the characteristics of the four SSF analyzed.

	SSF			
	C1	C2	D1	D2
Aspect ratio	2:1	2:1	2.5:1	2.5:1
Medium size (mm)	10	3.5	10	3.5
Porosity (%)	39	40	39	40
Water depth (m)	0.5	0.5	0.27	0.27

9.2 MATERIALS AND METHODS

9.2.1 PILOT PLANT SYSTEM

The pilot HSSF system used in this study treats part of the urban wastewater generated by the Can Suquet housing development in the municipality of Les Franqueses del Vallès (Barcelona, north-east Spain). A detailed description of the plant can be found in García et al., 2004 and García et al., 2005. The wastewater was previously screened and flowed to an Imhoff tank. After primary clarification, the effluent was equally divided and flowed to eight parallel HSSF constructed wetlands with a surface area of 54–56 m² each. The system had a valve and a flowmeter that allowed the adjustment and monitoring of the flow pumped to the wetlands. The plant began to operate in March 2001, when the beds were planted with *Phragmites australis.*

The aspect ratio of the HSSF constructed wetlands varied in pairs. Pair A had an aspect ratio of 1:1, B of 1.5:1, C of 2:1 and D of 2.5:1. Furthermore, the size of the granular medium within each pair also differed. Thus, type 1 contained a coarse granitic gravel (D60 = 10 mm, Cu = 1.6) while type 2 had fine granitic gravel (D60 = 3.5 mm, Cu = 1.7). For example, the HSSF constructed wetland A1 presented an aspect ratio of 1:1 and contained coarse gravel, while the HSSF constructed wetland A2 contained fine gravel and had the same aspect ratio. HSSF constructed wetland from types A, B, and C had an average water depth of 0.50 m, and D of 0.27 m.

All HSSF wetlands had 2 perforated tubes (0.1 m in diameter) inserted into the middle part of the gravel and uniformly distributed throughout the length of the bed, so that intermediate samples could be obtained. These tubes were perforated all along their depth, and they were installed at the bottom of the beds. They were referred to as P1 (near the inlet, at 1/4 of the length) and P2 (near the outlet, at 3/4 of the length). It should be taken into account that these perforated tubes had a minor impact on the flow field because of their small size in relation to the width of the HSSF wetlands.

9.2.2 SAMPLING

Influent and effluent water samples for evaluation of the removal efficiency of microbial indicators were taken once a month in types C and D HSSF constructed wetlands, from January 2003 to February 2005. Types A and B did not present significant differences in performance with type C, while type D differed in removal efficiency from the rest (García et al., 2004 and García et al., 2005). Thus, experiments were conducted in four HSSF wetlands: C1 and C2 were chosen as representatives of the deep beds, and D1 and D2 were chosen as shallow ones. The four wetland systems differ in terms of aspect ratio (length to width), granular medium size and water depth (Table 1). A 36 mm/day flow rate was used and all HSSF received the same flow and therefore they operated with the same hydraulic loading rate (HLR). The nominal hydraulic retention time (HRT) was 5.6 days for type C and 3 days for the type D.

In November 2004, samples from the perforated tubes (P1 and P2) and from deep (C1, C2) and shallow (D1, D2) wetland types were obtained for biofilm analysis. Gravel was quickly removed and placed in a container full of water in order to avoid drying of the biofilm and brought to the laboratory, where the contents were aseptically removed and analyzed. Prior to the analyses, gravel were rinsed with 200 mL of saline solution (NaCl 0.9%), to remove non-attached deposits and introduced into a plastic container with 15 mL of saline solution. Attached microorganisms were detached by sonication (3 min, 40 W), according to the procedures of the European Biofilm Workgroup, AGHTM Biofilm Group (1999). Details of the procedure of biofilm analyses can be found in Morató et al. (2005). For molecular analysis, aliquots of detached microorganisms were removed and centrifuged, and the pellets were then stored at − 20 °C for DNA extraction. For epifluorescence microscopy (Nikon Optiphot, Barcelona), aliquots of samples were treated after sonication with a 2% v/v solution of formaldehyde, filtered through a Nuclepore filter (0.2 mm pore and 13 mm diameter) and examined to evaluate the number of total cells and viable microbial counts using Live/Dead (Molecular Probes, Inc.).

Heterotrophic plate counts (HPC) were carried out using the spread plate method with PCA agar (Merck). The plates were incubated for 72 h at 22 °C. The presence of fecal bacterial indicators such as total coliform bacteria (TC) and *Escherichia coli*, fecal enterococci (FE) and *Clostridium perfringens* and other microbial groups such as *Pseudomonas* and *Aeromonas* were assessed by membrane filtration according to standardized methods (APHA, 1995).

9.2.3 STATISTICAL ANALYSIS

Statistical procedures for the evaluation of the effect of season, bed type (which included aspect ratio and water depth), and medium size on wetlands microbial removal performance were carried out using the SPSS statistical software package. One-way and three-way ANOVA methods were used to evaluate the influence of each factor considered for every microbial indicator parameter of the effluent and also to assess interactions between factors. For all ANOVA tests it was verified that the variables were distributed normally. Otherwise, the variables were log-transformed.

9.2.4 MOLECULAR ANALYSIS OF BACTERIAL DIVERSITY

Bacterial DNA from gravel biofilm was extracted as described by Massana et al. (1997). Gravel samples were suspended in 2 mL of lysis buffer (50 mM Tris–HCl, pH 8.3; 40 mM EDTA, pH 8.0; 0.75 M sucrose). 0.5-mm-diameter sterile glass beads were added to the cultures, and vortexed three times during 30 s. DNA was extracted using the lysis/phenol extraction method as described below. Lysozyme (1 mg/mL final concentration) was added and samples were incubated at 37 °C for 45 min in slight movement. Then, sodium dodecyl sulfate (1% final concentration) and proteinase K (0.2 mg/mL final concentration) were added and samples were incubated at 55 °C for 60 min in slight movement. Nucleic acids were extracted twice with phenol–chloroform–isoamyl alcohol (25:24:1, vol:vol:vol), and the residual phenol was removed once with chloroform–isoamyl alcohol (24:1, vol:vol). Nucleic acids were purified, desalted and concentrated with a Centricon-100 concentrator (Millipore). DNA integrity was checked by agarose gel electrophoresis, and quantified using a low DNA mass ladder as a standard (Invitrogen).

Fragments of the 16S rRNA gene suitable for denaturing gradient gel electrophoresis (DGGE) analysis were obtained by using the bacterial specific primer set 358f-907rM (Sánchez et al., 2007). Polymerase chain reaction (PCR) was carried out with a Biometra thermal cycler using the following program: initial denaturation at 94 °C for 5 min; 10 touchdown cycles of denaturation (at 94 °C for 1 min), annealing (at 63.5–53.5 °C for 1 min, decreasing 1 °C each cycle), and extension (at 72 °C for 3 min); 20 standard cycles (annealing at 53.5 °C, 1 min) and a final extension at 72 °C for 5 min.

PCR mixtures contained 1–10 ng of template DNA, each deoxynucleoside triphosphate at a concentration of 200 µM, 1.5 mM MgCl$_2$, each primer at a concentration of 0.3 µM, 2.5 U Taq DNA polymerase (Invitrogen) and PCR buffer supplied by the manufacturer. BSA (Bovine Serum Albumin) at a final concentration of 600 µg/mL was added to minimize the inhibitory effect of humic substances (Kreader, 1996). The volume of reactions was 50 µL. PCR products were verified and quantified by agarose gel electrophoresis with a low DNA mass ladder standard (Invitrogen).

The DGGE was run in a DCode system (Bio-Rad) as described by Muyzer et al. (1988). A 6% polyacrylamide gel with a gradient of 40–80% DNA-denaturant agent was cast by mixing solutions of 0% and 80% denaturant agent (100% denaturant agent is 7 M urea and 40% deionized for-mamide). One thousand nanograms of PCR product was loaded for each sample and the gel was run at 100 V for 18 h at 60 °C in 1 × TAE buffer (40 mM Tris [pH 7.4], 20 mM sodium acetate, 1 mM EDTA). The gel was stained with SYBR Gold (Molecular Probes) for 45 min, rinsed with 1 × TAE buffer, removed from the glass plate to a UV-transparent gel scoop, and visualized with UV in a Gel Doc EQ (Bio-Rad).

Prominent bands were excised from the gels, resuspended in milli-Q water overnight and reamplified for its sequencing. Purification of PCR products from DGGE bands and sequencing reactions were performed by Macrogen kit (South Korea) using primer 907rM and the Big Dye Termi-nator version 3.1 sequencing kit. Reactions were run in an automatic ABI 3730XL Analyzer-96 capillary type.

Sequences were subjected to a BLAST search (Altschul et al., 1997) to obtain an indication of the phylogenetic affiliation. Eighteen 16S rRNA gene sequences were sent to the EMBL database (http://www.Ebi.ac.uk/embl) and received the following accession numbers: from FN429742, FN429743, FN429744, FN429745, FN429746, FN429747, FN429748, FN429749, FN429750, FN429751, FN429752, FN429753, FN429754, FN429755, FN429756, FN429757, FN429758 and FN429759.

Digitized DGGE images were analyzed with Quantity One software (Bio-Rad). Bands occupying the same position in the different lanes of the gels were identified. A matrix was constructed for all lanes, taking into ac-count the presence or absence of the individual bands. This matrix was used to calculate a distance matrix using hierarchical cluster analysis with the statistical software SPSS. Finally, a dendrogram comparing samples for each HSSF constructed wetland was obtained utilizing the unweighted-pair group method with average linkages (algorithm: Dice) and SPSS.

9.3 RESULTS AND DISCUSSION

9.3.1 BACTERIAL REMOVAL

All HSSF constructed wetlands received the same wastewater flow with the same quality. According to García et al., 2004 and García et al., 2005, the shallow pair (D1, D2) had the lowest effluent concentrations of chemical oxygen demand (COD), biochemical oxygen demand (BOD_5), ammonia and dissolved reactive phosphorus (DRP), and therefore it was more efficient in removing organic matter and nutrients. Thus, differences in performance between the shallow pair and the rest of the HSSF wetlands were not due to aspect ratio, but rather to water depth. The overall removal efficiency was approximately 60% (types A–C), and 75% (type D) for COD with an inlet value of 170 ± 55 mg/L; 60% and 80% for BOD5 with an inlet value of 140 ± 54 mg/L; 30% and 50% for ammonia with an inlet value of 36.8 ± 11 mg/L; and 0% and 10% for DRP with an inlet value of 5.6 ± 1.9 mg/L, respectively. According to these conclusions, evaluation of the removal efficiency of microbial indicators, as well as molecular analyses, was performed in deep (C1, C2) and shallow (D1, D2) wetland types as representatives of different water depths.

Table 2 shows the overall averages and the standard error of the different microbial groups determined in the influent and the effluent water samples of deep and shallow wetlands (C and D types, respectively). The overall microbial inactivation ratio ranged between 1.4 and 2.9 log-units for heterotrophic plate counts (HPC), from 1.2 to 2.2 log units for total coliforms (TC) and from 1.4 to 2.3 log units for *E. coli*. The inactivation of *Pseudomonas* and *Aeromonas* ranged between 1.4 to 2.1 log units and 0.4 to 1.9 log units, respectively. For fecal enterococci (FE) and *Clostridium* it ranged between 1.4 to 2.2 log units and 1.2 to 1.6 log units respectively. The shallower HSSF wetlands, especially D1, showed a lower efficiency in the removal of the classical bacterial indicators such as TC, *E. coli* and FE. A similar trend was observed in the *Pseudomonas* group and with lesser differences, in the *Aeromonas* group. On the other hand, the shallower wetlands showed a higher efficiency in the removal of *Clostridium* spores.

Table 3 shows the ANOVA analysis of data from water from the outlet of the treatment plant, performed in order to assess the effect of the treatments (HSSF wetland type, granulometry and season) on microbial removal. The presence of fine granulometry strongly influenced the removal of each bacterial group. This effect was significant for TC ($p = 0.009$), *E. coli* ($p = 0.004$), FE ($p = 0.012$) and *Clostridium* spores ($p = 0.036$). Removal of the last bacterial groups including *Clostridium* spores was higher in summer season, and this effect was significant for TC (p = 0.011), *E. coli* (p = 0.028) and *Clostridium* spores (p = 0.001).

TABLE 2: Overall averages[a] and standard error (n = 17) of the different microbial groups determined in the influent and the effluent of SSF types C (deep) and D (shallow).

	Influent	C1	C2	D1	D2
Heterotrophic plate count (HPC)	1.73E+ 07	5.73E+ 05	1.90E+ 04	4.84E+ 05	4.70E+ 05
	9.24E+ 06	5.65E+ 05	5.54E+ 03	3.66E+ 05	4.35E+ 05
Total coliforms (TC)	4.40E+ 07	2.52E+ 05	2.30E+ 05	2.27E+ 06	4.88E+ 05
	1.62E+ 07	3.30E+ 04	5.87E+ 04	9.31E+ 05	1.66E+ 05
Escherichia coli	6.32E+ 06	4.13E+ 04	4.26E+ 04	2.17E+ 05	3.17E+ 04
	3.43E+ 06	9.35E+ 03	1.40E+ 04	8.63E+ 04	1.82E+ 04
Fecal enterococci (FE)	3.16E+ 05	7.04E+ 03	3.70E+ 03	1.22E+ 04	1.97E+ 03
	6.95E+ 04	3.02E+ 03	1.29E+ 03	3.35E+ 03	3.35E+ 03
Clostridium perfringens spores	1.56E+ 05	9.72E+ 03	7.20E+ 03	9.50E+ 03	4.00E+ 03
	6.54E+ 04	5.19E+ 03	2.09E+ 03	7.68E+ 03	1.80E+ 03
Pseudomonas	1.61E+ 08	1.19E+ 06	3.31E+ 06	6.73E+ 06	1.18E+ 06
	4.96E+ 07	2.38E+ 05	1.35E+ 06	2.70E+ 06	5.02E+ 05
Aeromonas	1.75E+ 07	2.21E+ 05	3.02E+ 05	6.26E+ 06	3.07E+ 05
	6.57E+ 06	6.86E+ 04	9.45E+ 04	5.55E+ 06	1.20E+ 05

[a]*CFU/mL for HPC and CFU/100 mL for all other microbial parameters.*

Although no clear relationship could be found between microbial removal and HSSF constructed wetland type, ANOVA interaction terms indicated that this fact was largely dependent on granulometry. In this sense, shallow HSSF wetlands were more efficient for removing total coliforms (p = 0.011) and *E. coli* (p = 0.013) when fine granulometry was used. Also, shallow HSSF wetlands were more effective for removing *Clostridium* spores (p = 0.039) but, in this case, this effect was independent of granulometry.

Granular media type and its granulometry can be identified as another key factor in the microbial removal process (Polprasert and Hoang, 1983). On average, smaller granular media improve the microbial inactivation ratio between 1 and 2 log units for both fecal coliforms and somatic coliphages (Ottová et al., 1997 and García et al., 2003). Garcia et al. (2005) reported that the slightly higher removal efficiency observed in HSSF constructed wetlands with a fine medium occurs in conjunction with a clearly greater macrophytes development (1800 and 600 g/m^2 of dry weight for aerial and underground biomass, respectively) than that observed in HSSF constructed wetlands with a coarse gravel (540 and 270 g/m^2, respectively). As a consequence, in shallow HSSF wetlands, a larger fraction of the water volume will be in contact with the root system of the macrophytes, maybe favoring the microbial removing process.

TABLE 3: Probabilities of the ANOVA test on the effects of the factors (one-way) and their interactions (three-way) on the bacterial indicators from the outlet of the treatment plant (log CFU/mL or CFU/100 mL).

Factor	HPC 22 °C	Total coliforms	*Escherichia coli*	Fecal enterococci	*Clostridium* spores
Type of CWs	*Deep*	*Deep*	*Deep*	*Deep*	*Shallow*
	0.005	*0.041*	*0.047*	*0.049*	*0.039*
Granulometry	Fine	*Fine*	*Fine*	*Fine*	*Fine*
	0.16	*0.009*	*0.004*	*0.012*	*0.036*
Season	Winter	*Summer*	*Summer*	Summer	Summer
	0.98	*0.011*	*0.028*	0.61	0.001
CWsType * granulometry	*D-1/2*	*Sh-2*	*Sh-2*	Sh-2	Sh-2
	0.047	*0.011*	*0.013*	0.052	0.572
CWsType * season	–	*Sh-Summer*	–	–	–
	0.79	*0.011*	0.17	0.22	0.126

Coarse gravel (1) and fine gravel (2) Significant values (p < 0.05) are in bold and italics.

In order to assess the effect of the different treatments (HSSF wetland type, granulometry and season) on microbial removal, an ANOVA data analysis from the gravel biofilm of the two perforated tubes (P1, near the inlet; P2, near the outlet) was performed (Table 4). Similarly to the results obtained with water samples, the presence of fine granulometry strongly influenced microbial removal. This effect was significant for all bacterial groups analyzed, with the exception of *Clostridium* spores. The higher specific surface area available for microbial attachment in the fine medium, which is considered the main mechanism for phosphorus removal in constructed wetlands (Vymazal, 2003), could explain the better performance for those with fine gravel.

TABLE 4: Probabilities of the ANOVA test on the effects of the factors (one-way) and their interactions (three-way) on the bacterial indicators from the gravel biofilm (log CFU/cm2).

Factor	HPC 22 °C	Total coliforms	*Escherichia coli*	Fecal enterococci	*Clostridium* spores
Type of CWs	Deep	Deep	***Shallow***	Shallow	Shallow
	0.185	0.088	***0.009***	0.527	0.662
Granulometry	*Fine*	*Fine*	*Fine*	*Fine*	Fine
	0.001	*0.001*	*0.001*	*0.001*	0.278
Season	*Winter*	Winter	***Summer***	Summer	–
	0.003	0.258	*0.048*	0.076	
Inlet–outlet	*Outlet*	*Outlet*	*Outlet*	*Outlet*	*Outlet*
	0.035	*0.001*	*0.001*	*0.001*	*0.035*
CWsType * Granulometry	*Deep-2*	Deep-2	Shallow-2	Shallow-2	Shallow-2
	0.036	0.778	0.206	0.533	0.472
Granulometry * Inlet–outlet	2-Outlet	2-Outlet	2-Outlet	*2-Outlet*	2-Outlet
	0.07	0.283	0.365	*0.006*	0.301

Significant values (p < 0.05) are in bold and italics.

Samples obtained from the P2 sampler (near the outlet) showed less bacterial charge for all groups, indicating a significant effect of the HSSF constructed wetland length. Interestingly, it seems that shallow wetlands were more efficient to remove *E. coli* (p = 0.009) on the biofilm, and a similar trend was observed for FE and *Clostridium* spores. On the other hand, the removal of HPC (22 °C) and TC attached to the biofilm was higher in deep HSSF wetlands, especially those with fine granulometry. Concerning to the seasons effect, a higher removal of *E. coli* (p = 0.048) and FE (p = 0.076) was observed in summer, while HPC (22 °C) (p = 0.003) removal was higher in winter. A clear decrease of the HPC in winter as a consequence of the temperature effect could increase the number of the bacterial indicators—*E. coli* and FE—or can suppress its detection, due to the reduction of some coliform antagonists belonging to the HPC (LeChevallier and McFeters, 1985), that could cause injury to the coliform population.

Finally, with the aim to assess the effect of the treatments (HSSF wetland type, granulometry and season) on total and viable microbial counts (Live/Dead), another ANOVA data analysis from the gravel biofilm of the two perforated tubes (P1, near the inlet; P2, near the outlet) was carried out (Table 5). Samples obtained from the P2 sampler (near the outlet) showed lower total cell numbers (p = 0.012), indicating a significant effect of the HSSF constructed wetland length. Similar to plate cultures, shallow HSSF wetlands were more efficient for microbial removal, showing a significant decrease of the total cell numbers (p = 0.001) and, as a consequence, the live (p = 0.002) and dead (p = 0.001) cell counts.

TABLE 5: Probabilities of the ANOVA test on the effects of the factors (one-way) and their interactions (three-way) on the microbial cell counts from the gravel biofilm ANOVA of data from the gravel biofilm (log CFU/cm²).

Factor	TCN Total	TCN Live	TCN Dead
Type of CWs	*Shallow*	*Shallow*	*Shallow*
	0.001	*0.002*	*0.001*
Inlet–outlet	*Outlet*	Outlet	*Outlet*
	0.012	0.58	*0.001*
CWsType * inlet–outlet	Shallow-Outlet	Shallow-Outlet	*Shallow-Outlet*
	0.059	0.905	*0.001*

Significant values (p < 0.05) are in bold and italics.

The oxidation and reduction potential (E_H) profiles carried out in summer 2001 and winter 2003 in the perforated tubes of all the HSSF constructed wetlands used in the present study indicate that the water inside the wetlands was under reducing conditions. Nevertheless, the E_H values were higher in the shallower HSSF wetlands (on average, they ranged from −144 to −131 mV) compared to the deep ones (from −183 to −151 mV), and therefore shallow wetlands had more oxidized conditions. Rivera et al. (1995) speculate that the redox status of the constructed wetlands affects the microbial removal efficiency as it occurs in other wastewater treatment systems. Differences in redox status in HSSF constructed wetlands at different depths can be explained because the mass transfer coefficient of oxygen from the atmosphere to the bulk water is inversely proportional to water depth (Kadlec and Knight, 1996). Therefore, it can be concluded that a lower water depth promotes more energetically favorable biochemical reactions, with more oxidized conditions that give in turn more efficiency as it has been previously described for the same system (García et al., 2004 and García et al., 2005).

9.3.2 BACTERIAL COMMUNITY COMPOSITION

Analysis of the bacterial community composition by PCR-DGGE was performed on gravel biofilm samples from perforated tubes P1 and P2 collected in November 2004 for deep and shallow HSSF constructed wetlands (types C and D, respectively). Banding patterns for the 16S rRNA DGGE-PCR amplicons are presented in Fig. 1. The number of bands per lane varied from 13 to 25. Some differences could be observed in band position, intensity, and number of bands present in the different samples for each HSSF wetland analyzed, demonstrating different bacterial community developments.

The dendrogram based on the DGGE banding pattern (Fig. 2) separates the samples according to the different water depth designs. Thus, samples split in two main clusters corresponding to deep (type C) and shallow wetlands (type D), confirming that it was the main factor affecting microbial communities. For deep HSSF wetlands (type C), gravel size and sample position (P1 or P2 perforated tubes) in the two systems analyzed (C1 and C2) did not seem to exert a strong effect on bacterial community structure. On the contrary, samples from shallow wetlands (D1 and D2) were separated according to gravel size, although gravel position (P1 or P2) did not have any influence on the bacterial assemblage composition.

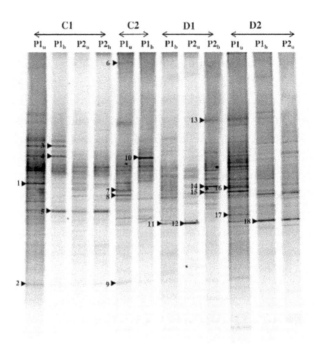

FIGURE 1: DGGE fingerprints from samples of gravel biofilm obtained from different positions (u: upper part; b: bottom part) of P1 (near the inlet) and P2 (near the outlet) perforated tubes of constructed wetlands C1, C2, D1 and D2.

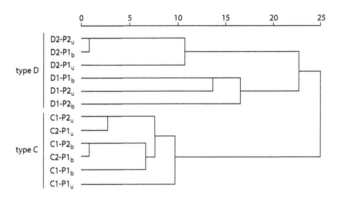

FIGURE 2: Dendrograms generated from the DGGE profiles of the different samples analyzed for each wetland, determined by the unweighted-pair group method using average linkages. The scale bar is linkage distance (P1, near the inlet; P2, near the outlet; u: upper part; b: bottom part).

The number of DGGE bands is a representation of the diversity of phylotypes present in the different microbial assemblages. For type C and D2 wetlands, it was observed a higher band number (between 20 and 25) near the upper part of the inlet (samples P1u) compared to the outlet (samples P2) and the bottom part of the inlet (samples P1b), indicating that sample position had some effect on diversity. Unfortunately, we could not obtain enough DNA for sample D1P1u.

A total of 38 band positions were excised and sequenced in order to determine their phylogenetic affiliation, although only 18 bands produced DNA sequences with enough quality. The closest matches (and percentages of similarity) for the sequences retrieved were determined by a BLAST search (Table 6). The number of bases used to calculate each similarity value is also shown in Table 6 as an indication of the quality of the sequence.

TABLE 6: Phylogenetic affiliation of sequences obtained from DGGE bands, with closest uncultured and cultured matches. Number of bases used to calculate the sequence similarity is shown in parentheses in the third column.

Band	Closest match	Similarity (%) (no bases)	Taxonomic group	Acc no. (GenBank)	Cultured closest match (% similarity)
FRV03-1	Uncultured Catellibacterium sp.	97.5 (505)	αproteobacteria	EU887785	Rhodobacter sp. (96.5)
FRV03-2	Uncultured bacterium	93.5 (488)	Acidobacteria	AB291529	Pelobacter carbinolicus (80.4)
FRV03-3	Uncultured bacterium	86.3 (441)	Synergistetes	EU864485	Synergistes sp. (81.8)
FRV03-4	Uncultured bacterium	89.3 (460)	Firmicutes	FJ390537	Clostridium ruminantium (88.3)
FRV03-5	Uncultured Deferribacter sp.	99.6 (518)	Deferribacteres	EU887800	Aminobacterium colombiense (88.5)
FRV03-6	Unidentified bacteria	96.5 (517)	Bacteroidetes	AJ224942	Owenweeksia hongkongensis (90.7)
FRV03-7	Desulfococcus biacutus	93.7 (463)	δproteobacteria	AJ277887	

TABLE 6: *Cont.*

Band	Closest match	Similarity (%) (no bases)	Taxonomic group	Acc no. (GenBank)	Cultured closest match (% similarity)
FRV03-8	Uncultured bacterium	92.8 (505)	β-proteobacteria	DQ836763	*Denitratisoma oestradiolicum (89.9)*
FRV03-9	Uncultured *Acidobacteria* bacterium	99.6 (525)	Acidobacteria	DQ383312	*Desulfuromonas alkaliphilus (85.9)*
FRV03-10	Uncultured *Thiobacillus* sp.	98.9 (539)	β-proteobacteria	AY082471	*Thiobacillus sayanicus (97.4)*
FRV03-11	*Candidatus Nitrospira defluvii*	100 (532)	Nitrospirae	EU559167	
FRV03-12	*Candidatus Nitrospira defluvii*	99.8 (532)	Nitrospirae	EU559167	
FRV03-13	Uncultured bacterium	95.7 (517)	Bacteroidetes	EU234211	*Flexibacter sp. (85.2)*
FRV03-14	Uncultured β-proteobacterium	96.1 (519)	β-proteobacteria	AB113610	*Siderooxidans ghiorsii (90.4)*
FRV03-15	Uncultured bacterium	95.4 (520)	β-proteobacteria	DQ836763	*β-proteobacterium G5G6 (93.6)*
FRV03-16	Uncultured bacterium	96.5 (526)	β-proteobacteria	DQ836763	*β-proteobacterium G5G6 (93.6)*
FRV03-17	Uncultured bacterium	93.0 (412)	β-proteobacteria	EU925882	*Dechloromonas sp. (80.8)*
FRV03-18	*Candidatus Nitrospira defluvii*	100 (533)	Nitrospirae	EU559167	

Different phylogenetic group sequences could be retrieved from deep and shallow wetlands (types C and D, respectively), which differed in water depth. Table 6 shows that sequence identity of most of the bands is closely associated with different species of bacteria from sludge or wastewater environments.

Some of them belonged to Beta-proteobacteria, and had uncultured closest matches; one of these sequences, although with a low similarity, was associated to *Denitratisoma oestradiolicum*, a denitrifying bacterium isolated from activated sludge of a wastewater treatment plant able to use 17-beta-oestradiol as the sole carbon and energy source (Fahrbach et al.,

2006). Another sequence corresponded to an uncultured *Thiobacillus* sp. related to *Thiobacillus sayanicus*, an autotrophic sulfur-oxidizing bacterium able to utilize thiosulfate and hydrogen sulfide. In general, sequences from Beta-proteobacteria are recurrent in this kind of environments (Sánchez et al., unpublished).

Other sequences, particularly abundant in the shallower HSSF wetlands (type D), were related with a high similarity (near 100%) to a microorganism from the nitrogen cycle, *Candidatus Nitrospira defluvii*. This nitrite-oxidizing bacterium was described as selectively enriched from a nitrifying activated sludge from a wastewater treatment plant, and is capable to further oxidize nitrite to nitrate (Spieck et al., 2006). The abundance of this kind of organisms involved in nitrogen removal in wastewater treatment plants usually accounts for 1–10% of the total bacterial population (Daims et al., 2001 and Juretschko et al., 2002).

Acidobacteria was another taxonomic group observed in our study. Sequences belonging to this widespread distributed phylum have been found traditionally in soil, aquatic environments and wastewater treatment plants (Ludwig et al., 1997 and Juretschko et al., 2002). On the other hand, the cluster Alpha-proteobacteria was represented by *Catellibacterium* sp., a strictly aerobic microorganism originally isolated from activated sludge belonging to the 'Rhodobacter group' characterized by the absence of photosynthetic activity (Tanaka et al., 2004).

Uncultured members of other groups, such as Bacteroidetes, Synergistetes, Firmicutes and Deferribacteres were also retrieved. One of these sequences had a high similarity (99.6%) with *Deferribacter* sp., a genus usually found in hydrothermal vents, petroleum reservoirs and gas fields (Takai et al., 2003 and Mochimaru et al., 2007). A sequence belonging to the Delta-proteobacteria and related to *Desulfococcus biacutus* was also detected, although similarity was low (93.7%). This strictly anaerobic sulfate-reducing bacterium, able of growing with acetone, was originally isolated from anaerobic digester sludge of a wastewater treatment plant (Platen et al., 1990).

In general, most of the sequences retrieved from deep HSSF wetlands (type C) gravel biofilm (bands 2, 3, 4, 5, 7, 8 and 9) corresponded to cultured closest matches represented by anaerobic microorganisms, in accordance with lower redox potential (E_H) measurements found in this type of systems. In contrast, closest matches of shallower HSSF wetlands (type D) were mainly related to aerobic bacteria.

9.4 CONCLUSION

Water depth and gravel granulometry are the two most important key design factors controlling the efficiency of subsurface flow constructed wetlands for wastewater treatment. During the period of activity evaluated, the HSSF wetland analyzed in this study with a water depth of 0.27 m and a granular medium with a size of 3.5 mm proved to be more effective than those designed with a water depth of 0.50 m and a granular medium size of 10 mm for the removal of TC ($p = 0.011$), *E. coli* ($p = 0.013$), FE ($p = 0.052$) and total cell numbers ($p = 0.001$). Microbial removal in all HSSF wetlands analyzed occurs mainly near the inlet by a combination of biological and physical mechanisms (i.e., sedimentation and filtration). Microbial removal effectiveness in HSSF wetlands with a fine medium can be partially explained by the fact that a larger fraction of the water volume is in contact with the root system of the macrophytes.

Water depth is an important parameter that should be taken into account for HSSF wetland design and for predictive performance models, although models currently available in the scientific and technical literature still do not include water depth as a variable. Measurements of E_H in different periods have shown that shallower HSSF wetlands have more oxidized conditions. Furthermore, microbial communities from constructed wetlands appear to be affected by water depth, being the communities from shallow HSSF wetlands more related to aerobic microorganisms, in accordance with the higher values of E_H found in this type of natural treatment systems.

Concerning to the season effect, a significant higher removal for bacterial indicators was observed in summer, while the HPC (22 °C) removal was higher in winter as a clear effect of the temperature decreasing.

On the other hand, DNA-based molecular tools showed that the HSSF constructed wetlands analyzed in our study contain significant hidden diversity of unknown and uncultured microorganisms that have the potential to act as degraders of environmental pollutants. Therefore, further attempts to isolate the key microorganisms involved in these processes will be essential in order to explore the degradation capacity of the microbial communities developed in the different designed wetlands.

REFERENCES

1. AGHTM Biofilm Group. Standard method to evaluate aquatic biofilms. C.W. Keevil (Ed.), Biofilm in the aquatic environment, Royal Society of Chemistry, London (1999), pp. 210–219
2. S.F. Altschul, T.L. Madden, A.A. Schäffer, J. Zhang, Z. Zhang, W. Miller, et al.. Gapped BLAST and PSI-BLAST: a new generation of protein database search programs. Nucl Acids Res, 25 (1997), pp. 3389–3402
3. APHA. American Public Health Association. Standard methods for the examination of water and wastewater. (19th ed.) (1995) [Washington DC]
4. C.A. Arias, A. Cabello, H. Brix, N.H. Johansen. Removal of indicator bacteria from municipal wastewater in an experimental two stage vertical flow constructed wetland system. Water Sci Technol, 48 (2003), pp. 35–41
5. W. Armstrong, J. Armstrong, P.M. Beckett. Measurement and modelling of oxygen release from roots of Phragmites australis. P.F. Cooper, B.C. Findlater (Eds.), The use of constructed wetlands in water pollution control, Pergamon Press, Oxford (1990), pp. 41–51
6. T. Asano, D. Levine. Wastewater reclamation, recycling and reuse: an introduction. T. Asano (Ed.), Wastewater reclamation and reuse, Technomic Publishing, Lancaster (1998), pp. 1–56
7. P.E. Axelrood, A.M. Clarke, R. Radley, S.J.V. Zemcov. Douglas-fir root-associated microorganisms with inhibitory activity towards fungal plant pathogens and human bacterial pathogens. Can J Microbiol, 42 (1996), pp. 690–700
8. J.C. Block. Biofilms in water distribution systems. L.F. Melo, T.R. Bott, M. Fletcher, B. Capdeville (Eds.), Biofilms—science and technology, Kluwer Academic Publishers, Dordrecht (1992), pp. 469–485
9. J.C. Block, K. Haudidier, J.L. Paquin, J. Miazga, Y. Le'vi. Biofilm accumulation in drinking water distribution systems. Biofouling, 6 (1993), pp. 333–343
10. H. Brix. Do macrophytes play a role in constructed treatment wetlands?. Water Sci Technol, 35 (1997), pp. 11–17
11. G. Burger, G. Weise. Untersuchungen zum einflux limnischer makrophyten auf die absterbegeschwindigkeit von Escherichia coli im wasser. Acta Hydrochim Hydrobiol, 12 (1984), pp. 301–309
12. J. Coleman, K. Hench, K. Garbutt, A. Sexstone, G. Bissonnette, J. Skousen. Treatment of domestic wastewater by three plant species in constructed wetlands. Water Air Soil Pollut, 128 (2001), pp. 283–295
13. P.F. Cooper, G.D. Job, M.B. Green, R.B.E. Shutes. Reed beds and constructed wetlands for wastewater treatment. WRc, Swindon (1996), p. 206

14. H. Daims, U. Purkhold, L. Bjerrum, E. Arnold, P.A. Wilderer, M. Wagner. Nitrifica-
    tion in sequencing biofilm batch reactors: lessons from molecular approaches. Water
    Sci Technol, 43 (2001), pp. 9–18

15. O. Decamp, A. Warren. Bacterivory in ciliates isolated from constructed wetlands
    (reed beds) used for wastewater treatment. Water Res, 32 (1998), pp. 1989–1996

16. O. Decamp, A. Warren. Investigation of *E. coli* removal in various designs of subsur-
    face flow wetlands used for wastewater treatment. Ecol Eng, 14 (2000), pp. 293–299

17. O. Decamp, A. Warren, R. Sánchez. The role of ciliated protozoa in subsurface flow
    wetlands and their potential as bioindicators. Water Sci Technol, 40 (1999), pp.
    91–97

18. M. Fahrbach, J. Kuever, R. Meinke, P. Kämpfer, J. Hollender. *Denitratisoma oestra-
    diolicum* gen. nov. sp. nov., a 17β-oestradiol-degrading, denitrifying betaproteobac-
    terium. Int J Syst Evol Microbiol, 56 (2006), pp. 1547–1552

19. U. Friedrich, H. Van Langenhove, K. Altendorf, A. Lipski. Microbial community
    and physicochemical analysis of an industrial waste gas biofilter and design of 16S
    rRNA-targeting oligonucleotide probes. Environ Microbiol, 5 (2003), pp. 183–201

20. J. García, J. Vivar, M. Aromir, R. Mujeriego. Role of hydraulic retention time and
    granular medium in microbial removal in tertiary treatment reed beds. Water Res, 37
    (2003), pp. 2645–2653

21. J. García, P. Aguirre, R. Mujeriego, Y. Huang, L. Ortiz, J.M. Bayona. Initial con-
    taminant removal performance factors in horizontal flow reed beds used for treating
    urban wastewater. Water Res, 38 (2004), pp. 1669–1678

22. J. García, P. Aguirre, J. Barragán, R. Mujeriego, V. Matamoros, J.M. Bayona. Effect
    of key design parameters on the efficiency of horizontal subsurface flow constructed
    wetlands. Ecol Eng, 25 (2005), pp. 405–418

23. R.M. Gersberg, R. Brenner, S.F. Lyon, B.V. Elkins. Survival of bacteria and viruses
    in municipal wastewaters applied to artificial wetlands. K.R. Reddy, W.H. Smith
    (Eds.), Aquatic plants for water treatment and resource recovery, Magnolia Publish-
    ing Inc., Orlando (1987), pp. 237–245

24. R.M. Gersberg, S.R. Lyon, R. Brenner, B.V. Elkins. Fate of viruses in artificial wet-
    lands. Appl Environ Microbiol, 53 (1987), pp. 731–736

25. R.M. Gersberg, R.A. Gearhart, M. Yves. Pathogen removal in constructed wetlands.
    D.A. Hammer (Ed.), Constructed wetlands for wastewater treatment; municipal, in-
    dustrial and agricultural, Lewis Publisher, Chelsea (1989), pp. 231–446

26. R.M. Gersberg, S.R. Lyon, R. Brenner, B.V. Elkins. Integrated wastewater treatment
    using artificial wetlands: a gravel marsh case study. D.A. Hammer (Ed.), Construct-
    ed wetlands for wastewater treatment; municipal, industrial and agricultural, Lewis
    Publishers, Chelsea (1989), pp. 145–152

27. P.D. Hiley. Wetlands treatment revival in Yorkshire. P.F. Cooper, B.C. Findlater
    (Eds.), Constructed wetlands in water pollution control, Pergamon Press, Oxford
    (1990), pp. 279–288

28. P.D. Hiley. The reality of sewage treatment using wetlands. Water Sci Technol, 32
    (1995), pp. 329–338

29. A.M. Ibekwe, C.M. Grieve, S.R. Lyon. Characterization of microbial communities
    and composition in constructed dairy wetland wastewater effluent. Appl Environ
    Microbiol, 69 (2003), pp. 5060–5069

30. S. Juretschko, A. Loy, A. Lehner, M. Wagner. The microbial community composition of a nitrifying–denitrifying activated sludge from an industrial sewage treatment plant analyzed by the full-cycle rRNA approach. Syst Appl Microbiol, 25 (2002), pp. 84–99

31. R.H. Kadlec, R.L. Knight. Treatment wetlands. (1st ed.)CRC Press, Florida (1996)

32. R. Kickuth, G. Kaitzis. Mikrobizid wirksame aromaten aus Scirpus lacustris L. Umweltschutz (1975), pp. 134–135

33. C.A. Kreader. Relief of amplification inhibition in PCR with bovine serum albumin or T4 gene 32 protein. Appl Environ Microbiol, 62 (1996), pp. 1102–1106

34. M.W. LeChevallier, G. McFeters. Interactions between heterotrophic plate count bacteria and coliform organisms. Appl Environ Microbiol, 49 (1985), pp. 1338–1341

35. M.W. LeChevallier, C.D. Cawthon, R.G. Lee. Factors promoting survival of bacteria in chlorinated water supplies. Appl Environ Microbiol, 54 (1988), pp. 649–654

36. M.W. LeChevallier, N.J. Welch, D.B. Smith. Full-scale studies of factors related to coliform regrowth in drinking water. Appl Environ Microbiol, 62 (1996), pp. 2201–2211

37. W. Ludwig, S.H. Bauer, M. Bauer, I. Held, G. Kirchhof, R. Schulze, et al.. Detection and in situ identification of representatives of a widely distributed new bacterial phylum. FEMS Microbiol Lett, 153 (1997), pp. 181–190

38. R. Massana, A.E. Murray, C.M. Preston, E.F. Delong. Vertical distribution and phylogenetic characterization of marine planktonic Archaea in the Santa Barbara Channel. Appl Environ Microbiol, 63 (1997), pp. 50–56

39. H. Mochimaru, H. Yoshioka, H. Tamaki, K. Nakamura, N. Kaneko, S. Sakata, et al.. Microbial diversity and methanogenic potential in a high temperature gas field in Japan. Extremophiles, 11 (2007), pp. 453–461

40. A. Morales, J.L. Garland, D.V. Lim. Survival of potentially pathogenic human-associated bacteria in the rhizosphere of hydroponically grown wheat. FEMS Microbiol Ecol, 20 (1996), pp. 155–162

41. J. Morató, F. Codony, J. Mas. Utilisation of a packed-bed biofilm reactor for the determination of the potential of biofilm accumulation in water systems. Biofouling, 21 (2005), pp. 151–160

42. G. Muyzer, T. Brinkhoff, U. Nübel, C. Santegoeds, H. Schäfer, C. Wawer. Denaturing gradient gel electrophoresis (DGGE) in microbial ecology. A.D.L. Akkermans, J.D. van Elsas, F.J. Brujin (Eds.), Molecular microbial ecology manual, Academic Publishers, Dordrecht (1988), pp. 1–27

43. V. Ottová, J. Balcarová, J. Vymazal. Microbial characteristics of constructed wetlands. Water Sci Technol, 35 (1997), pp. 117–123

44. H. Platen, A. Temmes, B. Schink. Anaerobic degradation of acetone by Desulfococcus biacutus, spec. nov. Arch Microbiol, 154 (1990), pp. 355–361

45. C. Polprasert, H. Hoang. Kinetics of bacteria and bacteriophages in anaerobic filters. J Water Pollut Control Fed, 55 (1983), pp. 385–391

46. J. Pundsack, R. Axler, R. Hicks, J. Henneck, D. Nordmann, B. McCarthy. Seasonal pathogen removal by alternative on-site wastewater treatment systems. Water Environ Res, 73 (2001), pp. 204–212

47. D.J. Reasoner, J.C. Blannon, E.E. Geldreich, J. Barnick. Nonphotosynthetic pigmented bacteria in a potable water treatment and distribution system. Appl Environ Microbiol, 55 (1989), pp. 912–921

48. F. Rivera, A. Warren, E. Ramírez, O. Decamp, P. Bonilla, E. Gallegos, et al. Removal of pathogens from wastewaters by the root zone method (RZM). Water Sci Technol, 32 (1995), pp. 211–218

49. O. Sánchez, J.M. Gasol, R. Massana, J. Mas, C. Pedrós-Alió. Comparison of different denaturing gradient gel electrophoresis primer sets for the study of marine bacterioplankton communities. Appl Environ Microbiol, 73 (2007), pp. 5962–5967

50. K. Seidel. Macrophytes and water purification. Biological control of water pollution, Pennsylvania University Press, Pennsylvania (1976), pp. 109–121

51. E. Spieck, C. Hartwig, I. McCormack, F. Maixner, M. Wagner, A. Lipski, et al.. Selective enrichment and molecular characterization of a previously uncultured Nitrospira-like bacterium from activated sludge. Environ Microbiol, 8 (2006), pp. 405–415

52. O.R. Stein, P.B. Hook. Temperature, plants, and oxygen: how does season affect constructed wetland performance?. Ü. Mander, C. Vohla, A. Poom (Eds.), Constructed and riverine wetlands for optimal control of wastewater at catchment scale, Publicationes Instituti Geographici Universitatis Tartuensis, Tartu (2003), pp. 37–43

53. K. Takai, H. Kobayashi, K.H. Nealson, K. Horikoshi. Deferribacter desulfuricans sp. nov., a novel sulfur-, nitrate- and arsenate-reducing thermophile isolated from a deep-sea hydrothermal vent. Int J Syst Evol Microbiol, 53 (2003), pp. 839–846

54. Y. Tanaka, S. Hnada, A. Manome, T. Tsuchida, R. Kurane, K. Nakamura, et al.. *Catellibacterium* nectariphilum gen. nov., sp. nov., which requires a diffusible compound from a strain related to the genus Sphingomonas for vigorous growth. Int J Syst Evol Microbiol, 54 (2004), pp. 955–959

55. N. Tanaka, K.B.S.N. Jinadasa, D.R.I.B. Werellagama, M.I.M. Mowjood, W.J. Ng. Constructed tropical wetlands with integrated submergent-emergent plants for sustainable water quality management. J Environ Sci Health Part A, 41 (2006), pp. 2221–2236

56. B. Tunçsiper. Removal of nutrient and bacteria in pilot-scale constructed wetlands. J Environ Sci Health Part A, 42 (2007), pp. 1117–1124

57. US EPA. Constructed wetlands treatment of municipal wastewaters, EPA/625/R-99/010. United States Environmental Protection Agency, Office of Research and Development, Cincinnati (2000), p. 165

58. G. Vacca, H. Wand, M. Nicolausz, P. Kuschk, M. Kästner. Effect of plants and filter materials on bacteria removal in pilot-scale constructed wetlands. Water Res, 39 (2005), pp. 1361–1373

59. D. Van der Kooij, H.R. Veenendaal, C. Baars-Lorist, D.W. van der Klift, Y.C. Drost. Biofilm formation on surfaces of glass and teflon exposed to treated water. Water Res, 29 (1995), pp. 1655–1662

60. J. Vymazal. Removal mechanisms in constructed wetlands. V. Dias, J. Vymazal (Eds.), The use of aquatic macrophytes for wastewater treatment in constructed wetlands, articles of the first international seminar, FundacSo Calouste Gulbenkian, Lisboa (2003), pp. 219–264

61. J. Vymazal. Removal of enteric bacteria in constructed treatment wetlands with emergent macrophytes: a review. J Environ Sci Health Part A, 40 (2005), pp. 1355–1367

62. L. Wang, J. Peng, B. Wang, R. Cao. Performance of a combined eco-system of ponds and constructed wetlands for wastewater reclamation and reuse. Water Sci Technol, 51 (2005), pp. 315–323

63. J. Williams, M. Bahgat, E. May, M. Ford, J. Butler. Mineralisation and pathogen removal in gravel bed hydroponic constructed wetlands for wastewater treatment. Water Sci Technol, 32 (1995), pp. 49–58

PART III

TREATMENTS

An Application of Wastewater Treatment in a Cold Environment and Stable Lipase Production of Antarctic Basidiomycetous Yeast *Mrakia blollopis*

MASAHARU TSUJI, YUJI YOKOTA, KODAI SHIMOHARA, SAKAE KUDOH, AND TAMOTSU HOSHINO

10.1 INTRODUCTION

Drainage from dairy parlors and milk factories produced in the process of cleaning transport pipes and milking tanks pollute rivers and groundwater with detergents, bactericides, mucus and milk fat are contaminating rivers and underground water [1]. In low temperature conditions, the wastewater is treated by bio-filters [2] and a reed bed system [3], [4]. However, the system is not used widely because of the high running cost and the necessity of a large space. Instead, an activated sludge system is now widely used for industrial treatment of dairy parlor wastewater [5] due to its advantages in maintenance and running cost. However, there is a problem in

An Application of Wastewater Treatment in a Cold Environment and Stable Lipase Production of Antarctic Basidiomycetous Yeast Mrakia blollopis. © *Tsuji M, Yokota Y, Shimohara K, Kudoh S, and Hoshino T. PLoS ONE **8**,3 (2013). doi:10.1371/journal.pone.0059376. Originally published under a Creative Commons Attribution License, http://creativecommons.org/licenses/by/3.0/.*

this system of low temperature conditions in winter having adverse effects on microbial functions.

The use of microorganisms living in polar regions for the purpose of removing nitrogen and phosphorus compounds from wastewater under low temperature conditions has been reported by Chevalier et al. [6] and Hirayama-katayama et al. [7], but it has not yet been applied for milk fat.

In our previous work, we examined 305 isolates of fungi including eight Ascomycetous and six Basidiomycetous species collected from Antarctica and found that they included fungi of the genus *Mrakia*, in psychrophilic Basidiomycetous yeast, suggesting that *Mrakia* is a major mycoflora highly adapted to the Antarctic environment (Fujiu, 2010; master's thesis in Graduate School of Science, Hokkaido University). *Mrakia* spp. and *Mrakiella* spp. are also common fungal species frequently found in cold climate areas such as Arctic, Siberia, Central Russia, the Alps and Antarctica [8]–[10]. Therefore, we screened our *Mrakia* isolates for their ability to decompose milk fat under low temperature conditions and evaluated their potential for application to an active sludge system in a region with a cold climate. The results showed that 56 *Mrakia* spp. exhibited a clear zone according to fat decomposition. Antarctic yeast strain SK-4 had physiological characteristics similar to those of *Mrakia blollopis* [11].

Here we report that activated sludge containing yeast strain SK-4 has the potential to remove milk fat BOD_5. We also describe identification of yeast strain SK-4 and the purification and characterization of the lipase, considered as a major enzyme to degrade milk fat in wastewater.

10.2 MATERIALS AND METHODS

10.2.1 ETHICS STATEMENT

All necessary permits were obtained for the described field studies. Permission required for field studies was obtained from the Ministry of the Environment of Japan. Sample collection in Antarctica was performed with the permission of the Ministry of the Environment of Japan.

10.2.2 SAMPLE ISOLATION

Algal mat samples were collected from sediments of Naga-ike, a lake in Skarvsnes, located near Syowa station, East Antarctica. The isolate was inoculated on potato dextrose agar (PDA) (DifcoTM, BD Japan, Tokyo, Japan) at 4°C for 1 week. Yeast strain SK-4 was selectively picked for isolation on the basis of its morphology. Yeast strain SK-4 was maintained on PDA plates at 4°C and long-term storage was performed in 40% (w/v) glycerol at −80°C.

10.2.3 PHYLOGENETIC ANALYSIS

Phylogenetic analysis was done by sequencing the ITS region including 5.8S rRNA and D1/D2 domain of 26S rRNA. Cells were harvested from 2-weeks-old cultures. DNA was extracted with an ISOPLANT II kit (Wako Pure Chemical Industries, Osaka, Japan) according to the manufacturer's protocol. Extracted DNA was amplified by PCR using KOD-plus DNA polymerase (TOYOBO, Osaka, Japan). The ITS region was amplified by using the following primers: ITS1F (5′-GTA ACA AGG TTT CCG T) and ITS4 (5′-TCC TCC GCT TAT TGA TAT GC). The D1/D2 domain was amplified using the following primers: NL1 (5′-GCA TAT CAA TAA GCG GAG GAA AAG) and NL4 (5′-GGT CCG TGT TTC AAG ACG G). Sequences were obtained with an ABI prism 3100 Sequencer (Applied Biosystems, Life Technologies Japan, Tokyo, Japan) using an ABI standard protocol. The ITS region and D1/D2 domain sequences of yeast strain SK-4 are deposited in DNA Data Bank of Japan (BBDJ) (Accession numbers AB630315 and AB691134). Alignment was made using CLUSTAL W (http://clustalw.ddbj.nig.ac.jp/) and corrected manually. Phylogenetic analysis was performed using MEGA software version 4.0 [12] with neighbor-joining analysis of the ITS region containing 5.8S rRNA and maximum parsimony analysis of the D1/D2 domain of 26S rRNA. Bootstrap analysis (1000 replicates) was performed using a full heuristic search.

10.2.4 PHYSIOLOGICAL CHARACTERIZATION

Assimilation of carbon was performed at 15°C on modified Czapek-Dox agar composed by 6.7 g/L of yeast nitrogen base without amino acids (DifcoTM, BD Japan, Tokyo, Japan), 2.0 g/L of sodium nitrate (Wako Pure Chemical Industries, Osaka, Japan), 30 g/L of carbon source and 15.0 g/L of Agar (DifcoTM, BD Japan, Tokyo, Japan). Assimilation of nitrogen and other physiological tests were carried out according to the protocols described by Yarrow [13]. All tests were performed at 15°C after 2 and 4 weeks of inoculation.

10.2.5 PREPARATION OF ACTIVE SLUDGE AND MEASUREMENT OF BIOCHEMICAL OXYGEN DEMAND

Activated sludge (AS) was cultivated at room temperature with aeration by using cow's milk as the substrate. After one month, the sludge was divided into two parts. One part of the activated sludge was mixed with *M. blollopis* SK-4 (1.4 g/L, dry weight), and the other part was used as a control. Separated activated sludge was prepared with Mixed Liquor Suspended Solids (MLSS, 3000 mg/L) and cow's milk at 10°C with aeration. One week later, prepared activate sludge was added to cow's milk, and biochemical oxygen demand (BOD_5) of waste-treated water was measured after 24 hours. BOD_5 assay was carried out using a coulometer (Ohkura Electric, Saitama, Japan).

10.2.6 INOCULUM

M. blollopis SK-4 was grown in YPD liquid medium (1% yeast extract, 2% peptone, and 2% glucose) at 15°C for 96 hours at 120 rpm. After 96 hours, *M. blollopis* SK-4 was collected by centrifugation at 3500×g for 15 min at 4°C. The pellet was transferred to fresh cream liquid medium (0.5% peptone, 0.5% NaCl, 5% fresh cream, pH 7.0) and incubated at 10°C for 14 days at 90 rpm. The resulting culture was used as inoculum.

10.2.7 LIPASE PRODUCTION MEDIUM

Lipase production medium was composed of 0.2% KH_2PO_4, 0.29% Na_2PO_4, 0.02% NH_4Cl, 0.04% $CaCl_2$, 0.001% $FeCl_3$, 0.5% yeast extract, and 1% Tween 80. The yeast was cultivated at 10°C for 324 hours at 90 rpm. One mL samples were collected every 24 h and centrifuged at 4°C for 10 min at 20000×g, and then lipase activity was measured.

10.2.8 ASSAY OF LIPASE ACTIVITY

Lipase activity was measured by a colorimetric method using p-nitrophenyl-palmitate as a substrate [14]. Forty mL of 50 mM sodium phosphate buffer (pH 7.0) containing 50 mg gum arabic and 0.2 g TritonX-100 was mixed with 3 mL 2-propanol containing 1 mM p-nitrophenyl-palmitate. Eight hundred μL of prepared substrate was added to 200 μL of enzyme solution. The enzyme reaction was carried out at 30°C for 30 min. The released p-nitrophenol was measured at A_{410}. One unit of lipase activity was defined as the activity required to release 1 μmol of free fatty acids per minute at 30°C.

10.2.9 MEASUREMENT OF PROTEIN CONCENTRATION

Protein concentration was measured by BCA protein assay reagent (Thermo Fisher Scientific, Waltham, MA, USA) according to the manufacturer's instructions using bovine serum albumin as a standard.

10.2.10 PURIFICATION OF LIPASE

M. blollopis SK-4 lipase was purified by ultrafiltration and Toyopearl-butyl 650 M (Tosho, Tokyo, Japan) hydrophobic interaction chromatography. Four hundred mL of lipase production liquid medium was centrifuged at 4°C for 15 min at 3000×g. The supernatant was filtered through a 0.45-μm of membrane filter (Advantec, Tokyo, Japan). The filtered medium was concentrated by ultrafiltration using an ultracel YM-30 membrane (Millipore, Billerica, MA, USA). The concentrated sample was adsorbed to a Toyopearl butyl 650 M column (2.5×20 cm) containing 1 M sodium chloride and eluted with a linear gradient from 750 mM to 100 mM sodium chloride in 20 mM Tris-HCl buffer (pH 8.5) at a flow rate of 60 mL/h. Fractions of high lipase activity was pooled and concentrated and then stored at 4°C until use. Protein molecular weight was estimated by SDS-PAGE according to Laemmli [15] and stained with CBB R-250. Precision plus protein unstained standards (Bio-Rad Laboratories Japan, Tokyo, Japan) were used as protein molecular weight makers.

10.2.11 CHARACTERIZATION OF LIPASE

Substrate specificity was determined by using substrate as different p-nitrophenyl esters (C_4–C_{18}). For determining the effects of metal ions and EDTA on lipase activity, residual lipase activity assays were carried out under standard assay conditions with final concentrations of 1 mM of various bivalent metal ions and EDTA. Lipase activity assay in the absence of metal ions and EDTA was carried out as a control. Optimum pH was measured at 30°C for 30 min and determined at various pH values of 50 mM buffer as follows: sodium citrate (pH 3.0–5.0), sodium phosphate (pH 6.0, 7.0 and 8.0), Tris-HCl (pH 7.5 and 8.5), glycine-NaOH (pH 9.0) and sodium carbonate (pH 9.3, 9.5 and 10.0).

Optimum temperature was measured in 50 mM sodium phosphate buffer (pH 7.0) for 30 min. To determine the pH stability of lipase, the enzyme was preincubated in various buffers for 15 h at 30°C and then adjusted to pH 8.5. The residual enzyme activity was measured by using p-nitrophenyl-palmitate as a substrate at 65°C for 30 min. For determining thermostability, lipase was preincubated for 30 min at different temperatures and the residual activity was measured at 65°C for 30 min in 50 mM Tris-HCl (pH 8.5). Effects of organic solvents on lipase activity were determined at 65°C for 30 min in 50 mM Tris-HCl (pH 8.5) containing various organic solvents at final concentration of 5% (v/v). Lipase activity assay in the absence of organic solvents was carried out as a control.

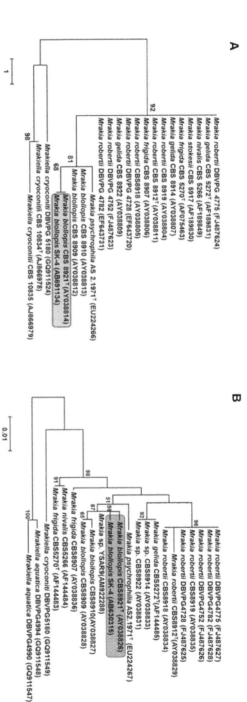

FIGURE 1: Phylogenetic tree of *Mrakia biollopis* SK-4 and related species. (A) Maximum parsimony analysis of the D1/D2 domain of the 26 rRNA gene sequence. Bootstrap percentages from 1000 replications are shown on the branches. *Mrakiella cryoconiti* CBS 10834T and *Mrakiella cryoconiti* CBS 10835 were used as an out group. (B) Neighbor-joining tree of the ITS region containing the 5.8 S rRNA gene sequence. Bootstrap percentages from 1000 replications are shown on the branches. *Mrakiella aquatica* DBVPG4994 and *Mrakiella aquatica* DBVPG4990 were used as an out group.

10.3 RESULTS

10.3.1 PHYLOGENETIC ANALYSIS

As a result of phylogenetic analysis of the ITS region and D1/D2 domain, yeast strain SK-4 was grouped with the clade of *Mrakia blollopis* CBS 8921[T] (Fig. 1 A and B). By comparison of the ITS region sequence containing 5.8 S rRNA, yeast strain SK-4 showed high homologies (>99.6%) with *M. blollopis* CBS8921[T]. The D1/D2 domain sequence showed no variation with *M. blollopis* CBS 8921[T].

TABLE 1: Comparison of physiological characteristics of *Mrakia blollopis* SK-4 and other *Mrakia* species.

Characteristic	*M. blollopis* SK-4	*M. blollopis* CBS8921[T]	*M. psychrophhila* AS2.1971[T]	*M. frigida* CBS5270[T]
Maximum growth temperature	22°C	20°C	18°C	17°C
Assimilation of				
Lactose	+	W	+	v
Inositol	+	w/+	+	v
D-arabinose	+	w/–	+	v
Ethanol	w/–	+	+	+
Growth on 50% glucose	w/–	–	+	–
Growth on vitamin-free medium	+	W	+	–
Fermentation of				
Ga lactose	+	–	–	W
Lactose	+	–	–	–
Raffinose	+	–	–	W
Maltose	+	–	–	W

Main physiology test results for characteristics of M. blollopis SK-4 and related species are shown. Physiological data were taken from Fell et al. (1969), Xin and Zhou (2007), Thomas-Hall et al. (2010) and this study. +, positive; w, weak; - , negative; v, variable; nd, no data.

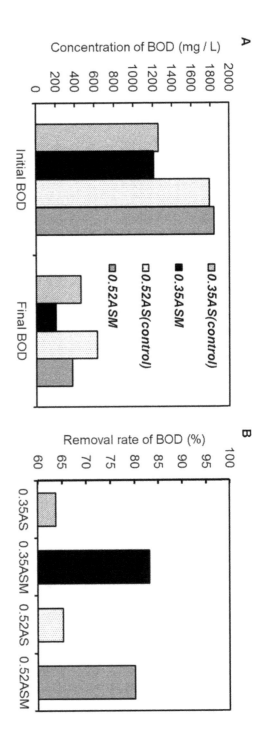

FIGURE 2: Performance of activated sludge system treating model milking parlor wastewater. (A) Results of measurement of initial BOD and final BOD. (B) Effect of BOD removal rate by ASM treatment. (0.35AS) for 0.35 BOD loading rate of model parlor wastewater treated by activated sludge (control), (0.35ASM) for 0.35 BOD loading rate of model parlor wastewater treated by activated sludge containing *M. blollopis* SK-4, (0.52AS) for 0.52 BOD loading rate of model parlor wastewater treated by activated sludge (control) and (0.52ASM) for 0.52 BOD loading rate of model parlor wastewater treated by activated sludge containing *M. blollopis* SK-4. Abbrebiations for Figure 3 are biochemical oxygen demand (BOD), activated sludge (AS) and activated sludge containing *Mrakia blollopis* SK-4 (ASM).

10.3.2 PHYSIOLOGICAL CHARACTERIZATION

Results of assimilation of carbon compounds and other physiological tests of *M. blollopis* SK-4 are shown in Table 1 with its type strain and related species. Test data for *M. blollopis* SK-4 are compared with those for *M. blollopis* CBS8921^T [8], *M. psychrophila* AS2.1971^T [16], and *M. frigida* CBS5270^T [17]. Maximum growth temperature of *M. blollopis* SK-4 was 22°C. Maximum growth temperatures of other related species were lower than 20°C. *M. blollopis* SK-4 differed from the other strains in substrate utilization as well. The strain could thrive well on lactose, D-arabinose, and inositol medium. Unlike other strains, this strain also grew on vitamin-free medium. A comparison of fermentabilities showed that *M. blollopis* SK-4 could ferment typical sugars such as glucose, sucrose, galactose, maltose, lactose, raffinose, trehalose and melibiose, while other related species were not able to strongly ferment such as various sugars (Table 1).

10.3.3 ASSESSMENT OF MILK FAT DECOMPOSITION IN MODEL WASTEWATER

Model wastewater containing cow's milk as a substrate of milk fat was prepared as an equivalent of BOD sludge loading in standard waste water treatment (0.35 kg-BOD/kg-MLSS·day). Activated sludge containing *M. blollopis* SK-4 had a BOD removal rate of 83.1%, higher than that in the control (63.8%, Fig. 2A). When BOD volume load was adjusted to 1.5 fold of standard wastewater treatment (0.52 kg-BOD/Kg-MLSS·day), BOD removal rate by activated sludge containing *M. blollopis* SK-4 was 80.1%, higher that in the control (65.2%, Fig. 2B). Regardless of BOD volume load, activated sludge containing *M. blollopis* SK-4 had a 1.25-fold higher BOD removal rate than that of the control.

10.3.4 PRODUCTION OF LIPASE
FROM MRAKIA BLOLLOPIS SK-4

Many microorganisms are known to produce lipase using Tween 80 as a substrate [18], [19]. Basidiomycetous yeast *M. blollopis* SK-4 also produced the enzyme. It is known that the production of lipase from Candida rugosa increased when yeast extract was used as a nitrogen source [20]. The same results as those for *M. blollopis* SK-4 were obtained. When 0.5% (w/v) yeast extract was used as a nitrogen source in the medium, lipase production by *M. blollopis* SK-4 lipase increased dramatically after 180 h. Maximum lipase activity of 0.695 U/mL was obtained with 0.5% (w/v) yeast extract, as compared to 0.059U/mL in its absence, after 324 h of inoculation, after which the accumulation of lipase markedly decreased (Spp. Fig. S1).

M. blollopis SK-4 morphology was observed by the fluorescence in situ hybridization (FISH) method during secretion of lipase. Therefore, all of the morphology of *M. blollopis* SK-4 during secretion of lipase was yeast form (data not shown).

10.3.5 PURIFICATION OF LIPASE

Lipase production medium was centrifuged at 3000×for 15 min at 4°C. The supernatant was filtered through a 0.45-μm membrane filter. The filtered solution was concentrated by ultrafiltration. After ultrafiltration, 72.3% of total lipase activities were recovered and 2.2-fold lipase specific activity was obtained. Then, enzyme solution was applied on a Toyopearl butyl-650 M column (2.5×20 cm) and purified by single-step hydrophobic interaction chromatography. Finally, 9.4% of the enzyme was recovered and 20.1-fold of specific activity, compared to crude sample, was obtained with a specific activity of 51.7 U/mg (Table 2). The purified enzyme showed a single band on SDS-PAGE with a molecular mass of 60 kDa (Supp. Fig. S2).

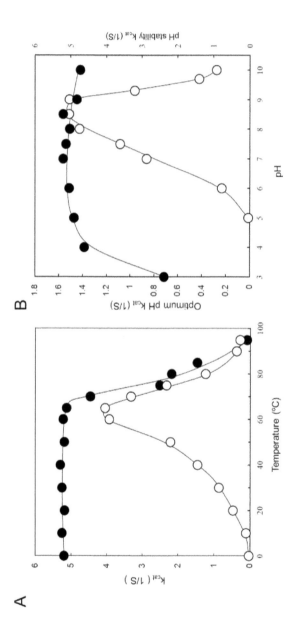

FIGURE 3: Effects of temperature and pH on SK-4 lipase. (A) Effects of temperature on lipase activity (○) and thermostability of the lipase (●). For effect of temperature, lipase activity assay was performed at various temperatures in 50 mM sodium phosphate buffer (pH 7.0) for 30 min. For thermostability, lipase was preincubated at various temperatures for 30 min. Remaining lipase activity was examined at 65°C for 30 min in 50 mM Tris-HCl buffer (pH 8.5). The line was fitted by using the Eyring-Arrhenius equation. (B) Effects of pH on lipase activity (○) and pH stability of lipase (●). For effect of pH, lipase activity assay was performed with various buffers at 30°C for 30 min using p-nitrophenyl-palmitate as a substrate. For pH stability, lipase was preincubated in various pH buffers at 30°C for 15 h and then pH of the buffer was adjusted to 8.5. The remaining enzyme activity was examined at 65°C for 30 min using p-nitrophenyl-palmitate as a substrate. The line was fitted by Henderson- Hasselbalch equation.

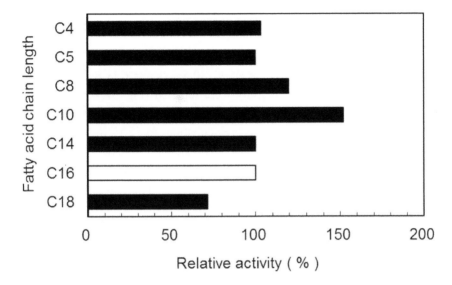

FIGURE 4: Substrate specificity of lipase form *Mrakia blollopis* SK-4. Substrates were p-nitrophenyl esters. Relative activity for each substrate (•) is expressed as the percentage of activity toward p-nitrophenyl-palmitate (□).

TABLE 2: Purification of lipase from *Mrakia blollopis* SK-4.

Purification Step	Total Protein (mg)	Total Activity (U)	Specific Activity (U/mg)	Recovery (%)	Fold
Supernatant of culture medium	180.0	278.0	2.5	100.0	1.0
Ultrafilter concentration	32.2	200.9	6.2	72.3	2.2
Toyopearl Butyl-650M	0.5	26.2	51.7	9.4	20.1

10.3.6 CHARACTERIZATION OF LIPASE

Optimum temperature of lipase activity was 60–65°C (k_{cat} = 3.93 and 4.04 S^{-1}). At temperatures of 80°C and 95°C, 30.5% (k_{cat} = 1.23 S–1) and 6.8% (k_{cat} = 0.27 S^{-1}) of enzyme activity were retained (Fig. 3A). The enzyme showed thermo-stability up to 65°C with 98.3% (k_{cat} = 5.12 S^{-1}) of residual enzyme activity even after 30-min preincubation. At 75°C, 80°C and 85°C, 48.2% (k_{cat} = 2.51 S^{-1}), 41.8% (k_{cat} = 2.18 S^{-1}) and 28.03% (k_{cat} = 1.46 S^{-1}) of enzyme activity remained after 30-min preincubation (Fig. 3A). Optimum pH range of lipase activity was between pH 8.0 and pH 9.0 (k_{cat} = 1.43 and 1.51 S^{-1}), whereas 57.0% (k_{cat} = 0.86 S^{-1}), 63.5% (k_{cat} = 0.96 S^{-1}) and 27.8% (k_{cat} = 0.42 S^{-1}) of the enzyme activity was retained at pH 7.0, pH 9.3 and pH 9.7 compared to 100% at pH 8.5 (Fig. 3B). The enzyme was quite stable over a wide pH range (4.0–10.0) and 45.5% (k_{cat} = 2.38 S^{-1}) of the enzyme activity remained at pH 3.0 even after preincubation for 15 h at 30°C (Fig. 3B). Enzyme activity was affected by metal ions at a concentration of 1 mM, retaining relative activity higher than 80%. There was little inhabitation of lipase activity in the presence of Cu^{2+} and Pb^{2+} ions. The metal-chelating agent EDTA did not affect lipase activity (Table 3). One of the most important characteristics of lipase was substrate specificity toward various p-nitrophenyl esters (C_4–C_{18}). The substrate specificity was determined in the presence of 1 mM p-nitrophenyl esters as a substrate with 50 mM sodium phosphate (pH 7.0) at 30°C for 30 min. Relative activity toward C_4–C_{14} was more than 100% compared to 100% toward p-nitrophenyl-palmitate. C_{18} had relative activity of 71.6% (Fig. 4). Various organic solvents (ethanol, methanol, diethyl ether, dimethyl sulfoxide, hexane and N, N- dimethyl formamide) enhanced SK-4 lipase acidity. Solvents such as acetone and chloroform however, had only a slight inhibitory effect in the activity, with relative activities of 77.9% and 70.4%, respectively (Table 3). Organic solvents are known to be severely toxic to most enzymes including lipase [21].

10.4 DISCUSSION

The Antarctic yeast strain SK-4, which we identified as *M. blollopis*, showed several uniques characteristics; for example, it can use various carbon sources as nutrition, it prefers relatively high temperature conditions among allied species and it can be activated even in vitamin-free conditions. These results suggested that strain SK-4 is a potential candidate for a biological agent to decompose various sugars in milk parlor wastewater under low temperature conditions. This expectation was further reconfirmed by the application of strain SK-4 in AS; namely, addition of SK-4 to AS in the model parlor wastewater improved the BOD removal rate.

TABLE 3: Comparison of the effects of various metal ions, EDTA and various organic solvents on *M. blollops* SK-4 and *Cryptococcus* sp. S-2 lipase activity.

Divalent cation or EDTA or Organic solvents	Relative activity (%)	
	M. blollopis SK-4	*Cryptococcus* sp. S-2
None	100.0	100.0
EDTA	106.7	ND
Metal salts		
$MgCl_2$	102.2	> 70
$MnCl_2$	101.6	> 70
$FeCl_2$	90.0	> 70
$ZnCl_2$	83.6	ND
$CoCl_2$	85.9	ND
$CaCl_2$	112.8	> 70
$PbCl_2$	61.6	ND
$CuCl_2$	67.2	< 70
Ethanol	107.5	ND
Methanol	112.3	53.3
Diethyl ether	113.1	116.7
2- Propanol	95.5	ND
Dimethyl sulfoxide (DMSO)	134.3	121.4
Hexane	104.9	95.2
Acetone	77.9	83.3
Chloroform	70.4	74.8
N, N- dimethylformamide (DMF)	122.3	92.4

Improved BOD removal rate of the activated sludge was attributed to lipase activity of SK-4 added. As expected, the lipase purified from *M. blollopis* SK-4 showed clearly weaker activity at lower temperatures but stronger activity in middle to high temperature conditions compared to activities of those from the other psychrophilic fungi [22]. *M. blollopis* SK-4 lipase, in addition, was quite stable in wide ranges of temperature and pH conditions and was not affected by the existence of EDTA, various metals ions, or organic solvents. *M. blollopis* SK-4 is thought to have acquired stable lipases by growing in extreme environments such as the Antarctica.

Comparison of the lipase from *M. blollopis* SK-4 with that from Cryptococcus sp. S-2, which has actually been used in wastewater treatment [23], [24], revealed that the former was superior to the latter both in thermo-stability and pH stability (Table 4); i.e., the lipase from *M. blollopis* SK-4 retained 71.1% of the enzyme activity after 30 min at 60°C and was stable for 6 hrs at 30°C in the pH range from 5.0 to 9.0.

TABLE 4: Comparison of *Mrakia blollopis* SK-4 lipase characteristics with other yeast lipase characteristics.

Yeasts	MW in KDa	Optimum pH	pH Stability Range	Optimum Temperature
Mrakia blollopsis SK-4	60	8.0~9.0	4.0~10.0	60–65°C
Candida antarctaca	43 (lipaseA)	5.0	–	50°C
Candida rugosa ATCC1483	60	5.0	–	–
Cryptococcus sp. S-2	22	7.0	5.0–9.0	37°C
Geotrichum candidum	44	5.6~7.0	4.2~9.8	40°C
Yarrowia lipolytica	44 (lip)	8.0 (lip)	4.5~8.0	37°C
Kurtzmanomyces sp. L-11	49	1.9~7.2	below 7.1	75°C
Trichosporon asteroids	37	5.0	3.0~10.0	50°C

In conclusion, *M. blollopis* SK-4 has the ability to assimilate various carbon compounds and to use various sugars for fermentation. Moreover, the lipase is more tolerant in relatively higher temperature conditions and wider pH ranges, less sensitive to various metal ions and organic solvents, and highly reactive to various chain lengths of substrates. SK-4 lipase, therefore, is a promising biological agent for parlor wastewater treatment even in low temperature regions of the world.

REFERENCES

1. Healy MG, Rodgers M, Mulqueen J (2007) Treatment of dairy wastewater using constructed wetland and intermittent sand filter. Bioresour Technol. 98: 2268–2281. doi: 10.1016/j.biortech.2006.07.036
2. Shah SB, Bhumbla DK, Basden TJ, Lawrence LD (2002) Cool temperature performance of a wheat straw biofilter for treating dairy wastewater. J Environ Sci Health B. 37: 493–505. doi: 10.1081/pfc-120014879
3. Biddlestone AJ, Gray KR, Job GD (1991) Treatment of dairy farm wastewaters in engineered reed bed systems. Process Biochem. 26, 265–268.
4. Kato K, Ietsugu H, Koba T, Sasaki H, Miyaji N, et al.. (2010) Design and performance of hybrid reed bed systems for treating high content wastewater in the cold climate., 12th International Conference on Wetland Systems for Water Pollution Control. http://reed-net.com/pdf/katoICWS2010.pdf. Accessed 1August 2012.
5. Ying C, Umetsu K, Ihara I, Sakai Y, Yamashiro T (2010) Simultaneous removal of organic matter and nitrogen form milking parlor wastewater by a magnetic activated sludge (MAS) process. Bioresour Technol. 101: 4349–4353. doi: 10.1016/j.biortech.2010.01.087
6. Chevalier P, Proulx D, Lessard P, Vincent WF, de la Noue J (2000) Nitrogen and phosphorus removal by high latitude mat-forming cyanobacteria for potential use in tertiary wastewater treatment. J Appl Phycol. 12: 105–112.
7. Hirayama-Katayama K, Koike Y, Kobayashi K, Hirayama K (2003) Removal of nitrogen by Antarctic yeast cells at low temperature. Polar Bioscience. 16: 43–48.
8. Thomas-Hall SR, Turchetti B, Buzzini P, Branda E, Boekhout T, et al. (2010) Cold-adapted yeasts from Antarctica and Italian alps-description of three novel species: Mrakia robertii sp. nov., Mrakia blollopis sp. nov. and Mrakiella niccombsii sp. nov. Extremophiles. 14: 47–59. doi: 10.1007/s00792-009-0286-7
9. Bab'eva IP, Lisichkina GA, Reshetova IS, Danilevich VN (2002) Mrakia curviuscula sp. nov.: A New psychrophilic yeast from forest substrates. Microbiology. 71: 449–454. doi: 10.1023/a:1019801828620
10. Margesine R, Fell JW (2008) Mrakiella cryoconiti gen. nov., sp. nov., a psychrophilic, anamorphic, basidiomycetous yeast from alpine and arctic habitats. Int J Syst Evol Microbiol. 58: 2977–2982. doi: 10.1099/ijs.0.2008/000836-0
11. Shimohara K, Fujiu S, Tsuji M, Kudoh S, Hoshino T, et al. (2012) Lipolytic activities and their thermal dependence of Mrakia species, basidiomycetous yeast from Antarctica. J water waste. 54: 691–696.
12. Tamura K, Dudley J, Nei M, Kumar S (2007) Molecular Evolutionary Genetics Analysis (MEGA) software version 4.0. Mol Biol Evol. 24; 1596–1599.

13. Yarrow D (1998) Methods for the isolation and identification of yeasts. In: Kurtzman CP, Fell JW (eds) The yeast, a taxonomic study 4th edn,. Elsevier, Amsterdam, pp 77–100.

14. Berekaa MM, Zaghloul TI, Abdel-Fattah YR, Saeed HM, Sifour M (2009) Production of a novel glycerol-inducible lipase from thermophilic Geobacillus steaorothermophilus strain-5. World J Microbiol Biotechnol. 25: 287–294. doi: 10.1007/s11274-008-9891-3

15. Laemmli UK (1970) Cleavage of structural proteins during the assembly of the head of bacteriophage T4. Nature 227: 680–685. doi: 10.1038/227680a0

16. Xin M, Zhou P (2007) Mrakia psychrophila sp. Nov., a new species isolated from Antarctic soil. J Zhejiang Univ-Sc B 8(4): 260–265. doi: 10.1631/jzus.2007.b0260

17. Fell JW, Statzell AC, Hunter IL, Phaff HJ (1969) Leucosporidium gen. n., the heterobasidiomycetous stage of several yeasts of the genus Candida. Anton Leeuw Int J G 35: 433–462. doi: 10.1007/bf02219163

18. Li CY, Cheng CY, Chen TL (2001) Production of Acinetobacter radioresistens lipase using Tween 80 as the carbon source. Enzyme Microb Technol. 29: 258–26318. doi: 10.1016/s0141-0229(01)00396-9

19. Taoka Y, Nagano N, Okita Y, Izumida H, Sugimoto S, et al. (2011) Effect of Tween 80 on the growth, lipid accumulation and fatty acid composition of Thraustochytrium aureum ATCC 34304. J Biosci Bioeng. 111: 420–424. doi: 10.1016/j.jbiosc.2010.12.010

20. Fadilog?lu S, Erkmen O (2002) Effect of carbon and nitrogen sources on lipase production by Candida rugosa. Turk J Eng Env Sci. 26: 249–254.

21. Iizumi T, Nakamura K, Fukase T (1990) Purification and characterization of a thermostable lipase from newly isolated Pseudomonas sp. KWI-56. Agric Biol Chem. 54: 1253–1258. doi: 10.1271/bbb1961.54.1253

22. Hoshino T, Tronsmo AM, Matsumoto N, Sakamoto T, Ohgiya S, et al. (1997) Purification and characterization of a lipolytic enzyme active at low temperature from Norwegian Typhula ishikariensis groupIII strain. Eur J plant Pathol. 103: 357–361.

23. Iefuji H, Iimura Y, Obata T (1994) Isolation and characterization of a yeast Cryptococcus sp. S-2 that produces raw starch-digesting α-amylase, xylanase and polygalacturonase. Biosci Biotechnol Biochem. 58: 2261–2262. doi: 10.1271/bbb.58.2261

24. Kamini NR, Fujii T, Kurosu T, Iefuji H (2000) Production, purification and characterization of an extracellular lipase from the yeast, Cryptococcus sp. S-2. Process Biochem. 36: 317–324. doi: 10.1016/s0032-9592(00)00228-4

There are several supplemental files that are not available in this version of the article. To view this additional information, please use the citation on the first page of this chapter.

CHAPTER 11

Revealing the Factors Influencing a Fermentative Biohydrogen Production Process Using Industrial Wastewater as Fermentation Substrate

IULIAN ZOLTAN BOBOESCU, MARIANA ILIE,
VASILE DANIEL GHERMAN, ION MIREL, BERNADETT PAP,
ADINA NEGREA, ÉVA KONDOROSI, TIBOR BÍRÓ,
AND GERGELY MARÓTI

11.1 BACKGROUND

Our society today increasingly requires more energy to maintain overall ascending economic trends. Although the demand for energy is permanently growing, the reserves of our primary energy carriers will be depleted within a few decades [1]. In addition, our fossil fuel-based economy is dramatically accelerating the process of global warming with severe and permanent consequences for the environment [2]. Therefore, novel and safe energy carriers must be introduced. Hydrogen satisfies all the requirements for a clean and renewable fuel, producing only water as a by-product upon combustion or direct use in fuel-cell technology. Hydrogen has the highest energy content per unit weight of any known fuel (142 kJ/g or 61,000 Btu/lb) and can be transported for domestic/industrial consumption through conventional means [3],[4]. In addition to this, H_2 gas is safer to handle than domestic natural gas, and can be used directly in internal combustion engines or in fuel cells to generate electricity [3]. Hydrogen use in fuel cells is inherently more efficient than the combustion currently required for the conversion of other potential fuels to mechanical energy [5]. However, most hydrogen is currently produced by conventional chemical or electrolytic methods, which require high amounts of energy and expensive technologies [6].

In the last few years, attention has shifted towards novel and less energy-intensive technologies for producing hydrogen [7]-[9]. Among the various hydrogen production processes, the biological methods (direct and indirect photolysis, photo-fermentation, and dark fermentation) appear to be the most promising [10]-[12]. In addition, certain methods of producing biological hydrogen such as dark fermentation can utilize various organic wastes as a substrate for fermentative hydrogen production, thus coupling organic waste treatment with renewable energy generation [13]-[16]. Recently, reducing the cost of wastewater treatment and finding ways to produce useful products from wastewater has been gaining importance with regard to environmental sustainability. One way to address both issues is to simultaneously generate bioenergy in the form of hydrogen by utilizing the organic matter present in wastewater [17]. In addition, certain types of wastewater generated by various industrial processes are considered ideal substrates because they contain high levels of easily degradable organic material [18].

The complete oxidation of glucose could yield a theoretical maximum of 12 moles of H_2 per mole of glucose, but in this case no energy can be utilized to support the growth and metabolism of the hydrogen-producing organism [7],[19],[20]. Strictly anaerobic hydrogen-producing microbes are able to generate a maximum of 4 moles of H_2 per mole of degraded glucose [21]-[23]. Fermentative H_2 production using complex microbial communities therefore has the advantage of a high hydrogen production rate utilizing complex organic wastes as fermentation substrates with limited amounts of additional external energy input [24]. Bacteria and other microbes capable of hydrogen production exist widely in natural environments rich in organic nutrients such as soil, wastewater sludge, and compost [25]-[27]. Microbial populations sampled from these habitats can thus be used as cheap and highly efficient inocula for fermentative hydrogen production. In addition, dark hydrogen production processes using mixed microbial cultures as starting inocula are more efficient than those using pure cultures. The reason is that mixed cultures represent more simple systems to operate which are easier to control, and may be able to degrade a broader range of feedstock [28]. The use of mixed microbial cultures as starting inocula also allows the use of unsterile fermentation substrates, such as most types of wastewater. However, in a fermentative hydrogen production process using mixed cultures, the hydrogen produced by hydrogen-evolving bacteria can be consumed by hydrogen-consuming bacteria [29],[30]. Strategies for pretreatment of the parent inocula are therefore required to restrict or even terminate the methanogenic process to ensure that H_2 remains the end product in the metabolic flow [31]-[33].

One of the major impediments in developing biohydrogen (bioH2) production processes for commercialization is low hydrogen production yield. The biological production rate of hydrogen and its molar yield, like most other bioprocesses, are dependent on several parameters including the activity rate of hydrogen-producing and -consuming bacteria, substrates, inorganic nutrients, and operational conditions of the bioreactor, among other factors [34]. Identifying these influencing factors and optimizing the fermentation conditions, particularly the nutritional and environmental parameters, is thus of primary importance in the bioprocess development. The most widely used screening and optimization strategy is the design of experiment (DOE) method, by which certain factors are selected and deliberately varied in a controlled manner, in order to study their effects, facilitate process comprehension, and even to improve performance [35]-[43]. The use of such DOE methods for process optimization in fermentative hydrogen production processes is critical due to the dynamic and complex nature of these systems.

In the present study, a central composite experimental design was used to investigate the influence of the process variables involved in batch-mode biohydrogen production, as well as to optimize their response. The experiments were performed using a defined mixed microbial consortium as the starting inoculum and wastewater obtained from a beer-brewing factory as the fermentation substrate. High-throughput metagenomic microbial community assessments as well as analysis of substrate degradation rates were performed during the experiments to understand the fermentation mechanisms involved. The results of this study have the potential to propel us closer to achieving and optimizing an industrial-scale system which will serve the dual purpose of providing wastewater pretreatment coupled with biohydrogen production.

11.2 RESULTS AND DISCUSSION

Batch experiments were conducted using mixed microbial consortia able to degrade complex organic substrates like wastewater in addition to simultaneous biohydrogen production. The involvement of biological, chemical, and physical factors influencing the fermentative biohydrogen production process as well as the interactions of these parameters were assessed by experiments designed by statistics-based methods and metagenomic monitoring of this complex ecosystem.

11.2.1 EFFECT OF VARIOUS INFLUENCING FACTORS AND THE INTERACTIONS BETWEEN THEM ON THE BIOHYDROGEN PRODUCTION PROCESS

To assess the influence of various factors on the biohydrogen production process, as well as the level of interaction between these factors, a cybernetic representation of the dark biohydrogen fermentation process was developed, with an input-output structure (Figure 1). The input data are defined as influencing factors (IF) while the system output data are defined as objective functions (OF). By developing and applying this experimental modeling approach, connecting relationships (most frequently with a predefined polynomial shape) between the IF and the OF (in this case, H_2 production rates) can be identified for an experimental domain previously defined as being of interest. The data obtained by applying this experimental design strategy can be used to generate predictions regarding the behavior of the system under investigation. They can also provide deep insight into the degree of influence of the investigated variables on the objective functions. In addition, by applying advanced statistical approaches, the optimum region of a system for a specific response can be identified (in this case, biohydrogen production rate).

Because of the steep curve generated by the response surface approach during the investigation of an optimal area within a specific system, using linear methods to adequately identify the stationary point becomes obsolete. Therefore, depending on the number of investigated variables, further additional experimental points are required. One of the most recommended approaches to address this issue is the central composite experimental design method. By applying this strategy, one can use the information obtained through a first-order mathematical model, completing these insights by consecutively extending the experimental approach to explain a second-order mathematical model. Thus, a central composite design (CCD) consists of a standard first-order design with a 2^k or 2^{k-p} orthogonal factorial point, where k is the number of selected factors and p the number of interactions replaced with influencing factors. 2^k axial points are displayed in a "star pattern" at distance α from the center of the design with n_0 center points. Based on preliminary screening experiments, three IF were selected; fermentation temperature, starting pH value, and degradable substrate availability (glucose addition; Table 1). Each IF was defined between two physical values, low and high, coded as -1 and +1 respectively, with a center value coded as 0.

TABLE 1: Physical and coded values of the variables used in the central composite design

Coded symbol	Variables	Values of coded levels				
		(-1.28)	(-1)	(0)	(+1)	(+1.28)
X1	Operating temperature (°C)	23.28	25	31	37	38.72
X2	Initial value of fermentation pH	4.56	4.8	5.65	6.5	6.74
X3	Glucose addition (g/L)	3.56	5	10	15	16.44

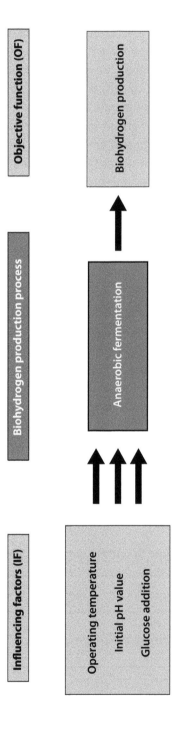

FIGURE 1: Cybernetic representation of the investigated dark biohydrogen fermentation process with an input-output structure. The anaerobic fermentation of the wastewater is considered a black-box system influenced by different variables, with consequences on the objective function (in this case, $bioH_2$ production).

The relationship between the physical and coded values is given by the following equation:

$$x_{jcod} = \frac{x_{jphys} - x_{j0phys}}{l_{jphys}}, j = 1,2,3 \qquad (1)$$

where:

- x_{jcod} is the coded value of the j factor
- x_{jphys} is the physical value of the j factor
- x_{j0phys} is the central level of the j factor
- I_{jphys} is the variation interval for the j factor; double this value gives the variation range D (Equation 2)

$$D = x_{jphys\ high} - x_{jphys\ low} = 2l_{jphys} \qquad (2)$$

The factorial portion of the CCD is a complete 2^3 factorial with eight runs, which contains all the possible combinations within the defined levels of the investigated variables (runs 1-8; Table 2). The additional experimental runs (9-14) represent supplementary axial points displayed in a "star pattern" around the center of the design, at distance α of 1.287 from the center. The design is completed with n_0=two observations at the experimental center (runs 15 and 16). Because of the different nature of the experimental design approaches (runs 1-8 and runs 9-14 together with runs 15 and 16), we will separately discuss the results obtained for these two main experimental groups. Biohydrogen production was monitored every 24 h for the duration of all experiments.

Even though the same starting microbial inocula and the same wastewater were used, notable differences in biohydrogen production rates were observed for the factorial portion of the CCD (runs 1-8; Figure 2A). This is the first indication that the selected variables manifest, beyond doubt, a clear influence on the OF in the investigated system. The highest biohydrogen production rates observed during the factorial portion of the CCD were measured during experimental run 4 (a total production of 25.2 mL

H_2 was measured at the end of the experiment), followed closely by experimental run 3 (a total production of 22.3 mL H_2 was measured at the end of the experiment). The other six experimental runs generated considerably smaller amounts of H_2, ranging from 14.6 mL (experimental run 7) to 8.2 mL (experimental run 5) H_2 at the end of the experiment (Figure 2A). Strong similarities in the H_2 production rates were observed between similar experimental conditions (Table 2). This was particularly true for most of the cases, where only the glucose addition variable (X_3) differed.

TABLE 2: Central composite experimental design matrix of the three investigated variables, with the total measured H_2 production for each of the experimental runs

Run number	Variable			Response
	X_1	X_2	X_3	Total hydrogen production mean (mL)
1	-1	-1	-1	8.66
2	-1	-1	1	9.62
3	-1	1	-1	22.34
4	-1	1	1	25.23
5	1	-1	-1	8.28
6	1	-1	1	8.35
7	1	1	-1	14.67
8	1	1	1	11.92
9	-1.28	0	0	17.67
10	1.28	0	0	10.43
11	0	-1.28	0	2.97
12	0	1.28	0	27.18
13	0	0	-1.28	14.95
14	0	0	1.28	25.92
15	0	0	0	21.08
16	0	0	0	21.66

Notable differences in volumetric biohydrogen production were also observed for the supplementary star and central points (experimental runs 9-16; Figure 2B). The highest biohydrogen production rates were obtained for experimental runs 12 and 14 (a total H_2 production of 27.1 mL and

25.9 mL was measured at the end of the experiments respectively), while the lowest biohydrogen production rate was obtained for experimental run 11 (a total H_2 production of 2.9 mL was measured at the end of the experiment). Regarding the fermentation conditions of the highest hydrogen-producing experimental run (12), the fermentation temperature (X_1) as well as the glucose addition (X_3) values were situated in the center of the experimental model, while the initial pH (X_2) was fixed at a value of 6.74, representing a distance of 1.287 from the experimental center (Table 2). These preliminary findings suggest that the initial value of the environmental pH has a strong influence on the biohydrogen production rate.

TABLE 3: ANOVA test performed on the second-order polynomial model used to discriminate between the significant linear (L) and quadratic (Q) effects of the investigated variables, as well as their interactions, for the investigated system and response (biohydrogen production)

	Sum of squares (SS)	Degree of freedom	Mean square	F-value	P-value
X1-L	270.935*	1*	270.935*	11.63704*	0.001725*
X1-Q	260.138*	1*	260.138*	11.17333*	0.002075*
X2-L	1314.436*	1*	1314.436*	56.45696*	0.000000*
X2-Q	185.342*	1*	185.342*	7.96069*	0.008033*
X3-L	61.992	1	61.992	2.66265	0.112237
X3-Q	0.236	1	0.236	0.01013	0.920420
X1X2	140.261*	1*	140.261*	6.02441*	0.019553*
X1X3	15.980	1	15.980	0.68634	0.413363
X2X3	0.291	1	0.291	0.01250	0.911647
Lack of fit	273.054	5	54.611	2.34561	0.062915
Pure error	768.309	33	23.282	-	-
Total SS	3290.973	47	-	-	-

Values considered statistically significant.

The analysis of variance (ANOVA) test performed on the central composite experimental design showed that both linear and quadratic effects of the temperature and pH, as well as their first-order interactions, are statistically significant parameters with regard to the OF (Table 3). However, the main influence of initial glucose addition (X_3), its quadratic effect, and first-order interactions on biohydrogen production are regarded as statistically insignificant. The lack of fit has a P-value slightly higher than $\alpha = 0.05$, which makes it marginally significant (meaning that it is significant at the 0.10 α - level).

The estimated effects of the IF on the OF together with their order of magnitude were revealed through additional data analysis (Figure 3A). The largest effect on biohydrogen production for the investigated system was caused by a switch in initial pH value from lower to higher values. Fermentation temperature also had a large effect on the OF, followed by the interaction between the fermentation temperature (X_1) and initial pH value (X_2). The interaction plot for X_1X_2 indicates a considerable increase in biohydrogen production at the point when initial pH values move from low to high levels, and temperature is low (Figure 3B). This demonstrates that understanding the interactions among the various process parameters in complex systems is crucial for developing and optimizing an efficient biohydrogen production process.

By analyzing the evolution of the main effects throughout the experimental process (120 h), an explicit shift in the direction and intensity of the effect of these variables on the biohydrogen production rate was detected (Figure 4). The main effects of initial pH (X_2) and temperature (X_1) tended to increase over time, while the effects of other variables, like glucose addition (X_3), tended to maintain a relatively constant value. These findings emphasize even more the dynamic characteristics of such a complex environment. Thus, the initial environmental conditions as well as the fermentation process settings greatly influenced the development of microbiological processes, and thereby the metabolic end products, in the investigated system. Because of the differences in the direction and intensity of the effects that investigated variables exerted on the objective functions, multiple optimal situations may be identified depending on the point selected in the fermentation development process. Consequently, application of these findings to an industrial-scale continuous biohydrogen fermentation system is crucial for establishing an optimum process development strategy with regard to influent feeding of the system among other factors. These findings particularly support the necessity for a complete exploration of the effects manifested over time on the biohydrogen production process by the different influencing factors.

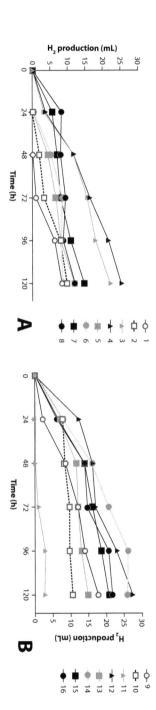

FIGURE 2: Biohydrogen evolution measured during the full factorial (A) and additional orthogonal central composite (B) multifactorial experiments. Clear differences can be observed between the different experimental lines, suggesting a strong influence of the investigated variables on the OF.

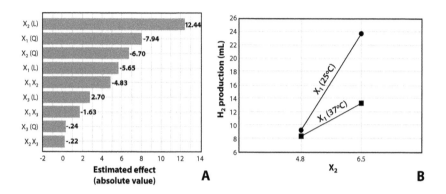

FIGURE 3: The effects and the interactions of the investigated variables on the biohydrogen production process. Pareto histogram depicting the estimated linear (L) and quadratic (Q) effects of each of the analyzed independent variables (in decreasing order of magnitude) on the biohydrogen production rate (A). Analysis of the effect of the interactions between the X_1 and X_2 variables on biohydrogen production rate (B).

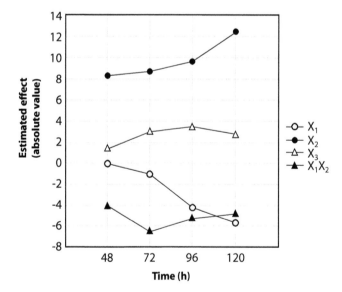

FIGURE 4: The evolution of the main effects of the investigated variables and their interactions on the biohydrogen production rate. The three investigated variables as well as the statistically significant interactions between the operating temperature and the initial pH are represented.

These crucial insights show that in complex biotechnological systems, both the effect of a particular variable at a specific time on the OF and the global effect of multiple influencing factors together with their different interactions must be considered. In addition, the evolution of the main effects of the IF throughout the biohydrogen fermentation process must be considered. Using these insights, an optimized industrial-scale biohydrogen production system can be successfully designed and operated in a feasible economic context.

11.2.2 IDENTIFICATION OF THE OPTIMUM CONDITIONS FOR BIOHYDROGEN PRODUCTION IN THE INVESTIGATED SYSTEM

Once the estimated effects on biohydrogen production rates and their magnitudes were established for each of the analyzed independent variables as well as for their interactions, a strategy to identify the optimum conditions of the investigated system to maximize hydrogen production rates was developed. To understand the specific hydrogen production conditions and to confirm the validity of the statistical experimental strategies applied, a response surface and contour plot methodology was developed. Prediction of H2 production at any tested parameter within the range of the applied experimental design is achieved by employing a second-order polynomial regression equation obtained from experimental data (Equation 3 and Figure 5). With respect to the coded factor levels, the second-order model used to fit the experimental data is:

$$Y = 20.95 - 2.83X_1 + 6.22X_2 + 1.35X_3 - 2.42X_1X_2 - 0.82X_1X_3 - 0.11X_2X_3 - 3.97X_1{}^2 - 3.35X_2{}^2 - 0.12X_3{}^2 \tag{3}$$

with 60% of the variation in biohydrogen production explained by the model ($R_{adj}{}^2 = 0.60$).

When considering the observed versus predicted values of biohydrogen production, it can be assumed that the statistical model developed is reasonably accurate. The response surface and contour plot analysis was developed for an initial glucose concentration of 10 g/L (Figure 6). The stationary point was found to be outside the investigated experimental domain. The canonical analysis of the response surface indicates a possible rising ridge, since two of the canonical coefficients are close to zero ($\lambda_1 = -0.002995$; $\lambda_2 = -0.099786$; $\lambda_3 = -4.65529$). In this type of ridge system, inferences about the true surface of the stationary point cannot be drawn because it is outside the region where the model has been fitted. The contour plots for three levels of X_3 (initial glucose concentration; low, medium, and high) display an optimum around $X_1 = 27°C$, $X_2 = 6.7$, which also indicates that the influence of X_3 on biohydrogen production in the investigated experimental domain is limited. Therefore, if the initial glucose addition is pooled out from the model (according to Equation 4), considering that it was found statistically insignificant in the investigated domain, the stationary point is found at $X_1 = 26.7°C$ and $X_2 = 6.66$, with a predicted optimal value of biohydrogen production of 25.57 mL with a ±95% confidence interval.

$$Y = -343.22 + 9.05X_1 + 74.42X_2 - 0.47X_1X_2 - 0.11X_1^2 - 4.64X_2^2 \qquad (4)$$

where the coefficients are calculated with uncoded (natural) factor values.

The stationary point was verified by carrying out experiments in triplicate. The mean biohydrogen production obtained was 28 mL, and this value was within the ±95% confidence interval for the predicted maximum. Therefore, the new empirical model, with a determination coefficient of $R_{adj}^2 = 0.62$ and from which all the terms containing the X_3 independent variable have been pooled out, was able to predict the behavior of the system within the experimental domain (Figure 7).

11.2.3 WASTEWATER DEGRADATION

During the dark fermentative biohydrogen production process, organic matter is converted from complex long-chain molecules to simple compounds.

By using wastewater as a fermentative organic substrate, at least partial biodegradation of this waste can therefore be achieved. Mass spectrometry (MS) was used to evaluate the wastewater biodegradation efficiency. Several organic compounds including lactose, glucose, acetic acid, propionic acid, and furfurol, among others, were identified and monitored during the wastewater degradation experiments. The results allowed direct comparison of the experimental runs by revealing the different degradation rates and metabolic pathways followed by the microbial communities involved.

The analysis of three different experimental situations, one with low hydrogen production (experimental run 11), one with medium hydrogen production (experimental run 13), and one with high hydrogen production (experimental run 14), revealed significant differences in the wastewater composition at the end of the biohydrogen fermentation process (Figure 8). As expected, comparison of low and high hydrogen-producing experimental runs revealed a general decrease in concentration of most of the measured components in the fermented wastewater with increasing hydrogen production. This tendency was especially marked for lactose, glucose, capric acid, maltose, lactic acid, furfurol, and caproic acid (Figure 8). These observations suggest that most of the macromolecules in the wastewater are metabolized, with biodegradation efficiency clearly correlating with the amount of hydrogen generated. Some components (glucose, maltose, and furfurol) were fully consumed during fermentation, suggesting that these compounds represent easily accessible energy sources which drive the hydrogen-producing fermentation pathways. At the same time, accumulation of propionic acid and galactose was observed concomitant with the increasing hydrogen yield (Figure 8).

The clear differences between experimental runs observed during the biohydrogen-producing wastewater degradation experiments indicate a strong correlation between the consumption of the organic constituents and hydrogen production rates. This suggests that the technological process developed here has the potential to generate biohydrogen using wastewater as an organic substrate.

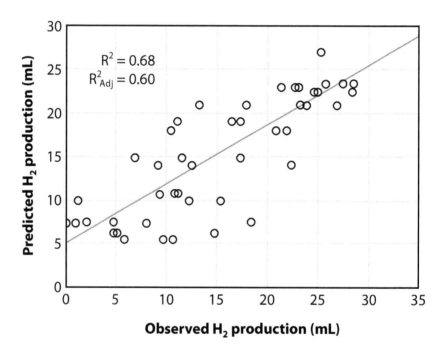

FIGURE 5: Modeling the biohydrogen production using beer-brewing wastewater as the fermentation substrate. Observed versus predicted values of the biohydrogen production process using a complex microbial consortium as the starting inoculum and beer-brewing wastewater as the fermentation substrate.

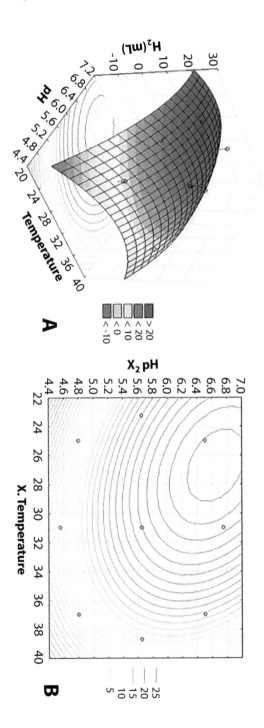

FIGURE 6: Prediction of the optimum area for the highest biohydrogen production yields. Response surface (A) and contour plot (B) analysis of H_2 production as a function of temperature and pH, with a constant glucose addition value of 10 g/L.

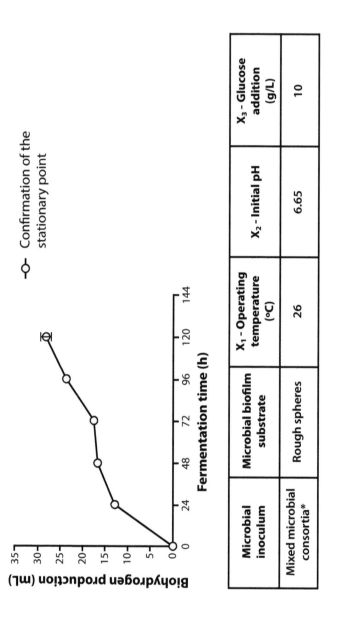

Microbial inoculum	Microbial biofilm substrate	X₁ - Operating temperature (°C)	X₂ - Initial pH	X₃ - Glucose addition (g/L)
Mixed microbial consortia*	Rough spheres	26	6.65	10

FIGURE 7: Confirmation of the predicted stationary point for the investigated system. The mean of the maximum biohydrogen production yield of 28 mL measured at the end of the confirmation experiments is within the ±95% confidence interval for the predicted optimum area. *The microbial population used as a starting inoculum during the wastewater degradation experiments was subjected to heat pretreatment prior to the inoculation.

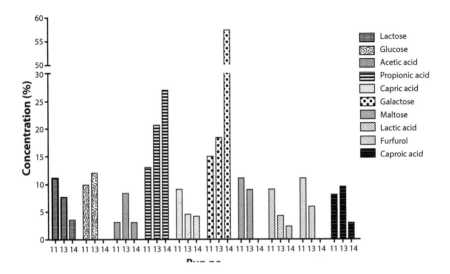

FIGURE 8: Wastewater chemical composition at the end of the fermentation period. Comparison between experimental runs which generated low (experimental run 11), medium (experimental run 13), and high (experimental run 14) hydrogen production rates.

11.2.4 ASSESSMENT OF MICROBIAL COMMUNITY COMPOSITION

The same mixed microbial consortium was used as a starting inoculum in all experimental runs and stages. This consortium was designed according to our previous research on the ability of various complex consortia to degrade complex organic substrates and simultaneously produce biohydrogen [44]. An organic nutrient-rich habitat (generated during the biological denitrification step at a communal wastewater treatment plant) was sampled in order to isolate the required microbial inoculum. To further increase the biohydrogen-generating abilities of the population, heat pretreatment was applied. As a result, the majority of the spore-forming hydrogen-producing bacteria survived the treatment, while the hydrogen-consuming methanogens were largely eliminated.

To better understand the hydrogen-producing dark fermentation processes, a sequencing-based metagenomic analysis was carried out on samples selected from different experimental runs. This approach allows for a detailed taxonomic and functional characterization of the selected eco-system as a function of time, which is essential when planning strategies for dark fermentation-based biohydrogen production using wastewater as a fermentative substrate.

To fully understand the experimental model, the complete factorial portion of the CCD was subjected to metagenomic analysis. Additional samples from the center as well as axial experimental points were ana-lyzed metagenomically, with the microbial composition of samples with the lowest and highest H_2-producing capabilities being investigated. This approach provided valuable insight into the transitions and structural rear-rangements occurring in the microbial communities under different fer-mentation conditions. The influence of these rearrangements on the OF (biohydrogen production rates) was also elucidated.

The metagenomes extracted from the full factorial experimental de-sign approach (experimental runs 1-8) revealed similarities in the micro-bial populations collected from the different experimental setups (Figure 9). Two major clusters formed at the genus level. Cluster A contained experimental runs 2, 3, 4, 6, and 7, while cluster B contained experimen-tal runs 1, 5, and 8. In addition, both clusters could be further divided into smaller units. Most of the cluster A microbial communities devel-oped under similar experimental conditions with regard to the initial pH value (X_2) and glucose addition (X_3). A notable difference between these clustered groups was the fermentation temperature (X_1). It seems that in the temperature range investigated (25 to 37°C) the microbial com-position in each experimental run was similar at the same levels of X_2 and X_3, regardless of X_1 levels. Interestingly, the microbial communities located at the closest Bray-Curtis distance (experimental runs 5 and 8), which are isolated from the experimental runs in cluster B, developed under the same temperature conditions (37°C) and different levels of X_2 and X_3. However, even in this cluster, the microbial community isolated from experimental run 1 developed at the same levels of X_2 and X_3 but at levels of X_1 that differed from the community in experimental run 5. These findings suggest a strong correlation between the microenviron-mental conditions inside the bioreactor and the different developmental pathways followed by the microbial community.

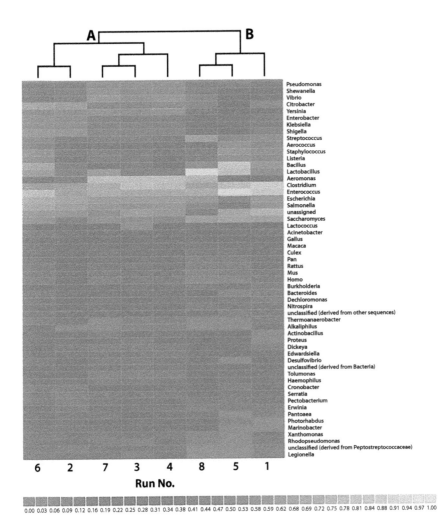

FIGURE 9: Partial heatmap calculated for samples from the full factorial experimental design approach (experimental runs 1-8). The heatmap was redrawn using normalized values clustered by genus using a Bray-Curtis distance metric (divided into clusters A and B for practical reasons).

The cluster formed by samples from experimental runs 3, 4, and 7 displayed the highest H_2 production potential (22.24 mL, 25.23 mL, and 14.67 mL, respectively). Certain key microorganisms identified in this cluster, such as *Aeromonas* spp., *Clostridium* spp., *Thermoanaerobacter* spp., and *Alkaliphilus* spp., were generally more abundant in experimental runs 3 and 4, while *Bacillus* spp. and *Lactobacillus* spp. were poorly represented in these runs (Figure 9).

Samples from the supplementary points (star and central points) in the experimental matrix (experimental runs 9-16) were also analyzed using metagenomics, with samples producing the lowest and highest amounts of H_2 being compared (Figure 2B and Figure 10). Strong correlations were observed between the structural changes within these microbial communities and their biohydrogen-producing potential. The microbial community with the lowest H_2 production capacity was dominated by *Lactobacillus* spp. and *Tetragenococcus* spp. (Figure 10A). The microbial consortium which produced moderate quantities of H_2 was dominated by *Citrobacter* spp. and *Aeromonas* spp. (Figure 10B). The microbial population which produced the highest amounts of H_2 was clearly dominated by *Clostridium* spp. (Figure 10C). A clear shift from a *Lactobacillus* spp.-dominated microbial population towards a *Clostridium* spp.-dominated population concomitant with increasing H_2 production rates was thus observed. Interestingly, the only factor that varied between these three experimental runs was the initial pH value (X_2; Table 2). Therefore, changing only one IF in the investigated system can result in significant changes in the OF (the biohydrogen production rate).

The results obtained from metagenomic investigations have led to a better understanding of the impact of each physicochemical parameter and the interactions between them, all of which are essential for the fermentative biohydrogen production process. To maximize the biohydrogen production potential of a microbial consortium during wastewater degradation, special emphasis has to be made not only on the careful selection and pretreatment of these populations but also on the optimal fermentation conditions for each system.

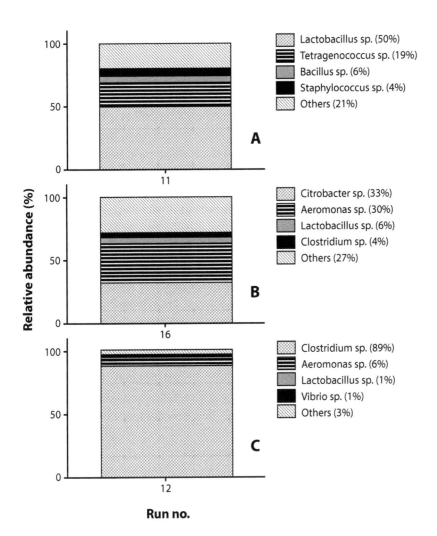

FIGURE 10: Microbial population shifts during the biohydrogen production experiments. Microbial community composition of samples from experimental runs 11 (A), 16 (B), and 12 (C), part of the central composite fractional factorial experimental design approach.

11.3 CONCLUSIONS

A central composite experimental design was used to underline the involvement of the various biological, chemical, and physical factors influencing the fermentative biohydrogen production process. The experiments were performed using a mixed microbial consortium as the starting inoculum and beer-brewing wastewater as the fermentation substrate.

It was demonstrated that the selected variables have a clear influence on the objective function in the investigated system. Both linear and quadratic effects of the fermentation temperature and initial pH, as well as their first-order interactions, were statistically significant with regard to the biohydrogen production rates for the system considered. The largest effect was caused by a change in initial pH value from its lower to its higher level. The fermentation temperature also had a strong effect on the objective function, followed by the interaction between the fermentation temperature (X_1) and initial pH value (X_2). Analysis of the evolution of these main effects over time during the experiment revealed a strong shift in the direction and intensity of the effect of these variables on the biohydrogen production rate. Because of this, further optimal situations can be identified depending on the point in the fermentation process being investigated. These crucial insights show that this complex biotechnological system is governed by the combined effect of several influencing factors as well as by their interactions with each other.

We successfully used response surface and contour plot methodology to understand the specific hydrogen production conditions of the system under study, and to confirm the validity of the statistical experimental strategies applied. In addition, confirmation experiments yielded a mean biohydrogen production of 28 mL, a value situated within the $\pm95\%$ confidence interval of the predicted maximum. The new empirical model was thus able to predict the system behavior within the experimental domain.

The mass spectrometry analyses performed on the degraded wastewater during each biohydrogen production experimental run revealed significant differences in the biochemical composition between runs, confirming a strong correlation between the consumption of most of the organic constituents and a high level of hydrogen generation. This indicates the potential of the developed technological process to generate biohydrogen by using wastewater as an organic substrate.

To better understand the microbiological factors driving the dark fermentative biohydrogen production processes in the system studied, high-throughput next generation metagenomic analyses were carried out on samples taken from different experimental runs. These analyses revealed strong correlations between the micro-environmental conditions inside the bioreactors and the developmental pathways taken by the microbial communities, even though the same microbial consortium was used as a starting inoculum in all experimental runs. By analyzing the metagenomes of the lowest and highest H_2-producing samples, population shifts in hydrogen production potential within the microbial consortia triggered by the different fermentation conditions can be traced. These results have led to a better understanding of the impact which a mixed microbial consortium has on the fermentative biohydrogen production process, and at the same time, the influence of the factors involved on the development of these populations over time.

Together the data generated by the present study reveal a strong interconnection between the investigated variables, as well as a permanent and shifting influence on the biohydrogen production process under the investigated conditions. Only by understanding these phenomena can an optimized industrial-scale biohydrogen production system be successfully designed and operated in a feasible economic context. This will bring us one step closer to a clean, fossil fuel-free future and a developed H_2-based economy.

11.4 METHODS

11.4.1 SEED INOCULA

Biological samples were collected from the denitrification step at a municipal sewage wastewater treatment plant. This ecosystem has a high microbial biodiversity consisting of naturally formed populations of microflora suitable for biodegradation of complex organic substrates. Once collected, the samples were stored at 4°C until inoculation.

Prior to inoculating the bioreactors, microbial samples were subjected to a heat pretreatment process at 70°C for one hour as described in our previous work [44]. The purpose of the pre-treatment strategy is to enrich with spore-forming hydrogen-producing microbes as well as to reduce the abundance of hydrogen-consuming microorganisms.

11.4.2 BIOREACTOR DESIGN AND OPERATION

The experimental setup was conducted using wastewater generated by the beer-brewing industry as the fermentation substrate in different experimental combinations as detailed in Table 2. The composition of the wastewater was: total COD—6558 mg/L, soluble COD—5066 mg/L, total N—75.8 mg/L, and total P—58.4 mg/L. The batch-mode experiments were conducted in 100-mL serum vials with 50 mL of wastewater and 10 mL pretreated sediment samples as a starting inoculum. A granular biofilm support material (sterilized river-bed rocks) was introduced into the bioreactors to increase the contact surface area of the microbial populations. The bottles were capped with rubber septum stoppers and aluminum rings. Incubation was performed at varying temperature conditions for a period of 120 h (Table 2). During the fermentation period, response parameters including biogas production and composition, metabolite concentration, substrate degradation, and total microbial community composition were monitored as described below. All batch experiments were performed in triplicate.

11.4.3 STATISTICAL EXPERIMENTAL DESIGN METHODS

A central composite experimental design approach was used to assess the IF and their effect on the OF (biohydrogen production rate). In the early stage of the experiments, a statistically based full factorial experimental design approach was utilized to determine the influence of the factors under study (fermentation temperature, starting pH value, and glucose addition) on the response (H_2 production; Table 1). Use of this strategy helped to avoid overlap of different effects and interactions among these variables. The screened factors were chosen based on our previous work and were tested at low, medium, and high levels, coded as -1, 0, and +1 (Table 1) [44]. The factorial portion of the design is a complete 2^3 factorial with eight runs, which contains all the possible combinations within the defined levels of the investigated variables (runs 1-8; Table 2). In addition, a second statistically based factorial experimental design method was applied in order to obtain a complete understanding of the investigated process. These experimental runs (9- 14) represent additional axial points displayed in a "star pattern" around the center of the design, at α distance of 1.287 from the center, a value that ensures its orthogonality. The design also contains two observations at the experimental center (runs 15 and 16; Table 2). This approach provided N_0 - 1 degrees of freedom in estimating the experimental error, and at the same time established the estimation precision of the OF (biohydrogen production rate) around the central experimental point. The biohydrogen production rate was monitored every 24 h for the duration of the experiments. Experimental design approaches were developed and analyzed using the Statistica 8 software suite (StatSoft Inc., USA).

11.4.4 ANALYTICAL METHODS

The quantity and composition of the biogas produced was directly measured by gas chromatography in 24-h intervals using an Agilent Technologies 7890A GC system equipped with a thermal conductivity detector and argon as a carrier gas. The temperatures of the injector, detector, and column were kept at 30°C, 200°C, and 230°C, respectively. An HP PLOTQ column (15 m × 530 mm × 40 mm) was used. Since a concentration gradient of H2 gas can form in the headspace, gas samples (0.5 mL) were taken out after mixing of the headspace gas by sparging several times with a gas-tight syringe.

Microbial degradation of wastewater was monitored at the end of the experimental runs by mass spectrometry conducted on a High Capacity Ion Trap Ultra mass spectrometer (HCT Ultra, PTM discovery; Bruker Daltonics, Bremen, Germany). All mass spectra were acquired in the mass range 100 to 3000 m/z, with a scan speed of 2.1 scans/s. Tandem mass spectrometry was carried out by collision-induced dissociation (CID) using He as the collision gas. For MS/MS sequencing, precursor ions were selected within an isolation width of 2 μm. The fully automated process was performed using a NanoMate 400 robot incorporating ESI chip technology (Advion BioSciences, Ithaca, USA) couplet on a High Capacity Ion Trap Ultra mass spectrometer (HCT Ultra, PTM discovery; Bruker Daltonics, Bremen, Germany). The robot was controlled and manipulated by ChipSoft software operated under the Windows system. The position of the electrospray chip was adjusted to the sampling cone potential to give rise to an optimal transfer of the ionic species into the mass spectrometer. To avoid contamination, a glass-coated microtiter plate was used in all experiments. Five microliter aliquots of working sample solutions were loaded onto a 96-well plate. The robot was programmed to aspirate the whole sample volume into the pipette tip followed by 2 μL of air, and then to deliver the sample to the inlet side of the microchip. Each nozzle had an internal diameter of 2.5 μm, and under the given conditions delivered a flow rate of approximately 200 nL/min. The nano-ESI process was initiated by applying voltages of 1.5 to 1.8 kV and a head pressure of 0.5 to 0.7 PSI. After spray initialization, infusion parameters (ESI voltage in the pipette tip, voltage, and desolvation gas flow) were optimized.

Values for ESI capillary, cone potential, and desolvation gas (nitrogen) were optimized to achieve an efficient ionization and to produce the optimum transfer of ions during MS. Measurement parameters were: capillary voltage of 1 kV, contraelectrode voltage (cone voltage) of 60 V, acquisition time 2 min, scan speed 2.1 scans/s, and a mass range of 100 to 3000 m/z. The NanoMate HCT MS system was tuned to operate in the positive ion mode. This technique was chosen because glucide ionization is highly efficient in this mode. The source block maintained at a constant temperature of 80°C provided optimal desolvation of the generated droplets without the need for desolvation gas. To prevent any cross-contamination or carry-over, the pipette tip was ejected and replaced after every sample infusion and MS analysis. All mass spectra were processed using Data Analysis 3.4 software (Bruker Daltonik, Bremen, Germany). The mass spectra were calibrated using sodium iodide. Accurate determination of the average mass was 20 ppm. Samples were dissolved in methanol at a concentration of approximately 5 pmol/μL. At the acquisition time of 2 min, the required volume of sample was approximately 2 pmol, a value indicating a highly sensitive analysis.

11.4.5 TOTAL DNA EXTRACTION FROM SAMPLES

DNA from the complex samples was extracted and purified according to described methods with some modifications [45]. Samples (0.5 g) were extracted with 1.3 mL extraction buffer (100 mM Tris-Cl pH 8.0, 100 mM EDTA pH 8.0, 1.5 M NaCl, 100 mM sodium phosphate pH 8.0, 1% CTAB). After thorough mixing, 7 μL of proteinase K (20.2 mg/mL) was added. After incubation for 45 min, 160 μL 20% SDS was added and mixed by inversion several times with further incubation at 60°C for 1 h with intermittent shaking every 15 min. Samples were centrifuged at 13,000 RPM for 5 min, and the supernatant was transferred into new Eppendorf tubes. The remaining soil pellets were treated three times with 400 μL extraction buffer and 60 μL SDS (20%) and kept at 60°C for 15 min with intermittent shaking every 5 min. Supernatants collected from all four extractions were mixed with an equal quantity of chloroform and isoamyl alcohol (25:24:1). The aqueous layer was separated and precipitated with 0.7 vol isopropanol. After centrifugation at 13,000 RPM for 15 min, the brown pellets were washed with 70% ethanol, dried at room temperature, and dissolved in TE (10 mM Tris-Cl, 1 mM EDTA, pH 8.0).

11.4.6 METAGENOMIC CHARACTERIZATION OF MICROBIAL COMMUNITIES

The total DNA from selected samples was prepared for high-throughput next generation sequencing analysis performed on the Ion Torrent PGM platform (Life Technologies). An average of 291.322 sequencing reads were generated for each sample, with a mean read length of 161 nucleotides. Bioinformatic analyses (taxonomic profiling and assessment of metabolic potential) were conducted using the public MG-RAST software package, which is a modified version of RAST (Rapid Annotations based on Subsystem Technology) [46]. The sequence data were compared to

M5NR using a maximum e-value of 1×10^{-5}, a minimum identity of 95%, and a minimum alignment length of 15, measured in amino acids for proteins and base pairs for RNA databases.

REFERENCES

1.	Matsakas L, Kekos D, Loizidou M, Christakopoulos P: Utilization of household food waste for the production of ethanol at high dry material content. Biotechnol Biofuels 2014, 7:4.
2.	Dziugan P, Balcerek M, Pielech-Przybylska K, Patelski P: Evaluation of the fermentation of high gravity thick sugar beet juice worts for efficient bioethanol production. Biotechnol Biofuels 2013, 6:158.
3.	Das D: Advances in biohydrogen production processes: an approach towards commercialization. Int J Hydrog Energy 2009, 34:7349-7357.
4.	Sevilla M, Mokaya R: Energy storage applications of activated carbons: supercapacitors and hydrogen storage. Energy Environ Sci 2014, 7:1250-1280.
5.	Levin DB, Chahine R: Challenges for renewable hydrogen production from biomass. Int J Hydrog Energy 2010, 35:4962-4969.
6.	De Vrije T, Budde MA, Lips SJ, Bakker RR, Mars AE, Claassen PA: Hydrogen production from carrot pulp by the extreme thermophiles Caldicellulosiruptor saccharolyticus and Thermotoga neapolitana. Int J Hydrog Energy 2010, 35:13206-13213.
7.	Ghosh D, Bisaillon A, Hallenbeck P: Increasing the metabolic capacity of Escherichia coli for hydrogen production through heterologous expression of the Ralstonia eutropha SH operon. Biotechnol Biofuels 2013, 6:122.
8.	Sarma SJ, Brar SK, Le Bihan Y, Buelna G, Soccol CR: Mitigation of the inhibitory effect of soap by magnesium salt treatment of crude glycerol - a novel approach for enhanced biohydrogen production from the biodiesel industry waste. Bioresour Technol 2014, 151:49-53.
9.	Li W-W, Yu H-Q, He Z: Towards sustainable wastewater treatment by using microbial fuel cells-centered technologies. Energy Environ Sci 2014, 7:911-924.
10.	Motte J-C, Trably E, Escudie R, Hamelin J, Steyer J-P, Bernet N, Delgenes J-P, Dumas C: Total solids content: a key parameter of metabolic pathways in dry anaerobic digestion. Biotechnol Biofuels 2013, 6:164.
11.	Roy S, Kumar K, Ghosh S, Das D: Thermophilic biohydrogen production using pretreated algal biomass as substrate. Biomass Bioenergy 2014, 61:157-166.
12.	Lin X, Xia Y, Yan Q, Shen W, Zhao M: Acid tolerance response (ATR) of microbial communities during the enhanced biohydrogen process via cascade acid stress. Bioresour Technol 2014, 155:98-103.
13.	Wicher E, Seifert K, Zagrodnik R, Pietrzyk B, Laniecki M: Hydrogen gas production from distillery wastewater by dark fermentation. Int J Hydrog Energy 2013, 38:7767-7773.

14. Fernández C, Carracedo B, Martínez EJ, Gómez X, Morán A: Application of a packed bed reactor for the production of hydrogen from cheese whey permeate: effect of organic loading rate. J Environ Sci Health A 2014, 49:210-217.
15. Mahmoud N, Tawfik A, Ookawara S, Suzuki M: Biological hydrogen production from starch wastewater using a novel up-flow anaerobic staged reactor. BioResources 2013, 8:4951-4968.
16. Deng L, Huang B, Zhao HY, Du JY, Gao L: Study on modified PVA-H3BO3 immobilization microorganism method for hydrogen production from wastewater. Adv Mater Res 2013, 634:280-285.
17. Sekoai PT, Gueguim Kana EB: Semi-pilot scale production of hydrogen from Organic Fraction of Solid Municipal Waste and electricity generation from process effluents. Biomass Bioenergy 2014, 60:156-163.
18. Qin Z, Qin Q, Yang Y: Continuous biohydrogen production with CSTR reactor under high organic loading rate condition. Adv Mater Res 2014, 864:225-228.
19. McKinlay JB, Oda Y, Rühl M, Posto AL, Sauer U, Harwood CS: Non-growing Rhodopseudomonas palustris increases the hydrogen gas yield from acetate by shifting from the glyoxylate shunt to the tricarboxylic acid cycle. J Biol Chem 2014, 289:1960-1970.
20. Cai G, Jin B, Monis P, Saint C: A genetic and metabolic approach to redirection of biochemical pathways of Clostridium butyricum for enhancing hydrogen production. Biotechnol Bioeng 2013, 110:338-342.
21. Cha M, Chung D, Elkins J, Guss A, Westpheling J: Metabolic engineering of Caldicellulosiruptor bescii yields increased hydrogen production from lignocellulosic biomass. Biotechnol Biofuels 2013, 6:85.
22. Maru BT, Bielen AAM, Constanti M, Medina F, Kengen SWM: Glycerol fermentation to hydrogen by Thermotoga maritima: proposed pathway and bioenergetic considerations. Int J Hydrog Energy 2013, 38:5563-5572.
23. Kumaraswamy GK, Guerra T, Qian X, Zhang S, Bryant DA, Dismukes GC: Reprogramming the glycolytic pathway for increased hydrogen production in cyanobacteria: metabolic engineering of NAD+-dependent GAPDH. Energy Environ Sci 2013, 6:3722-3731.
24. Varrone C, Van Nostrand JD, Liu W, Zhou B, Wang Z, Liu F, He Z, Wu L, Zhou J, Wang A: Metagenomic-based analysis of biofilm communities for electrohydrogenesis: from wastewater to hydrogen. Int J Hydrog Energy 2014, 39:4222-4233.
25. Abreu A, Karakashev D, Angelidaki I, Sousa D, Alves M: Biohydrogen production from arabinose and glucose using extreme thermophilic anaerobic mixed cultures. Biotechnol Biofuels 2012, 5:6.
26. Rafrafi Y, Trably E, Hamelin J, Latrille E, Meynial-Salles I, Benomar S, Giudici-Orticoni M-T, Steyer J-P: Sub-dominant bacteria as keystone species in microbial communities producing bio-hydrogen. Int J Hydrog Energy 2013, 38:4975-4985.
27. Ning Y-Y, Wang S-F, Jin D-W, Harada H, Shi X-Y: Formation of hydrogen-producing granules and microbial community analysis in a UASB reactor. Renew Energy 2013, 53:12-17.
28. Gadhe A, Sonawane SS, Varma MN: Kinetic analysis of biohydrogen production from complex dairy wastewater under optimized condition. Int J Hydrog Energy 2014, 39:1306-1314.

29. Masset J, Calusinska M, Hamilton C, Hiligsmann S, Joris B, Wilmotte A, Thonart P: Fermentative hydrogen production from glucose and starch using pure strains and artificial co-cultures of *Clostridium* spp. Biotechnol Biofuels 2012, 5:35.

30. Gadow SI, Jiang H, Hojo T, Li Y-Y: Cellulosic hydrogen production and microbial community characterization in hyper-thermophilic continuous bioreactor. Int J Hydrog Energy 2013, 38:7259-7267.

31. Park J-H, Lee S-H, Yoon J-J, Kim S-H, Park H-D: Predominance of cluster *Clostridium* in hydrogen fermentation of galactose seeded with various heat-treated anaerobic sludges. Bioresour Technol 2014, 157:98-106.

32. Penteado ED, Lazaro CZ, Sakamoto IK, Zaiat M: Influence of seed sludge and pretreatment method on hydrogen production in packed-bed anaerobic reactors. Int J Hydrog Energy 2013, 38:6137-6145.

33. El-Bery H, Tawfik A, Kumari S, Bux F: Effect of thermal pre-treatment on inoculum sludge to enhance bio-hydrogen production from alkali hydrolysed rice straw in a mesophilic anaerobic baffled reactor. Environ Technol 2013, 34:1965-1972.

34. Veeravalli SS, Chaganti SR, Lalman JA, Heath DD: Optimizing hydrogen production from a switchgrass steam exploded liquor using a mixed anaerobic culture in an upflow anaerobic sludge blanket reactor. Int J Hydrog Energy 2014, 39:3160-3175.

35. Guo L, Zhao J, She Z, Lu M, Zong Y: Statistical key factors optimization of conditions for hydrogen production from S-TE (Solubilization by thermophilic enzyme) waste sludge. Bioresour Technol 2013, 137:51-56.

36. Gadhe A, Sonawane SS, Varma MN: Optimization of conditions for hydrogen production from complex dairy wastewater by anaerobic sludge using desirability function approach. Int J Hydrog Energy 2013, 38:6607-6617.

37. Whiteman J, Kana EG: Comparative assessment of the artificial neural network and response surface modelling efficiencies for biohydrogen production on sugar cane molasses. BioEnergy Res 2013, 7:295-305.

38. Xiao Y, Zhang X, Zhu M, Tan W: Effect of the culture media optimization, pH and temperature on the biohydrogen production and the hydrogenase activities by Klebsiella pneumoniae ECU-15. Bioresour Technol 2013, 137:9-17.

39. Wu X, Lin H, Zhu J: Optimization of continuous hydrogen production from cofermenting molasses with liquid swine manure in an anaerobic sequencing batch reactor. Bioresour Technol 2013, 136:351-359.

40. Mullai P, Yogeswari M, Sridevi K: Optimisation and enhancement of biohydrogen production using nickel nanoparticles - a novel approach. Bioresour Technol 2013, 141:212-219.

41. Pakshirajan K, Mal J: Biohydrogen production using native carbon monoxide converting anaerobic microbial consortium predominantly Petrobacter sp. Int J Hydrog Energy 2013, 38:16020-16028.

42. Boonsayompoo O, Reungsang A: Thermophilic biohydrogen production from the enzymatic hydrolysate of cellulose fraction of sweet sorghum bagasse by Thermoanaerobacterium thermosaccharolyticum KKU19: optimization of media composition. Int J Hydrog Energy 2013, 38:15777-15786.

43. Jollet V, Gissane C, Schlaf M: Optimization of the neutralization of Red Mud by pyrolysis bio-oil using a design of experiments approach. Energy Environ Sci 2014, 7:1125-1133.

44. Boboescu IZ, Gherman VD, Mirel I, Pap B, Tengölics R, Rákhely G, Kovács KL, Kondorosi É, Maróti G: Simultaneous biohydrogen production and wastewater treatment based on the selective enrichment of the fermentation ecosystem. Int J Hydrog Energy 2014, 39:1502-1510.

45. Sharma PK, Capalash N, Kaur J: An improved method for single step purification of metagenomic DNA. Mol Biotechnol 2007, 36:61-63.

46. Meyer F, Paarmann D, D'Souza M, Olson R, Glass EM, Kubal M, Paczian T, Rodriguez A, Stevens R, Wilke A, Wilkening J, Edwards RA: The metagenomics RAST server – a public resource for the automatic phylogenetic and functional analysis of metagenomes. BMC Bioinformatics 2008, 9:386.

CHAPTER 12

Microbial Communities from Different Types of Natural Wastewater Treatment Systems: Vertical and Horizontal Flow Constructed Wetlands and Biofilters

BÁRBARA ADRADOS, OLGA SÁNCHEZ,
CARLOS ALBERTO ARIAS, ELOY BÉCARES, LAURA GARRIDO,
JORDI MAS, HANS BRIX, and JORDI MORATÓ

12.1 INTRODUCTION

Natural wastewater treatment systems such as constructed wetlands, biological sand filters and other decentralised solutions are becoming an increasingly relevant alternative to conventional systems when treating wastewater from small communities and dwellings due to its efficiency, low establishment costs and low operation and management requirements. In order to treat wastewater effectively, several factors have to be taken into account, e.g. the system's capacity, the plant species used, colonization characteristics of certain microbial groups, and the interactions of biogenic compounds and particular contaminants (wastewater components) with the filter bed material (Stottmeister et al., 2003). Although filtration

Microbial Communities from Different Types of Natural Wastewater Treatment Systems: Vertical and Horizontal Flow Constructed Wetlands and Biofilters. Adrados B, Sánchez O, Arias CA, Becares E, Garrido L, Mas J, Brix H, Morató J. Water Research **55** *(2014). doi:10.1016/j.watres.2014.02.011. Reprinted with permission from the authors. Reproduced with permission of ELSEVIER BV.*

is considered an important process in these removal mechanisms, additional interactions occur among media, plants and water. Many processes and relations between them take place: microbial-mediated processes, chemical networks, volatilization, sedimentation, sorption, photodegradation, plant uptake, transpiration flux and accretion (Kadlec and Wallace, 2009). The importance of microbial processes has been further studied as many reactions are microbiologically mediated (Stottmeister et al., 2003 and Kadlec and Wallace, 2009).

The most stable microbiota in these systems is found in the biofilm associated to the plant's roots and/or attached to the surface of the filter bed material. This complex microbial community created by interactions with wastewater, is mainly responsible for the degradation performance of the system (Sleytr et al., 2009). Furthermore, the diversity of microorganisms in this environment may be critical for its proper functioning and maintenance (Ibekwe et al., 2003). To improve the design of these systems, a detailed knowledge of the structure of these communities should be acquired in order to understand the biological processes that are taking place within them (Truu et al., 2009 and Dong and Reddy, 2010). Recently, several studies have characterized microbial populations in laboratory scale units, sand filters and full scale constructed wetlands under specific conditions (Ragusa et al., 2004, Vacca et al., 2005, Baptista et al., 2008, Calheiros et al., 2009, Krasnits et al., 2009, Sleytr et al., 2009, Zhang et al., 2010 and Dong and Reddy, 2010). However, there is a general lack of information on the diversity and changes of the microbial communities in long-term operation systems treating domestic wastewater at real time scale (Krastnits et al., 2009).

Increased removal efficiency of nitrogen from wastewater is one of the key issues for further development of constructed wetlands and other decentralised technologies. The diversity of microorganisms involved in the N-cycle is expected to be high in these systems. In fact, previous studies have suggested that archaeal nitrifiers, denitrifying fungi, aerobic denitrifying bacteria and heterotrophic nitrifying microorganisms may play an important role in nitrogen transformations in constructed wetlands (Truu et al., 2005). Most importantly, the effects of biofilms on nitrogen transformation and removal have not been adequately studied and modelled. As microorganisms affect processes like nitrification, denitrification, uptake, and sedimentation, they have to be taken into consideration when modelling the transformation and removal of nitrogen from wastewater (Mayo and Bigambo, 2005). Thus, a first step for establishing the role of biological communities in N-removal in constructed wetlands is to evaluate the diversity of microorganisms under different conditions and systems. With this purpose recent studies have introduced the characterization of bacterial communities by means of molecular methods based on 16S rRNA gene analysis (Sleytr et al., 2009).

The aim of this study was to compare the composition of microbial communities of three different types of domestic wastewater treatment systems used in Denmark: Horizontal Flow Constructed Wetlands (HFCW), Vertical Flow Constructed Wetlands (VFCW) and Biofilters (BF, with combined configurations of vertical or horizontal flow) using the PCR-DGGE based method. The systems were composed of different bed filling media, namely soil, sand and LWA (lightweight aggregate). In this work, we enlarged the microbial analysis by analyzing both the bacterial and archaeal populations, focussing in the possible influence of the water influent composition, the design and the bed filling of the treatment systems in the structure of these microbial communities.

12.2 MATERIAL AND METHODS

12.2.1 SITE DESCRIPTION

The wastewater treatment systems (WWTS) investigated were rural facilities used in Denmark for the treatment of domestic wastewaters. All the systems were built following Danish guidelines and comply with Danish wastewater discharge standards (for details see Brix and Arias, 2005). The layout of all the studied systems included a primary treatment step, using a sedimentation tank with a hydraulic residence time proportional to the number of people served and a minimum of 2 m^3. The second treatment step differs depending on the system chosen by the users among an array of technical possibilities approved by the Danish EPA.

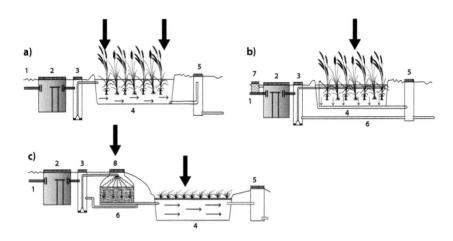

FIGURE 1: Schemes of the three types of systems studied; a) HFCW, b) VFCW, c) BF. 1) Inlet, 2) sedimentation tank, 3) pumping well, 4) bed, 5) outlet well, 6) recycling, 7) P removal system, 8) LWA dome biofilters. Arrows indicate the sampling sites of each system.

Three types of systems were selected for the study: two horizontal flow constructed wetlands (HFCW) with soil beds, two vertical flow constructed wetlands (VFCW) with sand bed and two LWA Biofilters (BF) fitted with a Filtralite-P® bed for the removal of phosphorous (Jenssen et al., 2010). The systems differed in flow configuration, operational and bed media characteristics.

The HFCWs studied have been operational for over 20 years. The systems were built following national guidelines (Miljøministeriet Miljøstyrelsen, 1990) and were composed of two soil filled beds operating in parallel with the necessary structures for distribution and collection of domestic water. After the treatment water was discharged to nearby watercourses (for details see Brix et al., 2010).

VFCWs were also built following the Danish design and construction guidelines (Miljøministeriet Miljøstyrelsen 2004). The domestic wastewater was pre-treated in a sedimentation tank; after that, water was loaded sequentially on the system surface at a rate of approx. 20 pulses/d to an unsaturated bed filled with sand, where it was homogeneously distributed in the surface trickling vertically. Once the water percolated through a one meter deep bed, it was collected at the bottom and evacuated. In order to improve the water quality, and enhance denitrification capacity, treated water was recycled back to the pumping well in one of the two systems studied, where conditions should favour the process (for details see Brix and Arias, 2005).

BFs are media filled systems that combine unsaturated conditions and a water saturated bed. The first section of the system operates unsaturated; it is housed in a fibreglass dome filled with a lightweight aggregate (LWA) from which wastewater is pumped at a rate of around 25 pulses/day. The second step of the treatment system involves the flow of water through a saturated bed filled with Filtralite-P® media, which is an LWA product chemically enriched, specifically engineered for phosphorus removal (see details in Jenssen et al., 2010). Different wastewater treatment systems studied are shown in Fig. 1 and their operational and design characteristics are shown in Table 1.

TABLE 1: Description of the systems evaluated. The averages of nitrification and total nitrogen removal percentages are based on six month sampling (n = 9).

Location	System	Area (m²)	PE served	Recycling	Years of operation	Hydraulic conditions	NH4–n (%)	Total N (%)	BOD$_5$ (mg/l)
Bjød-strup	HFCW1	470	80	No	>20	Saturated	60	64	103
Moes-gaard	HFCW2	520	80	No	>20	Saturated	23	34	–
Friland 1	VFCW1	90	30	Yes	1	Unsaturated	99	84	169
Tisset	VFCW2	15	2	No	4	Unsaturated	99	21	240
Friland 2	BF1	50	4	No	6	Both	59	44	290
Janne	BF2	50	6	Yes	6	Both	91	85	280

The flow conditions within the systems control the oxygen availability and therefore, anoxic conditions predominated in saturated HFCWs while oxic conditions prevail in VFCWs (Vymazal et al., 2006 and Brix and Arias, 2005).

On the other hand, because of the combination of two different modules, oxic conditions are found in the first section of BF systems, while anoxic conditions develop in the P removal bed.

12.2.2 SOIL AND WATER SAMPLING

Soil samples were taken in May 2010 from each system (Figs. 1 and 2), the two HFCW (HFCW 1 and HFCW 2), the two BF (BF 1 and BF2) and the two VFCW (1 and 2). When sampling HFCW, because of the horizontal flow, two separated zones were differentiated and samples were taken at the influent (I) and effluent (E) zone, and considered separately. In the case of BF, samples were also taken in two different parts of the system: in the first module (also represented as I) and in the main bed (E). Sampling points are shown with arrows in Fig. 1.

Three subsamples were collected in each sampling point at random by means of a core (1 m length, 2.54 cm diameter) and then mixed to yield one composite sample per point. Samples were stored at 4 °C, and processed within 24 h.

Grab water samples from influent and effluent were taken in three sampling campaigns, once a month between March and May 2010. Each campaign consisted of three consecutive sampling days. Samples were frozen at −20 °C until they were processed.

12.2.3 WATER ANALYSIS

The water quality parameters measured included in situ measurements of water temperature, oxygen saturation and electric conductivity as standard water control by means of calibrated electrodes. Additional water quality analysis included BOD5 determination using APHA5210B method, and nitrogen species such as total nitrogen (Kjeldhal Method), ammonia (APHA 4500 NH_3 D method), nitrite (APHA 4500 NO_2 B method) and nitrate (APHA 4500-NO_3^- F method).

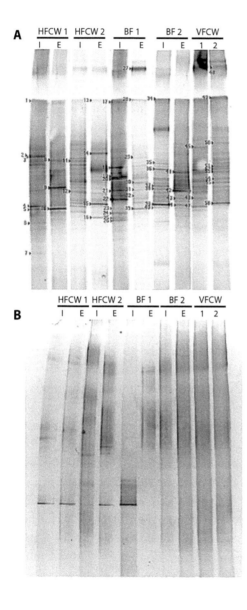

FIGURE 2: Negative images of DGGE gels with PCR products amplified with bacterial (A) and archaeal (B) primer sets from samples of the different systems: HFCW (Horizontal Flow Constructed Wetlands), BF (Biofilters) and VFCW (Vertical Flow Constructed Wetlands); 1 and 2 are replicates from each system; when applied, I: Influent zone, E: Effluent zone.

12.2.4 SOIL DNA EXTRACTION

A total of 100 g for each composite sample were collected in 100 ml of sterile saline solution (9% NaCl) and sonicated for 5 min in an ultrasonic water bath (Selecta, Barcelona, Spain). Samples were also vortexed 1 min to release the biofilm attached to the solution into the liquid phase. Subsequently, 10 ml were recovered and concentrated by centrifugation (5 min, 8000 g), and then samples were stored at −20 °C until further processing. DNA extractions were performed using the EZNA® Soil DNA kit (Omega Bio-Tek, Doraville, USA) following the manufacturer's recommendations.

12.2.5 PCR AMPLIFICATION, DGGE AND SEQUENCING OF 16S RRNA GENES

Amplification of 16S rRNA gene fragments for DGGE analysis was performed by using the bacterial specific primer set 358F with a 40 bp GC clamp, and the universal primer 907RM (Sánchez et al., 2007). Polymerase chain reaction (PCR) was carried out with a Biometra thermocycler using the following program: initial denaturation at 94 °C for 5 min, 10 touchdown cycles of denaturation (at 94 °C for 1 min), annealing (at 63.5 °C–53.5 °C for 1 min, decreasing 1 °C each cycle), and extension at 72 °C for 3 min. This procedure was followed by 20 additional cycles at an annealing temperature of 53.5 °C. During the last cycle of the program, the length of the extension step was 15 min at 72 °C.

Primers 344F-GC and 915R were used for archaeal 16S rRNA gene fragment amplification (Casamayor et al., 2002). The PCR protocol included an initial denaturation step at 94 °C for 5 min, followed by 20 touchdown cycles of denaturation (at 94 °C for 1 min), annealing (at 71 °C–61 °C for 1 min, decreasing 1 °C each cycle), and extension (at 72 °C for 3 min); 20 standard cycles (annealing at 55 °C, 1 min) and a final extension at 72 °C for 5 min.

PCR mixtures contained 1–10 ng of template DNA, each deoxynucleoside triphosphate at a concentration of 200 µM, 1.5 mM $MgCl_2$, each primer at a concentration of 0.3 µM, 2.5 U Taq DNA polymerase (Invitrogen) and PCR buffer supplied by the manufacturer. Bovine Serum Albumin (BSA) at a final concentration of 600 µg ml⁻¹ was added to minimize the inhibitory effect of humic substances (Kreader, 1996). The volume of reactions was 50 µl. PCR products were verified and quantified by agarose gel electrophoresis, with a low DNA mass ladder standard (Invitrogen).

The DGGE was run in a DCode system (Bio-Rad) as described by Muyzer et al. (1998). A 6% polyacrylamide gel with a gradient of 40–80% DNA denaturant agent was cast by mixing solutions of 0% and 80% denaturant agent (100% denaturant agent is 7 M urea and 40% deionized formamide). Seven hundred ng of PCR product were loaded for each sample and the gels were run at 100 V for 18 h at 60 °C in 1x TAE buffer (40 mM Tris [pH 7.4], 20 mM sodium acetate, 1 mM EDTA). The gel was stained with SybrGold (Molecular Probes) for 45 min, rinsed with 1x TAE buffer, removed from the glass plate to a UV-transparent gel scoop, and visualized with UV in a Chemi Doc EQ (Bio-Rad). Prominent bands were excised from the gels, resuspended in milli-q water overnight and reamplified for their sequencing.

Purification of PCR products from DGGE bands and sequencing reactions were performed by Macrogen (South Korea) with primer 907RM for Bacteria and primer 915R for Archaea. PCR products of the reamplified bands were used as DNA template in a sequencing reaction with the Big Dye Terminator version 3.1 sequencing kit in an automatic ABI 3730XL Analyzer-96 capillary type. Sequences were subjected to a BLAST search (Altschul et al., 1997) to obtain an indication of the phylogenetic affiliation.

Fifty-six 16S bacterial rRNA gene sequences were submitted to the EMBL database (http://www.ebi.ac.uk/embl) and received the following accession numbers: from HE716787 to HE716842.

12.2.6 ANALYSIS OF DGGE PATTERNS AND STATISTICAL ANALYSES

Digitalized DGGE images were analysed with the Quantity One software (Bio-Rad, Hercules, USA). Bands occupying the same position in the different lanes of the gels were identified. A matrix was constructed for all lanes, taking into account the presence or absence of the individual bands. Raup-Crick index was used for absence-presence data as this index utilizes a randomization procedure (Monte Carlo) comparing the observed number of species occurring in both samples in 200 pairs of random replicates of the pooled sample. The PAST program (Hammer et al 2008) was used for theses analyses.

DGGE banding data were used to calculate the Shannon–Weaver index as a measure of the diversity of microbial communities. It was calculated using the following function:

$$H' = -\sum_{i=1}^{i=n} p_i \ln p_i$$

where n is the number of bands in the sample and p_1 the relative intensity of the band.

12.3 RESULTS AND DISCUSSION

The aim of this study was to investigate the factors affecting the structure of prokaryotic communities established in three different types of natural wastewater treatment systems, each with different substrate and configuration. Analysis of bacterial and archaeal community composition from the substrate samples collected was performed by means of PCR-DGGE. The banding patterns for the 16S rRNA gene DGGE-PCR amplicons are presented in Fig. 2 for Bacteria and Archaea. Clear differences could be observed in both gels concerning band position, intensity and band number for the different samples, demonstrating that different bacterial and archaeal communities developed in the different systems.

In the bacterial DGGE, a high number of bands could be observed in all lanes (Fig. 2A). Band richness fluctuated from 31 in HFCW1I to 17 in the BF1E system (Table 2). Significant differences were found in total band richness among the influents and effluents ($p < 0.05$), influents harbouring higher richness than effluents (27 and 21 mean band richness for influents and effluents respectively). Similar results were found for Shannon diversity indexes (2.65 and 2.25 for influents and effluents respectively). On the other hand, although archaeal amplification was also found, the DGGE banding profile clearly revealed a lower diversity in comparison with the bacterial community (Fig. 2B).

TABLE 2: Shannon diversity index (H) and band richness calculated for each sample from bacterial data.

System	H	Band richness
HFCW 1 I	2.84	31
HFCW 1 E	2.83	24
HFCW 2 I	2.96	26
HFCW 2 E	2.33	18
BF 1 I	2.27	26
BF 1 E	2.02	17
BF 2 I	2.51	26
BF 2 E	1.81	23
VFCW 1	2.32	25
VFCW 2	2.24	27

Excision of prominent bacterial DGGE bands and subsequent sequencing allowed the characterization of the predominant microorganisms in the different systems studied. Informative sequences were obtained from 56 bacterial bands. The number of bases used to calculate each similarity value is also shown in Table 3, as an indication of the quality of the sequence. Unfortunately, bands recovered from the archaeal DGGE gel yielded sequences with a very poor quality that have not been included in this study. The most represented taxonomic groups in all samples belonged to the γ-Proteobacteria (26% of recovered bands) and Bacteroidetes (26%). Firmicutes (15%) were present in all systems with the exception of samples from VFCW. Members of the Actinobacteria group, although found in HFCW and VFCW, seemed to be more abundant in BF systems. Finally, some representatives of α, β and δ-Proteobacteria, Acidobacteria and Chloroflexi were also retrieved in some of the samples.

Most of the sequences corresponded to uncultured microorganisms (71% of the retrieved sequences), while others matched with a high percentage of similarity to cultured bacteria (29%). In general, typical bacteria from soil and wastewater environments were found in all the systems analyzed. For example, we could retrieve in HFCW typical soil bacteria such as sequences related to *Acinetobacter* sp. (γ-Proteobacteria), *Arthrobacter* sp (from the Actinobacteria group, also found in samples from VFCW and BF), and *Bacillus* sp. (Firmicutes), all of them potential denitrifying bacteria. Besides, other non-culturable matches corresponding to different groups were present. *Acinetobacter* sp. is commonly present in activated sludge (Snaidr el al. 1997) especially in those where enhanced biological phosphate removal is observed (Ivanov et al., 2005). On the other hand, *Arthrobacter* sp has been related to the nitrogen cycle, particularly to nitrogen fixation (Cacciari et al., 1971). The fact that some aerobic microorganisms have been found suggests that although HFCW systems are mostly all the time saturated, enough oxygen is present to allow proliferation of these microbial groups, with the subsequent possibility of nitrification in the system. Oxygen is present probably due to plant aeration and also because the upper part of the bed normally remains unsaturated.

Concerning the Bacteroidetes phylum, a group of chemoheterotrophic bacteria known by its ability to degrade complex organic matter, sequences with a high similarity at the species level were found. Thus, some of the retrieved sequences related to *Flavobacterium* sp., another potential denitrifying bacteria, and have been detected in VFCW and BF; it is a typical genus that can be found in activated sludge (Park et al., 2007). Another sequence similar to the denitrifying *Thauera terpenica* (cultured closest match 99.6% similarity) was also observed in VFCWs. Other species were also found in BF systems, such as sequences related to the γ-Proteobacteria *Xanthomonas* sp., *Dokdonella* sp., and some denitrifying bacteria such as *Rhodanobacter* sp. and *Stenotrophomonas* sp.

The application of molecular techniques (PCR-DGGE profiling) on different wastewater treatment systems has allowed the identification of some players and their potential role in the nitrogen removal processes. The diversity of N-cycling bacteria found in the analyzed systems is an indicator of the multiple possibilities of biological nitrogen transformations inside them. In addition, this profiling method is a useful tool to classify microbial community under different substrates by clustering and diversity analyses.

A cluster analysis of bacterial DGGE banding patterns based in band richness is shown in Fig. 3. Samples separated in two clusters; samples coming from VFCW and BFI, corresponding to unsaturated samples with a high organic load (Table 1), clustered together in one of the two main clusters, while all the other samples, corresponding to saturated systems with low organic load, clustered in another group. As there is almost no relation between the influent and effluent bacterial communities inside the same wetland, these results suggest that factors other than the influent wastewater, such as the organic load and the design of the treatment system, contribute to shape the microbial community.

Previous studies have shown that shifts in the structure of bacterial communities can be associated with changes in a number of soil properties, including soil texture and soil nitrogen availability (Dong and Reddy, 2010). The substrate is an important component since it supports plant growth (in case of planted wetland systems), as well as the establishment of a microbial biofilm, and it influences the hydraulic processes (Stottmeister et al., 2003). A porous matrix substrate such as LWA will probably favour the development of biofilms. Additionally, recent studies concluded that the type of substrate is one of the main factors influencing bacterial communities (Vacca et al., 2005 and Calheiros et al., 2009). However, none of these studies took place in real constructed wetlands; both of them consisted in different pilot systems, with the same influent water. In our study, no relation between the microbial assemblage and the substrate was found, as different communities were retrieved within systems with the same substrate. On the contrary, from the cluster analysis we did observe two separated groups that appeared to be influenced by factors such as the organic load, as well as for the absence/presence of oxygen, since one of the groups is composed only by samples from unsaturated samples, which receive a higher load of organic matter (VFCW and influent of BF), and the other group by saturated conditions with a lower load of organic matter (HFCW influent and effluent zone, and BF effluent zone). Since influent water is different for each system, the results suggest a community configuration more related with the design of the treatment system and its operational conditions. These results are in consonance with the work carried out by Baptista et al. (2008), who suggested that stochastic processes could play an important role in the microbial community assembly in engineered and natural systems.

Different authors, such as Ibekwe et al. (2003) and Calheiros et al. (2009) indicated that the diversity of the bacterial community in the constructed wetlands systems might influence the final effluent quality, and so the engineering should be directed to develop a higher diversity in order to enhance processes such as nitrification and denitrification (Ibekwe et al., 2003). The Shannon index obtained for our samples showed a very similar diversity for all the samples. Significant differences (p-value < 0.05) were only found between HFCW and BF.

On the other hand, despite we could not retrieve sequences directly affiliated to known nitrifiers, nitrogen removal occurred in all the systems evaluated, although the removal rates were different among systems (Table 1). Saturated systems did not reach high nitrification rates but they were able to denitrify almost all the nitrified ammonia. Unsaturated systems were capable of high nitrification rates but total nitrogen removal was lower than unsaturated CW.

The removal of nitrogen in constructed wetlands is usually limited by the nitrification process, and in order to reach high total nitrification rates is important that biological nitrification takes place. Additionally, in order to increase denitrification rates in the unsaturated systems, the establishment of recycling or an additional step is a must. In this sense, the application of molecular techniques in this study has revealed the presence of several groups of denitrifiers. Finally, the diversity for bacterial groups has proven to be higher than for archaeal representatives. Further studies are needed to assess the activity of these groups under different conditions, and to go deeper into the functional groups present in each system.

12.4 CONCLUSIONS

- The application of molecular techniques (PCR-DGGE profiling) on different wastewater treatment systems showed that there is no relation between the influent and effluent bacterial communities inside the same treatment system.

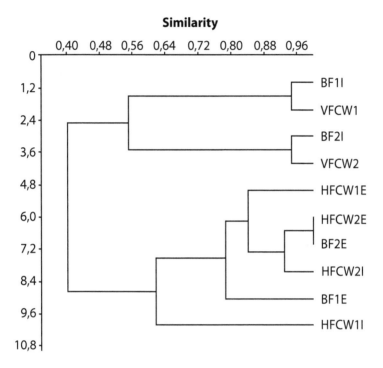

FIGURE 3: Cluster analysis of bacterial DGGE profiles, determined by the Raup-Crick method.

- Microbial community structure was related to the oxygen conditions (saturated or unsaturated) and organic matter load.
- High diversity of bacteria was found in all systems studied. A lower archaeal diversity was found in comparison with the bacterial population.

REFERENCES

1.   S.F. Altschul, T.L. Madden, A.A. Schäffer, J. Zhang, Z. Zhan, W. Miller, D.J. Lipman. Gapped BLAST and PSI-BLAST: a new generation of protein database search programs. Nucl. Acids Res., 25 (1997), pp. 3389–3402

2. J.C. Baptista, R.J. Davenport, T. Donnelly, T.P. Curtis. The microbial diversity of laboratory-scale wetlands appears to be randomly assembled. Water Res., 42 (2008), pp. 3182–3190

3. H. Brix, C.A. Arias. Danish guidelines for small-scale constructed wetland systems for onsite treatment of domestic seawage. Water Sci. Technol., 51 (2005), pp. 1–9

4. H. Brix, T. Koottatep, O. Fryd, C.H. Laugesen. The flower and the butterfly constructed wetland system at Koh Phi Phi – system design and lessons learned during implementation and operation. Ecol. Eng., 37 (2010), pp. 729–735

5. I. Cacciari, G. Giovannozzi-Sermanni, A. Grapelli, D. Lippi. Nitrogen fixation by Arthrobacter sp. Part I. Taxonomic study and evidence of nitrogenase activity of two new strains. Ann. Microbiol. Enzimologia, 21 (1971), pp. 97–105

6. C.S.C. Calheiros, A.F. Duque, A. Moura, I.S. Henriques, A. Correia, A.O. Rangel, P.M. Castro. Substrate effect on bacterial communities from constructed wetlands planted with Typha latifolia treating industrial wastewater. Ecol. Eng., 35 (2009), pp. 744–753

7. E.O. Casamayor, R. Massana, S. Benlloch, L. Ovreas, B. Díez, V.-J. Goddard, J.M. Gasol, I. Joint, F. Rodriguez-Valera, C. Pedrós-Alió. Changes in archaeal, bacterial and eukaryal assemblages along a salinity gradient by comparison of genetic fingerprinting methods in a multipond solar saltern. Environ. Microbiol., 4 (2002), pp. 338–348

8. X. Dong, G.B. Reddy. Soil bacterial communities in constructed wetlands treated with swine wastewater using PCR-DGGE technique. Bioresour. Technol., 101 (2010), pp. 1175–1182

9. O. Hammer, D.A.T. Harper, P.D. Ryan. PAST-paleontological Statistics. V 1.81 (2008)

10. A.M. Ibekwe, C.M. Grieve, S.R. Lyon. Characterization of microbial communities and composition in constructed dairy wetland wastewater effluent. Appl. Environ. Microbiol., 69 (2003), pp. 5060–5069

11. V. Ivanov, V. Stabnikov, W.Q. Zhuan, J.H. Tay, S.T. Tay. Phosphate removal from the returned liquor of municipal wastewater plant using iron-reducing bacteria. J. Appl. Microbiol., 98 (2005), pp. 1152–1161

12. P.D. Jenssen, T. Krogstad, A.M. Paruch, T. Maehlun, K. Adam, C.A. Aricas, A. Heistad, L. Jonsson, D. Hellström, H. Brix, M. Yli-Halla, L. Vrale, M. Valve. Filter bed systems treating domestic wastewater in the Nordic countries – performance and reuse of filter media. Ecol. Eng., 36 (2010), pp. 1651–1659

13. R.H. Kadlec, S. Wallace. Treatment Wetlands. CRC press, Boca Raton, FL, USA (2009)

14. E. Krasnits, E. Friedler, I. Sabbah, M. Beliavski, S. Tarre, M. Green. Spatial distribution of major microbial groups in a well established constructed wetland treating municipal wastewater. Ecol. Eng., 35 (2009), pp. 1085–1089

15. C.A. Kreader. Relief of amplification inhibition in PCR with bovine serum albumin or T4 gene 32 protein. Appl. Environ. Microbiol., 62 (1996), pp. 1102–1106

16. A.W. Mayo, T. Bigambo. Nitrogen transformation in horizontal subsurface flow constructed wetlands I: model development. Phys. Chem. Earth, 30 (2005), pp. 658–667

17. Miljøministeriet Miljøstyrelsen. Spildevandsforskning fra Miljøstyrelsen Nr 8 Spildevandsrensning i rodzoneanlæg. Copenhagen (In Danish) (1990)

18. Miljøministeriet Miljøstyrelsen. Retningslinier for Etablering Af Beplantede Filteranlæg op Til 30 PE, Økologisk Byfornyelse og Spildevandsrensning Nr. 52. Copenhaguen (In Danish) (2004)

19. G. Muyzer, T. Brinkhoff, U. Nübel, C. Santegoeds, H. Schäfer, C. Wawer. Denaturing gradient gel electrophoresis (DGGE) in microbial ecology. A.D.L. Akkermans, J.D. Van Elsas, F.J. De Bruijn (Eds.), Molecular Microbial Ecology Manual, Kluwer Academic Publishers, Dordrecht, Boston, London (1998), pp. 1–27

20. M. Park, S.H. Ryu, T.H. Vu, H.S. Ro, P.Y. Yun, C.O. Jeon. Flavobacterium defluvii sp nov, isolated from activated sludge. Int. J. Syst. Evol. Microbiol., 57 (2007), pp. 233–237

21. S.R. Ragusa, D. McNevin, S. Qasem, C. Mitchell. Indicators of biofilm development and activity in constructed wetlands microcosms. Water Res., 38 (2004), pp. 2865–2873

22. O. Sánchez, J.M. Gasol, R. Massana, J. Mas, C. Pedrós-Alió. Comparison of different denaturing gradient gel electrophoresis primer sets for the study of marine bacterioplankton communities. Appl. Environ. Microbiol., 73 (2007), pp. 5962–5967

23. K. Sleytr, A. Tietz, G. Langergraber, R. Haberl, A. Sessitsch. Diversity of abundant bacteria in subsurface vertical flow constructed wetlands. Ecol. Eng., 35 (2009), pp. 1021–1025

24. J. Snaidr, R. Amann, I. Huber, W. Ludwig, K.H. Schleiffer. Phylogenetic analysis and in situ identification of bacteria in activated sludge. Appl. Environ. Microbiol., 63 (1997), pp. 2884–2896

25. U. Stottmeister, A. Wiessner, P. Kuschk, M.K. Kappelmeyer, R.A. Bederski, H. Müller, H. Moormann. Effects of plants and microorganisms in constructed wetlands for wastewater treatment. Biotechnol. Adv., 22 (2003), pp. 93–117

26. M. Truu, J. Juhanson, J. Truu. Microbial biomass, activity and community composition in constructed wetlands. Sci. Total Environ., 407 (2009), pp. 3958–3971

27. M. Truu, K. Nurk, J. Juhanson, U. Mander. Variation of microbiological parameters within planted soil filter for domestic wastewater treatment. J. Environ. Sci. Health. Part Toxic/Hazard. Subst. Environ. Eng., 40 (2005), pp. 1191–1200

28. G. Vacca, H. Wand, M. Nikolausz, P. Kuschk, M. Kästner. Effect of plants and filter materials on bacteria removal in pilot-scale constructed wetlands. Water Res., 39 (2005), pp. 1361–1373

29. J. Vymazal, H. Brix, P.F. Cooper, M.B. Green, R. Haberl. Constructed wetlands for wastewater treatment, in: J.T.A. Verhoeven, B. Beltman, R. Bobbink, D.F. Wigham (Eds.), Wetlands as a Natural Resource, Wetlands and Natural Resource Management, vol 1, Springer Verlag (2006), pp. 69–96

30. C.-B. Zhang, J. Wang, W.-L. Liu, S.-X. Zhu, H.L. Ge, S. Chang, J. Chang, Y. Ge. Effects of plant diversity on microbial biomass and community metabolic profiles in a full-scale constructed wetland. Ecol. Eng., 36 (2010), pp. 62–68

Author Notes

CHAPTER 1

Acknowledgments

The authors thank the Hong Kong General Research Fund (HKU7197/08E) for the financial support of this study and Lin Ye thanks The University of Hong Kong for the postgraduate studentship.

CHAPTER 2

Funding

This work was supported by the Hong Kong General Research Fund (7198/10E). The funders had no role in study design, data collection and analysis, decision to publish, or preparation of the manuscript.

Competing Interests

The authors have declared that no competing interests exist.

Acknowledgments

We would like to thank the Hong Kong General Research Fund (7198/10E). Dr. Feng Guo and Lin Cai want to thank the University of Hong Kong for the postdoctoral fellowship. Mr. Feng Ju wished to thank the University of Hong Kong for the postgraduate studentship.

Author Contributions

Conceived and designed the experiments: FG TZ. Performed the experiments: FG LC. Analyzed the data: FG FJ TZ. Contributed reagents/materials/analysis tools: FG FJ TZ. Wrote the paper: FG FJ LC TZ.

CHAPTER 3

Funding

This work was funded by a research grant from FORMAS, The Swedish Research Council for Environment, Agricultural Sciences and Spatial Planning. The funders had no role in study design, data collection and analysis, decision to publish, or preparation of the manuscript.

Competing Interests

The authors have declared that no competing interests exist.

Acknowledgments

We thank the staff at Gryaab AB for assistance in obtaining samples. We thank the staff at the Genomics Core Facility platform at the Sahlgrenska Academy, University of Gothenburg, for assistance with the T-RFLP analysis. We also thank Frank Persson who provided valuable comments on the manuscript.

Author Contributions

Conceived and designed the experiments: NJF BMW. Performed the experiments: NJF BMW. Analyzed the data: NJF BMW MH. Contributed reagents/materials/analysis tools: NJF. Wrote the manuscript: NJF. Contributed with critical revisions of the manuscript: BMW MH.

CHAPTER 4

Data Availability

The authors confirm that all data underlying the findings are fully available without restriction. The sequences of amoA for cloning library construction have been deposited in GenBank under accession numbers KF720406 to KF720513. The pyrosequencing datasets have been deposited into the NCBI Short Reads Archive Database (accession number: SRX396800). The Illumina metagenomic datasets are available at MG-RAST under accession numbers 4494863.3, 4494888.3, 4494854.3 and 4494855.3.

Funding

This study was financially supported by National Natural Science Foundation of China (Grant No. 51378252, PI: BL, URL: http://www.nsfc.

gov.cn), National Science and Technology Major Project of China (No. 2011ZX07210-001-1, PI: AL, URL:http://www.nmp.gov.cn/) and Technology Support Project of Jiangsu Province (Grant No. BE2011722, PI: BL; Grant No. BE2013704, PI: XXZ; URL: www.jskjjh.gov.cn). The funders had no role in study design, data collection and analysis, decision to publish, or preparation of the manuscript.

Competing Interests
The authors have declared that no competing interests exist.

Author Contributions
Conceived and designed the experiments: XXZ BL CL AL. Performed the experiments: ZW XXZ XL YL. Analyzed the data: ZW XXZ XL BL. Contributed reagents/materials/analysis tools: XXZ BL AL. Wrote the paper: ZW XXZ XL YL AL.

CHAPTER 5

Acknowledgments
We thank the Bioinformatics Platform UAB (BioinfoUAB), Ramiro Logares and Guillem Salazar for help and support with sequence analyses.

Funding Information
This work was supported by the Spanish projects Consolider TRAGUA (CSD2006-00044) and CTQ2009-14390-C02-02.

CHAPTER 6

Conflict of Interest
The authors declare that the research was conducted in the absence of any commercial or financial relationships that could be construed as a potential conflict of interest.

Acknowledgments
This study was financed by the Swiss National Science Foundation, Grants no. 120536 and 138148. David Weissbrodt was supported for the research collaboration with UFZ Magdeburg by obtaining the PhD Mobility Award

of the EPFL Doctoral Program in Civil and Environmental Engineering in 2011. Jean-Pierre Kradolfer and Marc Deront (EPFL-LBE), and Jonathan May (master student from Agrosup'Dijon, France) are acknowledged for their excellent assistance on reactor designs and automation. We are further grateful to Julien Maillard (EPFL-LBE) for helpful comments on the manuscript.

CHAPTER 7

Acknowledgements
This work was supported by grants of the Israel Science Foundation (No. 1290/08 and 1583/12). The authors confirm that they have no conflicts of interest.

CHAPTER 8

Funding
This study was supported by National Natural Science Foundation of China funds (51078207 and 51178239) and the special fund of State Key Joint Laboratory of Environment Simulation and Pollution Control (12L03ESPC). This study made use of the GeoChip and associated computational pipelines funded by ENIGMA-Ecosystems and Networks Integrated with Genes and Molecular Assemblies through the Office of Science, Office of Biological and Environmental Research, the U. S. Department of Energy under Contract No. DE-AC02-05CH11231. The funders had no role in study design, data collection and analysis, decision to publish, or preparation of the manuscript.

Competing Interests
The authors have declared that no competing interests exist.

Author Contributions
Conceived and designed the experiments: X. Wen YY JZ. Performed the experiments: X. Wang YX. Analyzed the data: X. Wang. Contributed reagents/materials/analysis tools: X. Wang YX. Wrote the paper: X. Wang YX.

CHAPTER 9

Acknowledgments
This work was possible due to the contribution of the Consorci per a la Defensa dels Rius de la Conca del Besòs and Les Franqueses del Val-lès Town Council and through the grants awarded by the Spanish Department of Science and Technology, Research Projects CTM2005-106457-C05-05/TECNO, 2FD1997-1298-C02-01, REN2000-3162-E, REN2002-04113-C03-03, CONSOLIDER-TRAGUA CDS2006-00044and PET2008-0165-02; FPI grant from Ministry of Education and Science of Spain, and FI grant from the Comissionat per a Universitats i Recerca del Departament d'Innovació, Universitats i Empresa de la Generalitat de Catalunya i del Fons Social Europeu.

CHAPTER 10

Competing Interests
The authors have declared that no competing interests exist.

Acknowledgments
The authors thank Dr. Yoshitomo Kikuchi, researcher AIST and faculty member of the Graduate School of Agriculture, Hokkaido University, for useful technical advice. The authors are also grateful to Dr. SM Singh, Scientist D of the National Centre for Antarctic & Ocean Research, Ministry of Ocean Development (NCAOR) and Dr. Teruo Sano, Professor in the Faculty of Agriculture and Life Science, Hirosaki University for useful suggestions.

Author Contributions
Identification of yeast: MT KS. Measuring Biological oxygen demand and design of wastewater treatment system: YY. Biological characteristics of yeast, purification of lipase and determination of yeast lipase characteristics: MT. Collection of algal mat from East Antarctica: KS TH. Conceived and designed the experiments: TH. Performed the experiments: MT YY KS SK TH. Analyzed the data: MT YY. Contributed reagents/materials/analysis tools: MT YY SK TH. Wrote the paper: MT.

CHAPTER 11

Competing Interests
The authors declare that they have no competing interests.

Author Contributions
IZB participated in the conception, design, experimental work, and data collection and analysis, and also drafted the manuscript. MI participated in the design of the experiment approaches and mathematical modeling. VDG carried out the microbiological sampling and pretreatment procedures. IM conceived the design and operation of the bioreactors. BP elaborated and carried out the total DNA extractions as well as community composition assessments by the 16S rRNA method and metagenomic approaches. AN designed and performed the wastewater analysis. EK participated in the critical discussion of the results. TB contributed to the evaluation of the analytics data. GM performed the metagenomics experiments, designed the study, and participated in the critical discussion of the results. All authors read and approved the final manuscript.

Acknowledgments
This work was supported by the following international (EU) and domestic funding bodies: "SYMBIOTICS" ERC AdG EU Grant, "BIOSIM" PN-II-PT-PCCA-2011-3.1-1129 European Fund (Romania), and by PIAC_13-1-2013-0145 supported by the Hungarian Government and financed by the Research and Technology Innovation Fund.

CHAPTER 12

Acknowledgements
This research was supported by predoctoral scholarships, FI and FPI, from the Comissionat per a Universitats i Recerca del Departament d'Innovació, Universitats i Empresa de la Generalitat de Catalunya i del Fons Social Europeu, and Ministry of Education and Science of Spain, respectively. Financial support was provided by grants CTM2008-06676-C05-02/TECNO from the Ministry of Science and Innovation of Spainto Jordi Morató and by Consolider TRAGUA (CSD2006-00044), CTQ2009-14390-C02-02.

Index

A

abundance, xviii, xx–xxii, 4, 13, 16, 18–19, 22, 26, 34–35, 38–40, 43–44, 48, 58–59, 61–62, 72, 77, 84–85, 90, 94, 96, 101–102, 104–105, 107–108, 110–111, 113, 115–120, 122, 128, 133, 135, 149, 151–152, 170, 177, 179, 181, 183–193, 195–199, 201, 203, 216, 229, 249, 302

accession number, 6, 87, 91–92, 108, 135, 208, 239, 261, 319, 330

Accumulibacter, xxi–xxii, 28–29, 35, 142–143, 146, 149–152, 154, 157, 162–164, 166–167, 170

acetate, 144–145, 150, 152, 163, 169, 182, 239, 307, 319

Acidobacteria, xxiv, xxvi, 59, 89, 247–249, 321

Actinobacteria, xxi, xxvi, 13, 17–18, 70–71, 131, 212, 321–322

aeration, xxii, 5, 144–145, 150, 152, 162, 207, 262, 322

Alphaproteobacteria, xviii–xix, 13, 17, 59, 61, 70–72, 76–77, 79, 90, 93, 128, 152, 163, 171, 179

amino acids, 110, 262, 306

Aminobacterium, 110, 247

ammonia, xx–xxi, xxiii, 22, 25, 50–51, 102–103, 112, 116, 119–120, 122, 126–127, 133, 135–139, 172, 180–181, 190, 200, 206–207, 216–217, 222–229, 240, 316, 324

ammonia monooxygenase (*amoA*), xx–xxi, 22, 102, 104, 108, 111–112, 114–116, 118–119, 122, 125, 127, 129, 132–133, 135, 137–138, 330

ammonia-oxidizing, xx, 22, 50–51, 102, 119–120, 122, 126, 133, 135–136, 138–139, 172, 200, 225, 227–229

ammonium, xx, 22, 101, 112, 116, 122, 126, 138, 143, 151, 180, 191

ammonium-oxidizing organisms (AOO), 143, 151, 153

amplification, 6, 20, 32, 44, 57–58, 78, 102, 104, 127, 135, 253, 318, 321, 326

anaerobic, xx–xxii, 5–6, 16, 18–22, 96, 103, 109–112, 117–119, 121–123, 143–145, 150–152, 154, 162–164, 166–168, 170, 202–203, 207, 224, 249, 253, 278, 282, 306–308

anaerobic/anoxic/aerobic (A2O), 207, 224

analysis of similarity (ANOSIM), 64, 85

analysis of variance (ANOVA), 237, 240–244, 285

antibiotic resistance, xxiii, 120, 206, 220–221, 223

aquaculture, xxii, 133, 177, 179, 201–203

Archaea, xx, xxvi, 12–13, 21–22, 50–51, 102, 120, 122, 135–139, 202, 212, 229, 253, 319–320

Ascomycota, 212

B

bacteria, xvii–xviii, xx–xxii, xxvi,
3–5, 10–12, 17–19, 21–22, 25–26,
28–30, 32, 34–35, 39, 41, 45, 50, 52,
56–59, 76–78, 83, 90, 92–94, 97, 99,
102, 104, 108, 110–111, 117–118,
120, 122, 128, 131, 133, 135–139,
153, 166, 172–173, 177–179, 181,
183–185, 187, 189–191, 193, 195,
197–203, 212, 217, 225–229,
232–234, 237, 247–248, 250–255,
278–279, 295, 307, 312, 319–320,
322, 325–327
  ammonia oxidizing bacteria
    (AOB), xx, 25–26, 28–29, 40,
    102, 104, 108, 111–112, 114,
    117–118, 225–226
  Bdellovibrio and like organisms
    (BALO), xxii, 178–179, 181–184,
    189–191, 193, 195–199, 201, 203
  denitrifiers, xx, 25, 28, 101–103,
    105, 107, 109, 111, 113, 115,
    117–119, 121, 123, 136, 147, 151,
    163–164, 229, 324
  fecal coliform, 232
  glycogen accumulating organisms
    (GAOs), 28, 171–172
  Gram-positive bacteria, 58, 76–77
  nitrite oxidizing bacteria (NOB),
    25–26, 28–29, 102, 117–118
  polyphosphate accumulating
    organisms (PAOs), 25, 28, 172
bacterial community structure, 17, 21,
51, 56, 97, 102, 109–110, 127, 138,
164, 200, 228, 245
bactericides, 259
Bacteriovoracaceae, 179, 181–182,
184, 191, 196–197, 200

Bacteroidetes, xxi–xxii, xxiv, xxvi,
13, 18, 70–71, 127–128, 131, 136,
190–191, 212, 247–249, 321–322
Basidiomycota, 212
Bdellovibrio, xxii, 178–179, 187,
191–193, 199–203
Bdellovibrionaceae, xxii, 178–179,
181–183, 191, 196–197, 202–203
Betaproteobacteria, xix, 17–18, 59,
61, 70–71, 76–77, 79, 90, 93, 127,
133, 135
bias, 27–28, 33, 38, 40, 44, 46, 52, 76,
78, 196
biochemical oxygen demand (BOD),
xxiv, 240, 260, 262, 267–268,
273–274, 315–316
biofilm, xxi–xxii, xxiv, 99, 122,
141–145, 147, 149–153, 155–161,
163–174, 181, 185, 189–190, 195,
198, 232–233, 235–238, 242–246,
249, 251–254, 302, 307, 312, 318,
323, 327
biofilters (BF), xxvi, 169, 190–191,
195, 200, 202, 225, 229, 311,
313–314, 316–317, 321–324
biogeochemistry, 25
biological nutrient removal (BNR),
141–142, 144, 150, 162–163, 165,
170
biological treatment, xvii–xviii, xxv, 3,
103, 109, 121
biomass, xxii, 121, 141–142, 144–145,
147–152, 162, 166, 169, 172–173,
178, 198, 234, 242, 306–307, 327
bioreactor, 50, 77, 95–97, 111–112,
116, 120–121, 126, 133, 135,
138–139, 198, 203, 225, 227–229,
279, 296, 301–302, 308, 334
  aerobic bioreactors, 112
bubble column (BC-SBR), xxi, 143–
150, 152, 155–164, 166, 169, 229

C

canonical correspondence analysis (CCA), xxiii, 208–209, 221–222, 224, 226

capillary gel electrophoresis, 84

carbon, xxiii–xxiv, 5, 136, 142, 171, 180, 206, 212–213, 216, 223–225, 248, 262, 268, 273–274, 276, 308
  carbon cycling, 212–213

cell percentage, xviii, 11, 16–17, 19

cellulose, 212, 224, 308

central composite design (CCD), 281, 283, 296

chemical oxygen demand (COD), xxiii, 5, 121, 127,147, 168, 206–207, 209, 222–226, 229, 240, 302

chimera, 10–12, 20, 33, 52, 86, 98, 106, 169, 200

chitin, 212, 214, 224

classification, xix, 27–28, 33, 35, 38–39, 43–46, 51–52, 60, 73, 89, 92, 201
  classification consistencies, 38
  classification method, xix, 35, 38

Clostridium, xxiv–xxv, 30, 47, 237, 240–243, 247, 298, 307–308

coagulation, 103

coastlines, 177

combustion, 277–278

community dynamics, 57, 74, 79, 96–97, 121, 142, 146, 148, 226–227

community profiling, 28

Competibacter, xxi–xxii, 28–29, 35, 142–143, 149–152, 157, 163–164, 166–167, 170, 173

conductivity, 144–145, 168, 171, 207, 223, 303, 316

confidence threshold, 10, 12, 16, 31, 33–34, 39–40, 45, 48, 107

confocal laser scanning microscopy (CLSM), 93, 142–143, 149–150, 152–153, 155, 157, 160–161, 166

cost, xvii, 3–4, 26, 141, 232, 259, 278, 311

D

dairy parlor, 259

database, xviii, xx, 6, 10–11, 20–21, 27, 31–34, 40, 45–46, 52, 56, 59, 65–75, 78, 80, 83, 87, 90–95, 98–99, 106, 108, 111, 115, 118, 120–121, 132–133, 135, 148, 169, 185, 200, 239, 251, 319, 325, 330
  database construction, 34

degradation, xviii, xxiii, xxv, 121, 178, 189, 206, 212–213, 216–217, 219, 221, 223, 251, 253, 280, 290–291, 294, 298, 302–303, 312

denaturing gradient gel electrophoresis (DGGE), xxiv, xxvi, 3, 20–22, 26, 126, 128, 133, 136–137, 202, 225, 229, 235, 238–239, 245–247, 253–254, 313, 317–327

denitrification, xx, 76, 101–102, 108, 110–111, 117, 119–120, 123, 135, 138, 143, 163, 174, 178, 216, 223–224, 228, 295, 301, 313, 315, 324

design of experiment (DOE), 279

detergents, 259

digestion/sedimentation basin (DB), 96–97, 138, 180, 185, 187–191, 306

disinfection, 231

dissolved oxygen (DO), xxiii, 110, 122, 142, 181, 206, 229

diversity, xvii–xxiii, xxvi, 4, 12–13, 17–22, 28, 45, 50–52, 55, 57–59, 61, 63, 65, 67, 69, 71, 73, 75–81, 83, 85, 87, 89, 91, 93–99, 101–102, 106, 109–111, 117–122, 125–129, 131–133, 135–139, 148, 159, 164–165, 170, 173–174, 177, 179, 181, 183, 185, 187, 189–191, 193,

195, 197, 199–201, 203–204, 209,
211, 223–229, 235, 238, 247, 250,
253, 312–313, 320–327
DNA, xix, 4–6, 20, 32–33, 43,
51–52, 56, 59, 73, 77, 81, 83–84, 86,
102–105, 111, 115–117, 119, 127,
135, 148, 153, 156, 181–185, 187,
207–208, 226, 228, 230, 237–238,
247, 250, 261, 305, 309, 318–319,
334
DNA extraction, 5, 43, 52, 77, 83,
103, 105, 181, 185, 237, 305,
318
DNA sequencing, 51, 102,
104–105

E

economic, xxv, 172, 277, 289, 301
efficiency, xvii, xx, 3, 26, 33, 38–39,
105, 116, 185, 187, 233–234, 236,
240, 242, 245, 250, 252, 291,
311–312
removal efficiency, xx, 116, 236,
240, 242, 245, 312
effluent, xviii, xxiv, 4–5, 11–13,
15–17, 19, 26, 117, 141, 168, 180,
202, 209, 233, 235–237, 240–241,
252, 316–317, 323–324, 326
endoglucanase, 212
Escherichia coli, xxiv, 21, 86–87,
92, 183–184, 187, 237, 240–244,
250–252, 306
esters, 264, 271–272
Euryarchaeota, 212
evaporation, 178
exopolysaccharides, 143, 163–166,
168, 170, 172–173
extracellular polymeric substances
(EPS), 142, 168, 170–171, 173–174

F

fecal enterococci (FE), xxiv, 237,
240–241, 243–244, 250
fermentation, xxv, 266, 274, 277–282,
285–286, 289, 291–292, 295–296,
298, 300–302, 306–309
fermentation temperature, 281,
285–286, 296, 300, 302
filter, xvii, 5, 33, 93–94, 180–181,
237, 253–254, 259, 263, 269, 275,
311–312, 326–327
filter bed, 311–312, 326
filter bed material, 311–312
sand filters, xvii, 311–312
filtration, 50, 233–234, 237, 250, 311
mechanical filtration, 234
ultrafiltration, 263–264, 269
Firmicutes, xxi, xxiv, xxvi, 13, 18, 59,
70–71, 78, 110, 131, 212, 247, 249,
321–322
fish, xxii, 20, 26, 58–59, 61, 77,
83, 93, 99, 120, 126, 128, 133,
136, 142–143, 149, 153, 157, 160,
165–166, 172, 177–181, 190–191,
198, 200–203, 229, 269
fish basins (FBs), 177, 181
Flavobacteria, xxii, 190–191, 197,
199
flocs, xxi–xxii, 5, 19, 35, 52, 111, 143,
149, 152–153, 155, 157, 160–161,
165, 171–172, 174, 198
fluorescence, xvii, xix, xxii, 26, 58,
84, 93, 126, 142–143, 172, 174, 269
fluorescence in situ hybridization
(FISH), xix, xxii, 20, 26, 58–59,
61, 77, 83, 93, 99, 120, 126, 128,
133, 136, 142–143, 149, 153,
157, 160, 165–166, 172, 177–
181, 190–191, 198, 200–203,
229, 269

fluorescence lectin-binding
analysis (FLBA), xxii, 143, 149,
153, 156, 165, 174
fuel cells, 278, 306
fuel-cell technology, 277
functional groups, xix, 26–29, 31–32,
34–35, 38, 40, 44–46, 57–58, 76, 324
functional remediators, 26

G

Gammaproteobacteria, xviii, 13,
17–18, 59, 61, 70–72, 78–79, 93,
126–127, 131, 133
gel electrophoresis, xxiv, xxvi, 3, 20–
21, 26, 84, 104, 126, 137, 182–183,
202, 207, 225, 235, 238, 253–254,
319, 327
GenBank, 87, 89, 108, 114, 132–133,
135, 195, 247, 330
gene
carbon degradation genes, 212, 216
gene amplification, 102
gene array, xxiii, 206
gene arrays, 206
gene library, 57–59, 69, 72, 75–77,
79, 104
gene percentage, xviii, 4, 11, 16, 19
metal resistance genes, 219–221
nitrogen cycling genes, 216
phosphorus cycling genes, 217
sulfur cycling genes, 218
genetic diversity, 20, 50, 102, 121
genome, 19–21, 51–52, 94, 98, 105,
149, 185, 187, 189, 200, 228, 230
GeoChip, xxiii, 121, 205–209, 216,
222–224, 226–228, 332
global warming, 126, 277
granulation, xxi–xxii, 141–143, 152–
153, 157, 160, 162–168, 171, 173

granules, xxi–xxii, 141–144, 147,
149–150, 152–153, 155–158,
160–170, 173, 307
granulometry, xxiv, 240–244, 250
gravel, xxiv, 225, 235–238, 242–246,
249–250, 252, 255

H

habitats, 191, 197–198, 275, 279, 295
harvest, 180
hemicellulose, 214
heterotroph, 20–21, 76, 200
high throughput sequencing, xviii, 4,
26–28, 43, 46, 50, 52, 185
homogenization, 103
hybridization, xvii, xix, 3, 20, 26, 58,
93, 126, 142, 157, 207–208, 226, 269
hydraulic retention time (HRT), 116,
144–145, 162, 223, 236, 252
hydrogen, xxv, 168, 249, 277–279,
284–285, 289, 291, 295, 300–302,
306–308
biohydrogen, xxv, 277, 279–303,
305–309
hydrogen production, 278–279, 284,
289, 291, 295, 300–301, 306–308
hydrogen yield, 291

I

Illumina sequencing, 102, 105–106,
108, 111, 115, 118–119
inductively coupled plasma mass
spectrometry (ICP-MS), 207
influent, xviii, xxiv, xxvi, 4–5, 11–13,
15–19, 21, 103, 109, 112, 127, 131,
145, 209, 216–217, 223, 232–233,
236, 240–241, 286, 313, 316–317,
323–324

inoculation, 159, 262, 269, 294, 301
inoculum, xxv, 262, 279, 292,
    294–295, 300–302, 308
ions, xxv, 101, 145, 168, 264,
    272–274, 277, 303–305
irrigation, 120, 232

L

Lactobacillus, 214, 298
lectin, xxii, 143, 153, 156, 162, 165,
    168, 171, 173–174
libraries, xvii, xix, 6, 16, 26, 51,
    56–61, 64, 67–69, 72–79, 81, 83,
    85–90, 92–93, 95–96, 98, 104–106,
    108, 111–112, 126, 128, 130, 133,
    136, 330
LIBSHUFF, 64, 88–89
lipase, xxiv, 259–260, 263–264,
    269–274, 276, 333
    lipase activity, xxiv, 263–264,
        269–270, 272–274
lipid, 18, 180, 276
lowest common ancestor (LCA), xix,
    11, 27, 29, 33–36, 39–41, 43, 45–48

M

macrophytes, 233–234, 242, 250–251,
    254–255
mass spectrometry (MS), xxv, 207,
    291, 300, 303–304
mass transfer coefficient, 245
mass transfer theory, 234
mesocosms, 178
metagenomic, xx, xxv, 21, 51, 102,
    106, 108, 115–117, 119–121, 135,
    139, 171, 186, 228, 280, 295–296,
    298, 301, 305, 307, 309, 330, 334
metatranscriptomic, 52, 120, 135, 139
methanogens, 136, 295

Microbial Community Analysis
    (MiCA), 56, 94, 102, 207, 307
microbial community functional
    structures, xxiii, 121, 205–207,
    209–211, 213, 215, 217, 219,
    221–227, 229
microbial diversity, 4, 21–22, 51, 95,
    97, 119, 126, 132, 170, 174, 201,
    225, 229, 235, 253, 326
microbial ecology, xix, 26, 28, 50, 57,
    95, 98, 137, 165, 172, 203, 226, 253,
    327
milk, xxiv–xxv, 259–260, 262,
    267–268, 273, 275
mismatch, 33, 65–69, 72–75, 78, 83,
    91–93, 95, 185
molecular fingerprinting, 232
Monte Carlo permutations, 85, 208
Mrakia, xxiv, 259–260, 265–267, 269,
    271, 274–276
    Mrakia blollopis, xxiv, 259–260,
        262–263, 265–269, 271, 272–275
Mycobacterium, xviii, 16–17, 19, 28,
    30–31, 47, 213, 219

N

natural gas, 278
nitrate, 101–102, 111, 117–120, 126,
    151, 180–181, 216, 249, 254, 262,
    316
nitrification, xx–xxi, 22, 51, 101–102,
    108, 117, 122–123, 126, 131, 135–
    136, 138–139, 143, 145, 147, 151,
    174, 178, 191, 216, 224, 226–227,
    229–230, 252, 313, 315, 322, 324
nitrite, 25, 50, 101–102, 112, 116, 119,
    122–123, 126, 135, 137, 151, 181,
    191, 216, 249, 316
nitrogen, xx–xxi, xxiii, 5, 26, 50,
    97, 101, 111, 117, 119, 122, 127,

135–136, 147, 151, 163, 168–169, 177–178, 180, 206–207, 209, 214, 216–217, 223–224, 249, 260, 262, 269, 275–276, 304, 312–313, 315–316, 322–324, 326
  nitrogen availability, 323
  nitrogen removal, xx–xxi, 5, 26, 50, 101, 117, 119, 135, 147, 151, 163, 216, 249, 315, 322, 324
  nitrogen transformation, 312, 326
Nitrosomonas, xx–xxi, 29, 34, 40, 46–47, 111, 117–119, 123, 126, 133, 135, 151, 200
nitrous oxide reductase (*nosZ*), xx, xxiii, 102, 104, 108, 115–119, 215–217, 223–224
nutrient removal, xvii, xxi, 125, 141–142, 170, 173

O

ordination analysis, 64, 84, 224
oxidation and reduction potential (EH), 244–245, 249–250
oxygen, xxii–xxiv, 109–110, 112, 117, 122–123, 127, 142–143, 169, 180–181, 190, 205–206, 209, 229, 233–234, 240, 245, 251, 254, 262, 267, 315–316, 322–323, 325, 333
  dissolved oxygen, xxiii, 110, 122, 142, 181, 206, 229

P

Pareto-Lorenz evenness curve, 90
pathogen, xviii, 19–20, 26, 28, 30, 37, 41, 50, 179, 220, 232, 234, 251–255
pH, xxv, 50, 95–96, 137, 144–145, 149, 163, 169–171, 173, 181, 190, 207, 223, 238–239, 262–264, 270,

272, 274, 281, 285–286, 288, 293, 296, 298, 300, 302, 305, 308, 319
phage, 178, 197, 202, 226, 230
phosphate, 5, 149, 168–169, 189–190, 198, 263–264, 270, 272, 305, 322, 326
Phragmites australis, 235, 251
phylogenetic analysis, 22, 104, 111, 132, 139, 191, 202, 261, 266, 327
pollutant, xvii, xxii, 25, 110, 121, 137, 177, 206, 250
polymerase chain reaction (PCR), xix–xx, xxiv, xxvi, 3, 5–6, 10–11, 20–21, 26, 32–33, 38, 44, 46, 50–52, 55–57, 59, 61, 63, 65, 67, 69, 71, 73–75, 77, 79, 81, 83–87, 89, 91, 93–97, 99, 102, 104–106, 112, 120, 126, 137, 148, 181–185, 200, 203–204, 229, 232, 238–239, 245, 253, 261, 313, 317–320, 322, 324, 326
polymorphism, xx, 3, 21, 26, 51, 58, 94–95, 98, 126, 142, 170
  terminal restriction fragment length polymorphism (T-RFLP), 3, 22, 26, 58, 62, 64, 80, 83–84, 95, 97, 126, 142–143, 148, 151, 153, 159, 330
population dynamics, 138, 179, 229
predation, 178, 187, 191, 196, 199, 201–203, 226, 230, 232–233
predator, xxii, 177–179, 181–183, 185, 187–189, 191, 193, 195–204
prey, 178–179, 187, 189, 197–201, 203–204
primer, xix–xx, 4–6, 10, 12, 16, 27, 32–33, 38–39, 43–45, 51–52, 55–61, 63–81, 83–97, 99, 104–105, 127–128, 137, 148, 181–185, 238–239, 254, 261, 317–319, 327
  primer pair, xix–xx, 56–61, 64, 67–80, 83–85, 90, 92–93, 95

primer sites, 72, 92–93
universal primer, xix, 56–57, 76, 80, 128, 318
principal component analysis (PCA), 106–107, 109, 237
protein expression, 117
Proteobacteria, xx–xxi, xxiv, xxvi, 13, 17, 50, 77, 90, 99, 107, 110, 119, 127–128, 131, 136, 139, 212, 248–249, 321–322
protists, 178, 196, 201, 233
pyrosequencing, xviii, xx–xxi, 3–6, 8, 10, 18–22, 27–28, 32, 44–45, 50–52, 96, 102, 105–109, 117–119, 121, 125–128, 131–133, 136–139, 143, 148, 151, 173, 185, 198, 204, 224, 227–228, 330
Python, 10, 33–34, 49, 106, 108

Q

quantitative real time polymerase chain reaction (qPCR), xx, xxii, 26, 102, 104–105, 115–119, 182–185, 187–188, 191–193, 196, 232

R

rarefaction, xxi, 8, 10, 12, 106, 109, 128, 130, 132
reactor, xxi, 3, 5, 22, 26, 51, 57, 97, 103, 110–112, 121–122, 131, 133, 139, 141–142, 144–146, 148, 154, 159–160, 163, 166, 168–171, 173, 180–181, 195, 200, 225–229, 252–253, 307–308, 332
integrated anoxic/oxic reactor (A/O), 103
up-flow anaerobic sludge reactor (UASB), 103, 110, 112, 118, 121–122, 307

reclamation, 231, 251, 255
recycling, xxii, 177–179, 251, 314–315, 324
redox status, 245
refinery, 112, 122
Rhizobiales, 13, 72–73, 76, 78, 92–93, 152, 164, 191
Rhodospirillaceae, 152, 163
Ribosomal Database Project (RDP), xviii–xix, 6, 8, 10–12, 16, 20, 27, 29, 31–36, 38–41, 45–49, 56, 60, 65–66, 68–75, 78, 83, 89–93, 95, 99, 106–107, 120, 128, 186, 200
ribotypes, 197
richness, xix, 12, 52, 57–58, 66, 70–71, 73–74, 79, 81, 87–88, 106, 110, 120, 128, 148, 159, 164, 225, 320–321, 323
RNA, xviii–xix, 11, 21, 50, 56, 72, 92, 94, 99, 226, 306
root, 233—234, 242, 250–251, 254, 312
rRNA, xviii–xx, xxii, xxiv, 4, 6, 10–11, 15–17, 19–22, 25–29, 31, 33, 35, 37, 39–41, 43–47, 49, 51–53, 55–61, 76–77, 80–81, 83, 85, 94–99, 102, 105, 109, 115, 117–118, 120, 123, 125–127, 129–130, 132–133, 136–138, 148–149, 169–171, 173, 181–182, 184–185, 187–196, 200, 202, 230, 238–239, 245, 252–253, 261, 265–266, 313, 318–320, 334

S

salinity, xxii, 5, 18, 109, 112, 121, 179–180, 196, 198, 201, 326
Salmonella, 31, 232
sanitation, 232

season, xxiii, 177, 179, 181, 183, 185,
187–191, 193, 195, 197–199, 201,
203, 233–234, 237, 240–244, 250,
253–254
sedimentation, 19, 77, 112, 180, 190,
232–234, 250, 302, 312–314
sequencing batch reactors (SBR),
xxi–xxii, 97, 141, 143–152, 154–164,
166, 170–171, 225, 227
settling, 97, 103, 142, 144–145,
151–152, 162–164
sewage, xviii, xxiv, 4–5, 19, 21–22,
26, 50–51, 96, 109, 112, 120–121,
131, 138–139, 170, 204, 225, 228,
231, 252–253, 301
    saline sewage, xviii, 4–5, 112
    sewage treatment, 22, 26, 50–51,
    96, 120–121, 139, 170, 204, 228,
    231, 252–253
    urban sewage, 131
Shannon index, 109, 320, 324
shrimp, 133, 179, 203
SILVA ribosomal RNA gene database
project, 56, 72, 92, 94
sludge, xvii–xxi, xxiii–xxiv, 4–6,
11–13, 15–22, 25–29, 32, 34–35, 39,
44–46, 50–52, 57–62, 64–66, 70–71,
73, 76–80, 83, 85, 91, 93–97, 102–
104, 106–107, 109–112, 114–123,
125–128, 130–133, 135–139, 141–
146, 151, 153–155, 159, 162–164,
167–174, 191, 204–207, 209–212,
216, 223–229, 248–249, 253–254,
259–260, 262, 267–268, 274–275,
279, 308, 322, 327
    activated sludge, xvii–xxi, xxiii,
    4–5, 11–13, 15–22, 25–29, 32,
    34–35, 39, 44–46, 50–52, 57–62,
    64–66, 70–71, 73, 76–80, 85, 91,
    93, 96–97, 102–103, 111, 116,
    120–123, 125–128, 130–133,

    135–139, 141, 143–144, 151, 155,
    159, 162–164, 167, 169–174,
    204–207, 209–212, 216, 223–229,
    248–249, 253–254, 259–260, 262,
    267–268, 274–275, 322, 327
    aerobic granular sludge (AGS),
    xxi, 121, 141–143, 148–149, 151,
    163–165, 168–171, 173–174, 229
    aerobic sludge, 103, 109–112, 116,
    118–119, 173
    sludge biodiversity, 126
    sludge retention time (SRT),
    144–145, 151, 154, 162
    sludge sampling, 103
    sludge volume, 5
soil, xxii, 4–5, 17, 22, 32, 52, 55–56,
83, 95, 97–98, 102–103, 112,
116–117, 171, 179, 181, 196–197,
200–201, 203, 206, 225, 229–230,
249, 251, 276, 279, 305, 313–314,
316, 318, 322–323, 326–327
Sphingobacteriales, 13, 151–152, 164
standard deviation, 148, 208
substrate, xxv, 142, 158, 163,
165–167, 172, 201, 212, 262–264,
268–272, 274–275, 277–281,
291–292, 295–296, 300–302, 306,
320, 323, 326
    substrate specificity, 264, 271–272
sugar, 165, 268, 273–274, 306, 308
surface area, 234–235, 243, 302
sustainability, 278

T

tannery, xx, 101–103, 105, 107,
109–113, 115, 117–121, 123
taxonomic, xviii–xix, xxii, 10, 12–13,
16, 22, 25–28, 33–35, 38–40, 44–47,
51–52, 59, 72, 81, 95, 98, 104, 106,
118, 120, 123, 128, 132, 138, 148,

186, 199, 247, 249, 276, 296, 305, 321, 326

temperature, xxiii–xxv, 83–84, 95, 97, 138, 142, 144, 149, 163, 171, 181, 190, 198, 201, 206, 221–223, 225–226, 229, 232, 244, 250, 253–254, 259–260, 262, 264, 266, 268, 270, 272–276, 281, 285–286, 288, 293, 296, 300, 302, 304–305, 308, 316, 318

　temperature variations, 198

terminal restriction fragment (T-RF), xx, 3, 21, 51, 58, 62–64, 74–75, 79, 82–85, 95, 97, 126, 138, 170

Tetrasphaera, 28–30, 43, 47, 152, 163

Thermotogae, xviii, 12–13, 16–18, 21

total suspended solids (TSS), 145, 147

transcription, 111, 116, 135, 137

V

variance inflation factors (VIFs), 208, 221–222

variance partitioning analysis (VPA), xxiii, 208–209, 223, 225

Vibrio, xviii, 17, 19, 31, 179, 197

volatile fatty acids (VFA), xxii, 145, 150, 159, 162, 164, 167

W

wash-out, xxi, 142, 144, 147, 149–151, 159, 162–164, 167, 173

wastewater

　brewery wastewater, xxv, 110

　domestic wastewater, x, xxvi, 18, 224, 251, 312–314, 326–327

　tannery wastewater, xx, 101, 103, 105, 107, 109–111, 113, 115, 117–121, 123

　urban wastewater, xxiii, 121, 123, 235, 252

　wastewater quality, 109

wastewater treatment plant (WWTP), xvii–xxi, xxiii, 3–5, 7, 9, 11, 13, 15, 17–19, 21–23, 25–26, 35, 51, 55, 57–58, 76–77, 80, 83, 95–96, 101–103, 105, 107, 109–113, 115–123, 125–127, 129, 131, 133, 135–139, 144, 163, 198, 200, 205–207, 209–213, 215–217, 219–227, 229, 248–249, 295, 301

　freshwater plants, 126

　natural wastewater treatment systems, xvii, 311, 313, 315, 317, 319–321, 323, 325, 327

water

　drinking water, 102, 120, 251, 253

　freshwater, xxii, 18, 126, 131–132, 179–181, 190, 196–197, 203

　seawater, xxi, 5, 17–19, 125–127, 129, 131, 133, 135–137, 139, 196, 201, 203

　water depth, xxiii–xxiv, 234–237, 240, 245, 248, 250

　water purification, 178, 254

　water quality, 103, 178, 181, 254, 315–316

　water requirements, 178

　water shortage, xvii, 126, 231

wetlands, xvii, xxiii–xxiv, xxvi, 231–237, 239–255, 311–314, 317, 323–324, 326–327

　constructed wetlands, xvii, xxiii–xxiv, xxvi, 231–236, 239, 242–246, 250–255, 311–314, 317, 323–324, 326–327

　Horizontal Flow Constructed Wetlands (HFCW), xxvi, 311, 313–314, 316–317, 321–324

　horizontal subsurface flow (HSSF), xxiii–xxiv, 231–233, 235–236, 239–245, 249–250, 252, 326

Vertical Flow Constructed
   Wetlands (VFCW), xxvi, 313–
   314, 316–317, 321–323, 327
wetland design, 235, 250

X

Xanthomonadales, 72–73, 78, 92–93,
   152, 164
xylanase, 214, 224, 276

Z

zero discharge systems (ZDS),
   177–181, 187–192, 195–199
Zoogloea, xxi–xxii, 29, 35, 142,
   149–150, 152–153, 157–160, 162,
   164–165, 167, 170
zooplankton, 197